Multilayer Thin Films - Versatile Applications for Materials Engineering

Edited by Sukumar Basu

Published in London, United Kingdom

IntechOpen

Supporting open minds since 2005

Multilayer Thin Films - Versatile Applications for Materials Engineering
http://dx.doi.org/10.5772/intechopen.77490
Edited by Sukumar Basu

Contributors
Huseyn Mamedov, Parthasarathy Panchatcharam, Tania Tsvetkova, Cheng-An Tao, Jianfang Wang, Rui Chen, Surajit Kumar Hazra, P.B Barman, Shikha Sinha, Anuradha Kashyap, Supriyo Bandyopadhyay, Sajjad Habibzadeh, Ehsan Rahmani, Mohammad Reza Saeb, Mohammad Reza Ganjali, Jamal Chaouki, Miguel Angel Pasquale, Nicolás Muzzio, Omar Azzaroni, Sergio Moya, Arnab Hazra, Nagesh Samane, Sukumar Basu, Subhashis Gangopadhyay, Atanu Dutta, Nithya S, Tynee Bhowmick, Vibhav Amabardekar, Partha Pratim Bandyopadhyay, Sudip Nag, Subhasish B Majumder, Abhishek Ghosh, Moumita Dewan, Antonio Riul Jr., Maria Luisa Braunger, Rafael Hensel, Gabriel Gaál, Mawin Jimenez, Varlei Rodrigues

First published in London, United Kingdom, 2020 by IntechOpen
IntechOpen is the global imprint of INTECHOPEN LIMITED, registered in England and Wales, registration number: 11086078, 7th floor, 10 Lower Thames Street, London, EC3R 6AF, United Kingdom
Printed in Croatia

British Library Cataloguing-in-Publication Data
A catalogue record for this book is available from the British Library

Additional hard and PDF copies can be obtained from orders@intechopen.com

Multilayer Thin Films - Versatile Applications for Materials Engineering
Edited by Sukumar Basu
p. cm.
Print ISBN 978-1-78985-437-4
Online ISBN 978-1-78985-438-1
eBook (PDF) ISBN 978-1-83968-463-0

An electronic version of this book is freely available, thanks to the support of libraries working with Knowledge Unlatched. KU is a collaborative initiative designed to make high quality books Open Access for the public good. More information about the initiative and links to the Open Access version can be found at www.knowledgeunlatched.org

For EU product safety concerns:
IN TECH d.o.o., Prolaz Marije Krucifikse Kozulić 3, 51000 Rijeka, Croatia, info@intechopen.com or visit our website at intechopen.com.

Meet the editor

Prof. Sukumar Basu obtained his Ph.D. in solid-state chemistry in 1973 from the Indian Institute of Technology (IIT), Kharagpur, India. He undertook his post-doctoral research at the Institute of Physical and Theoretical Chemistry, University of Vienna (1972-1974), Austria and he was a Research Associate in the Max Planck Institute for Carbon Research, Muelheim a.d Ruhr (1975-1978). After returning to India, Prof. Basu joined IIT Kharagpur as a faculty member in 1979 and was promoted to full professor in 1995. He visited several international research institutions abroad e.g. visiting scientist in the Department of Physical Chemistry and Materials Science, State University of Milan, Italy (1995-1996) with ICTP fellowship and Japan Research Center for Atom Technology (JRCAT), Tsukuba, Japan (May-July 1999) with JSPS fellowship. He has attended a number of international and national seminars, conferences, and workshops in India and abroad and presented research papers on semiconductor materials, crystal growth and characterizations, solar energy conversions, spintronics, and chemical sensors. Prof. Basu has organized almost a dozen international and national seminars and conferences. He has taken part in joint SRC (Swedish Research Council) research project with the institute of Physics, Chemistry and Biology, Linkoping University, Sweden and IBSA (India-Brazil-South Africa) projects on the development of chemical gas sensors. His present research areas are graphene and graphene oxides for chemical gas sensors. Prof. Basu is associated with the Department of Physics and Materials Science, Jaypee University of Information Technology, Himachal Pradesh (HP), India as Professor Emeritus.

Contents

Preface

In the present scenario of using thin films to realize miniature devices, multilayer thin films have become important for applications. Although the single layer is the ideal technology, the multilayer thin films are a more realistic technology, from the price index and the device dimensions point of view. Mechanical, electrical and electronic, magnetic, and optical devices have shown better performance with large scale integration thereby saving lots of space. Ultra-thin films with multilayer configurations can function more efficiently. If we consider the physics, chemistry, and engineering aspects, multilayer thin films can be controlled more precisely. It is also relatively easier to fabricate and study the multilayer films than the single layer films. With these views, technologists prefer multilayer thin films for recent applications in solar cells, sensors, and spintronic devices.

This edited book consists of thirteen chapters. The authors have presented different aspects of multilayer thin films. Repeated research has confirmed the vital use of multilayer thin films with optimal results. Except for some rigorously precise gadgets where single layer thin films are superior, the multilayer films are largely used and make miniature devices more popular. The chapters discuss the different materials and their thin films for various applications. Of late, solar cells and sensors technology have replaced many conventional textures of thin films and are using multilayer thin films. Especially in the sensor devices with graphene technology with multilayer structures shown as the most efficient detectors. The authors have described the use of multilayer graphene thin films as gas/vapor sensors for monitoring environmental pollution. The composite thin films with different combinations of nanomaterials have also been highlighted with a score of advantages.

The chapters are evenly distributed on different aspects of materials, devices, and the mechanism of ultrathin layers. With the development of multilayer thin films, a new type of binary and ternary composite sensors has been evaluated and been found to be quite efficient as multilayer devices. The material part is very relevant for the formation of nanolevel thin films. In this book, most of the chapters discuss the materials technology. The device configurations are also given importance. Finally, the applications have been elaborated on in some chapters. It is envisaged that this book will lead to more awareness of the multilayer thin films, its importance, and the advantages over the single layer films for uses in the modern technology.

Finally, it is expected that the book may act as a map for the scientists and technologists for carrying on more research and development work on multilayer thin films for varied applications. This development will show how a material with some intrinsic defects in the language of science and technology can have more advantages than the so-called more perfect materials. The successful technology does not only mean the theoretical perfections and available at much higher costs but it means the relatively low-cost materials with appreciably high output. In the present trend, the functional materials mean the upliftment of so-called inferior materials through different modifications such as thermal annealing, laser annealing, and chemical treatment of the multilayer thin films. In this book we aim to bring attention on these directions.

I owe thanks to the help of many important personalities in my field of research and others. At first, I am thankful to IntechOpen book personnel for day-to-day interactions. The initiative taken by Ms. Ana Pantar, the commissioning editor, is highly praiseworthy. Mr. Josip Knapic, the author service manager, always helped me when I was in need of his assistance. His friendly co-operation made it possible for me to complete the work in time. Dr. Surajit Hazra, Assistant Professor (Department of Physics and Materials Science, Jaypee University of Information Technology, Himachal Pradesh, India) provided technical assistance. I appreciate his friendly co-operation. I am especially thankful to Dr. Arnab Hazra (Assistant Professor, Department of Electrical and Electronics Communication Engineering, Birla Institute of Technology and Science (BITS), Pillani, Rajasthan, India) for helping me in literature collections, computer system arrangement, and other important technical jobs. I am also thankful to Dr. (Ms.) Sukanya Basu (ex-faculty member, University of Michigan, USA) for being associated with me during the preparation of the project. I am indebted to the authors from the different parts of the world for kindly responding to my request and contributing their invaluable chapters to enrich the quality of the edited book. I thankfully remember the help and co-operation of the technical group of IntechOpen publishing group. Last but not the least; I am extremely grateful to my family for their kind understanding and absence from the daily domestic activities for the final completion of the book.

Dr. Sukumar Basu
Department of Physics and Materials Science,
Jaypee University of Information Technology (JUIT),
Solan, India

Chapter 1

A Review on Metal Oxide-Graphene Derivative Nano-Composite Thin Film Gas Sensors

Arnab Hazra, Nagesh Samane and Sukumar Basu

Abstract

Most of the available commercial solid-state gas/vapor sensors are based on metal oxide semiconductors. Metal oxides (MOs) change their conductivity while exposed to gas or vapors ambient can be utilized as gas or vapor sensing materials. In recent days, graphene has attracted tremendous attention owing to its two-dimensional structure with an extremely high surface to volume ratio, electron mobility, and thermal conductivity. However, intrinsic graphene is relatively inefficient for the adsorption of gas/vapor molecules. In this regard, graphene oxide (GO) and reduced graphene oxide (rGO), which are graphene species functionalized with different oxygen groups that offer a higher amount of adsorption sites improving the sensitivity of the film. Up to now, many research groups across the globe have reported the promising performance towards gas detection using various GO/rGO-metal oxide nanocomposites. This chapter reviews the composites of graphene oxide or reduced graphene oxide and metal oxides in nanoscale dimensions (0-D, 1-D, 2-D, and 3-D) for gas sensing applications considering two specific focus areas, that is, synthesis of nanocomposites and performance assessment for gas/vapor sensing.

Keywords: nanoscale metal oxide, graphene derivatives, nanocomposites, efficient gas sensing

1. Introduction

In today's world, gas/vapor sensors have received significant attention because of their important applications in numerous areas such as environmental monitoring at industry and domestic area [1], disease diagnosis [2], agriculture [3], industrial wastes [4], food quality monitoring, etc. The detection of gases like NO, NO_2, NH_3, CO, CO_2, SO_2, H_2S, etc. is essential in many fields especially in environmental monitoring due to their toxicity and the related risk to the ecosystem [1–4]. Detection of volatile organic compounds (VOCs) is of great importance in environmental safety, supervision of human health, and food quality monitoring [1–3]. The detection of frequently used VOCs like acetone [5], formaldehyde [6], methanol [7], etc. is essential because they produce toxic effects, even in low concentrations, on human health. Detection of ethanol in human breath is important to restrict the drunken driving-related issue [8]. Timely detection of released VOCs from stored vegetables and fruits is important to monitor their quality and freshness [9]. So, simple and reliable detections of gases and VOCs are important in everyday life.

Most of the existing commercial gas/vapor sensors are based on metal oxide (MO) semiconductors and polymer materials. However, the limitations of these gas sensors can be one or more as follows: costly, low sensitivity in lower ppm or ppb level, poor selectivity, limited lifetime, poor repeatability, difficult to miniaturization high power consumption [4, 10, 11], etc. As an alternative, nanostructured material-based gas/vapor sensors have gained significant importance due to many promising electrical, thermal, and optical characteristics combined with very high effective surface area, high sensitivity, fast response and recovery, selectivity, repeatability and stability [11], etc. Different carbon nanomaterials, such as graphene, graphene oxide (GO), carbon nanotube (CNT), charcoal, etc. have been shown to be promising gas/vapor sensing behavior due to the simple modifying their sensitivity by easy chemical treatments [12–14].

The limitations of intrinsic graphene are: (i) difficult to synthesize in large scale, (ii) it has almost no functional groups that can use for the adsorption of gas/vapor molecules, and (iii) it has metallic behavior with almost zero band gap [4, 13]. The prime performance enhancement methods in graphene-based sensors are found to be suitable impurity doping, composite formation, functionalization, implementation in field-effect transistor (FET) structure, etc. In this situation, reduced graphene oxide (rGO), which is graphene functionalized with different oxygen groups that provide enhanced adsorption sites, is more favorable for improving sensitivity. Besides very high thermal stability, the rGO sample contains many dangling bonds which can act as adsorption sites for gas analytes [15, 16].

Although many literatures suggested that the gas sensing performance can be improved by the structural and morphological variations, this is an insufficient approach for the growing demands of the gas/vapor sensing device performance. Single component transition metal oxide and carbon-based materials still suffer from some limitations arising from their inadequate physical and chemical characteristics that may hinder their large scale applications for high-performance gas/vapor sensors. Owing to their variable chemical conformation, synergistic properties, heterostructured nano-hybrids components, and nanocomposites are expected to show more admirable gas/vapor sensing performance [15, 17].

Metal oxide nanostructures are frequently hybridized with (i) noble and transition metals like Pd, Pt, Au, Ag, Ni, Nb, and so on, (ii) other metal oxides, (iii) carbon-based nanomaterials like CNT, graphene, and graphene-derivatives like GO and rGO to improve the gas sensing performance. Among all these functionalized materials, graphene and its derivatives attract tremendous attention for hybridizing with nanostructured metal oxides for promising gas/vapor sensing applications. Improvement of gas sensing properties of graphene/metal oxides hybrids principally depends on the following four factors:

(i) graphene derivative like GO or rGO supplies more dangling bonds and active interaction sites for gas/vapor molecule adsorption/reaction; (ii) its large effective surface area also enhance the gas sensing performance [15, 16]; (iii) metal oxide nanostructures have been extensively discovered as gas/vapor sensors due to the relatively high sensitivity of their electrical conductance to the target adsorbents. Thus the presence of rGO layers on metal oxide surface, electrical properties exhibit large and fast changes in the occurrence of gases/vapors improving overall sensing performance of the sensor; (iv) while GO and rGO show ambipolar behavior in the electron and hole concentration, they show hole-dominant p-type conducting properties owing to the adsorbed water and oxygen molecular species. Also, a nanocomposite of p-type rGO with an n-type transition metal oxide form a p-n heterojunctions and the resulting complex nanostructure may exhibit better sensing performances than those of the individual materials. Numerous research has confirmed that the p-n heterojunction formed by p and n-type materials can play a positive role in the sensing mechanism [18–20].

However, a wide variety of nanostructured metal oxides and its composite with GO and rGO have been reported for efficient gas/vapor sensing applications in last one decade. In this chapter, we have categorized the graphene nanocomposites based on the morphology of metal oxides, that is, zero-dimensional (0-D like nanoparticles, quantum dots, etc.), one dimensional (1-D like nanorods, nanotubes, nanofibers, etc.), two dimensional (2-D like nanosheets, nanoplates, etc.), and three-dimensional (3-D like nanoflower, nanospheres, etc.). Synthesis, fabrication of graphene/nanoscale metal oxides nanocomposites and their performance assessment for gas/vapor sensing application are the main objective of the article.

2. Synthesis nanoscale metal oxides and graphene derivatives composite

In this section, the synthesis of graphene and its derivatives like graphene oxide (GO) and reduced graphene oxides (rGO) is described in the first sub-section. Then the synthesis of nanoscale metal oxides, as well as the nanohybrid formation, is described in the next sub-section.

2.1 Synthesis of graphene and graphene-derivatives

Graphene is considered as the parent of all graphitic forms [21]. The purest form of graphene is named as pristine graphene (with no heteroatomic contamination) where 'scotch tape method' widely accepted for producing the highest quality of graphene [22]. Graphene produced from micromechanical cleavage, that is, adhesive tape method can isolate only a small amount of graphene, hence this method is used to isolate graphene for research purposes. For large scale production of graphene, various methods have been reported in the literature which can be broadly classified into two categories: top-bottom approach and bottom-up approach [23].

Top-bottom methods mainly involve breaking of the van der Waals bonds which hold layers of graphene to form graphite [22]. Top-bottom approach involves electrochemical exfoliation, exfoliation of graphite intercalation compounds (GIC), micromechanical cleavage, solvent-based exfoliation of graphite oxide, arc discharge, etc. [23]. Among these methods, exfoliation of graphite oxide has received great attention as graphite oxide is easily produced by oxidation of graphite as reported in the Hummers method. Graphite oxide is exfoliated to obtain graphene oxide which is reduced to form reduced graphene oxide (rGO). Reduction process can be thermal, chemical, or UV-based method [24]. Bottom-up approach involves forming of large-area graphene sheet via growth over the substrates and one of the most potential methods is chemical vapor deposition (CVD) [23].

Along with graphene, researchers have also worked on the synthesis of graphene oxide (GO) as well as reduced graphene oxide (rGO) in recent years. rGO nanoparticles was prepared by thermal reduction of GO which is again obtained from Hummer's method [25]. However, the required quality of graphene and graphene derivatives (rGO, GO) depends on its applications and based on that the methods of production are decided. Till date, CVD [26, 27] and modified Hummer's method [28–30] are most suitable for the synthesis of graphene and GO, respectively, in context of the formation of metal oxide/graphene, metal oxide/rGO, metal oxide/GO nanocomposite.

2.2 Synthesis metal oxide nanostructures and graphene derivatives composites

Synthesis of hybrid graphene with different nanoscale metal oxides are classified in four categories, that is, graphene/0-D metal oxides, graphene/1-D metal oxides, graphene/2-D metal oxides, and graphene/3-D metal oxides.

2.2.1 Synthesis of graphene/0-D metal oxides composites

Synthesis of metal oxide nanoparticles (NPs) and GO/rGO composites which was used for efficient gas sensing applications is described in this section. Among all the metal oxides, SnO_2 was reported mostly to synthesize nano composites with graphene and its derivatives (GO and rGO). At the same time, nanoparticles of metal oxides were preferred majorly to prepare the monohybrids with GO and rGO. Different chemical synthesis techniques were followed to develop the nanocomposites of metal oxide/rGO like hydrothermal, solvothermal, flame spray pyrolysis, etc. [31, 32].

Hydrothermal is one of the commonly reported techniques for preparing metal oxide nanoparticles-rGO composites. Among different metal oxides, SnO_2 nanoparticles were reported extensively to prepare nano-hybrid with rGO for efficient gas sensing application [33–42]. SnO_2/rGO [33–35] nano-hybrid was prepared by facile hydrothermal treatment where precursor was prepared with mixture of $SnCl_4$, HCl, H_2O, and GO (or rGO). Heating temperatures were reported as 120°C [33] and 180°C [34, 35] whereas the heating time was 12 h, consistent for all the reports. Different weight% (0.5–5 wt.%) of Au was added in the SnO_2/rGO nanocomposite by using $HAuCl_4$ salt to study the effect of Au concentration on the sensitivity of SnO_2/rGO gas sensors [34]. Scanning electron micrograph (SEM) of SnO_2/rGO films which was used for promising gas sensing application are represented in **Figure 1(a** and **b)**. Mishra et al. reported rGO/SnO_2 nanocomposite by surfactant-assisted hydrothermal method, in which hexamethyldisilazane (HDMS) was used as a surfactant [36]. Ghosh et al. [37] reported SnO_2 nanoparticle synthesis by hydrothermal method and SnO_2/rGO film synthesis by mixing of SnO_2 nanoparticles with GO. The GO-SnO_2 mixture was then ultrasonicated to obtain uniform dispersion. Then the sample was drop cast on the platinum electrode and heated at 160°C to reduce GO and get SnO_2/rGO hybrid sensing layer [37]. The hydrothermal method was also used for the synthesis of SnO_2/rGO hybrid with

Figure 1.
SEM image of hydrothermally grown SnO_2 nanoparticles and rGO composites reported by (a) Zhang et al. [33] and (b) Peng et al. [41].

a high concentration of oxygen vacancy [40, 42] and Pt-activated SnO_2 nanoparticles-rGO hybrid [41]. Transmission electron micrograph (TEMs) of SnO_2 quantum dot decorated on rGO surface is represented in **Figure 2(a–c)**.

NiO/rGO nanohybrid [39] was prepared via two-step hydrothermal treatment. NiO nanoparticles powder was prepared by hydrothermal method using $NiCl_4{\cdot}6H_2O$ as the source of Ni and then calcined at 400°C. NiO nanoparticle powder was then mixed with rGO solution and treated by hydrothermal method with a various ratio of NiO/rGO as 2:1, 4:1, and 8:1 (**Figure 3**).

Undoped and Ni-doped SnO_2 nanoparticle and graphene composites were developed by flame spray pyrolysis (FSP) method as reported in references [32, 43], respectively. About 0.1–2 wt.% Ni-doped SnO_2 nanoparticles were synthesized by FSP technique and graphene was produced from graphite by the electrolytic exfoliation technique. Then, a paste was prepared by mixing Ni-doped SnO_2 and graphene powder and finally spin coating method was used to deposit a film for gas sensing application. Bright field (BF) TEM images of 0.5 wt.% SnO_2 NPs loaded graphene composites and 2 wt.% Ni doped SnO2 NPs loaded graphene composites are represented in **Figure 4(a)** and **(b)**, respectively.

ZnO/rGO composite was prepared by the solvothermal method for low-temperature acetylene sensing as reported by Iftekhar Uddin et al. [44, 45]. ZnO powder was prepared through the solvothermal method by using $Zn(NO_3)_2$ and NaOH in ethanol at 120°C. Ag-loaded ZnO/GO hybrid was synthesized by chemical route. $AgNO_3$ was added to the ZnO/GO solution with 2:1 ratio, then stirred continuously for 30 min. Hydrazine hydrate was then added to the mixer to reduce GO

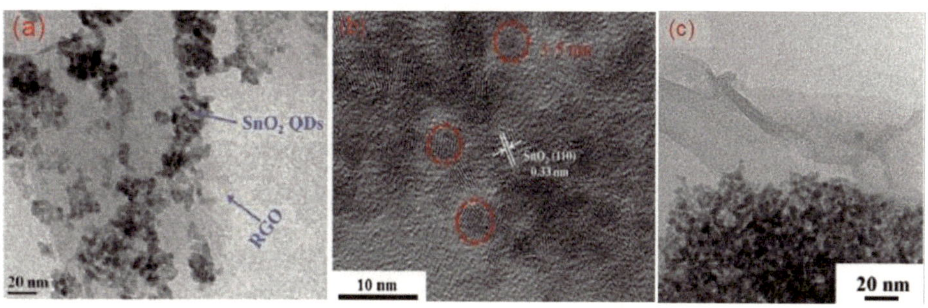

Figure 2.
TEM images of hydrothermally grown SnO_2 nanoparticles and rGO composites (a) SnO_2 quantum dot on rGO film surface [36], (b) high resolution (HR) TEM image of SnO_2 NPs on rGO [40], and (c) dense SnO_2 NPs on rGO [42].

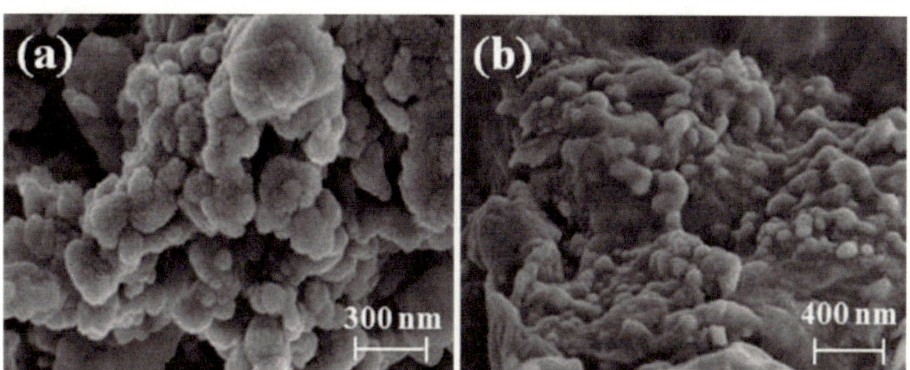

Figure 3.
SEM of hydrothermally grown (a) NiO NPs and (b) NiO/rGO nanocomposites with 2:1 ratio [39].

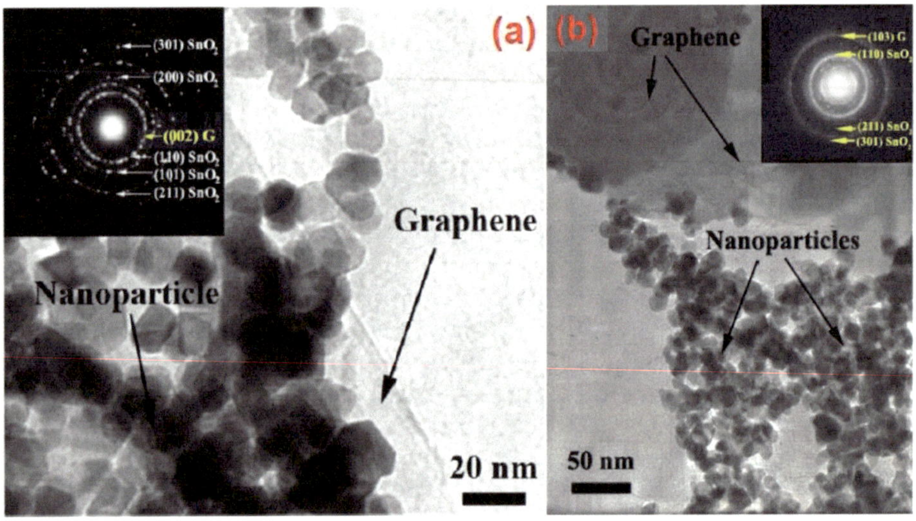

Figure 4.
BF TEM images of 0.5 wt.% SnO_2 NPs loaded graphene composites and 2 wt.% Ni-doped SnO_2 NPs loaded graphene composites. Inset: Corresponding selected area electron diffraction (SAED) pattern [32, 43].

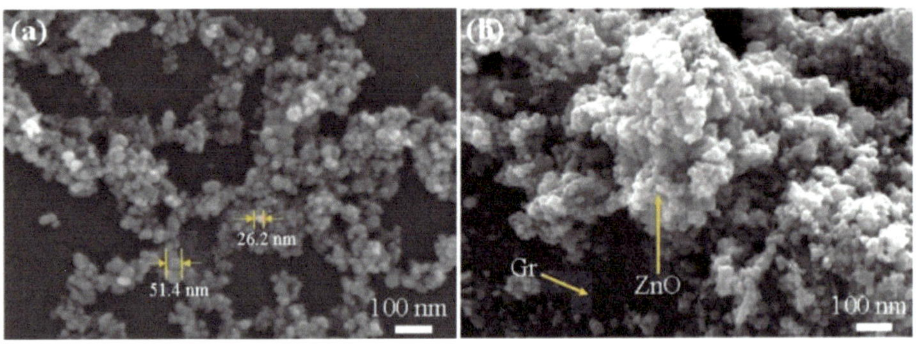

Figure 5.
Plane-view FESEM micrographs of (a) pure ZnO nanoparticles and (b) ZnO nanoparticle rGO hybrids [44].

at 110°C for 8 h [44]. Morphology of ZnO NPs and rGO nanocomposite is shown in **Figure 5(a** and **b)**.

ZnO quantum dots (QDs) decorated on graphene nanosheets were synthesized by facile solution-processed method (**Figure 6(a)**). ZnO QDs were nucleated and grown on the surface of graphene by controlling the distribution density by reaction time and precursor concentration [46]. ZnO-rGO hybrid was prepared by wet chemical method followed by deposition of Au using HAuCl4, which was added to the ZnO-rGO dispersion. Finally, the addition of $NaBH_4$ through sonication process completed the formation of ZnO QD [47]. To understand the impact of particle size on gas sensing performance, Tung et al. [48] prepared rGO-Fe_3O_4 nanoparticle hybrid with different particle sizes (5, 10, and 20 nm) via in situ chemical reduction of GO in presence of poly-ionic liquid (PIL) (**Figure 6(b)**). Kamal [49] prepared graphene-NiO nanoparticles composites by decomposition of nickel benzoate dihydrazinate complex used for hydrogen sensing application.

Graphene oxide was synthesized from natural graphite flakes by Hummers' method which was further used to prepare rGO-$CuFe_2O_4$ nanocomposite by combustion method [50]. In this process, sonicated GO was dissolved with 1:2 ratio of

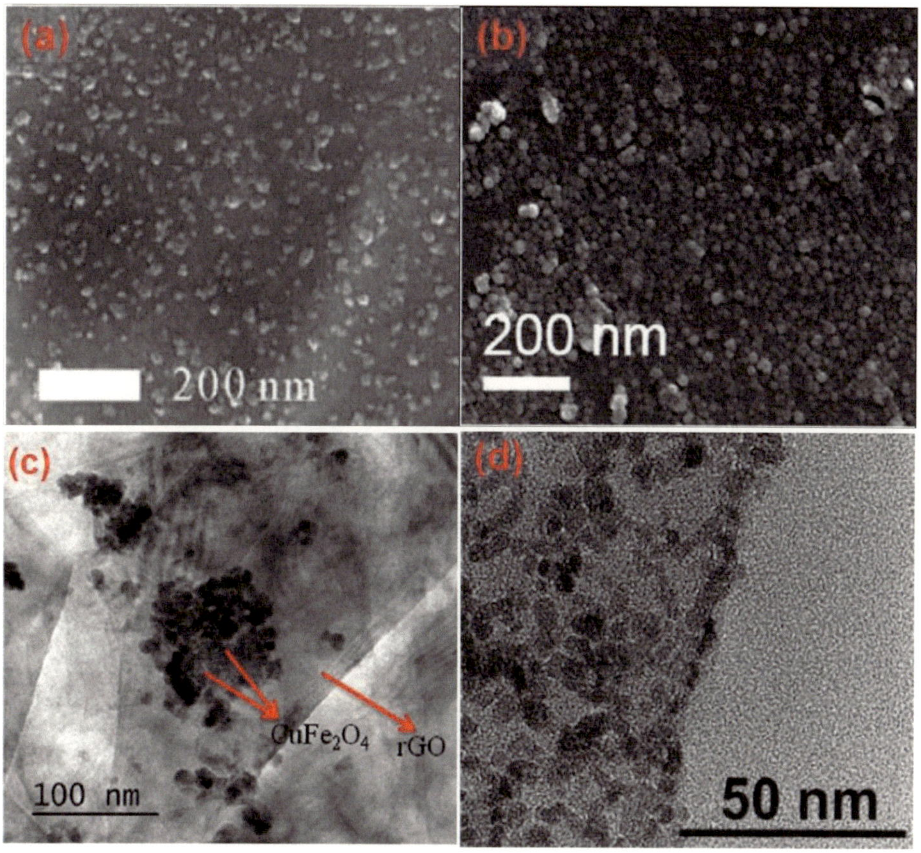

Figure 6.
SEM images of (a) ZnO QDs/graphene nanocomposites [46], (b) rGO-Fe_3O_4 nanoparticles [48] synthesized by facile solution-processed method. TEM images of (c) rGO-$CuFe_2O_4$ by combustion method [50] and (d) microwave assisted rGO-SnO_2 nanparticles composites [51].

Cu^{2+} to Fe^{3+} salts and distilled water. The resulting mixture was stirred at 100°C to get a viscous solution which was further heated around 450°C in a muffle furnace. Finally, the mixture was frothed and it gave a foamy powder of nanocomposite (**Figure 6(c)**). The one-pot microwave-assisted non-aqueous sol-gel method was used to synthesize pure SnO_2 nanoparticles and SnO_2/rGO nanocomposite (**Figure 6(d)**) [51]. Kim et al. [52] reported the microwave-assisted the formation of SnO_2/graphene nanocomposite in which mixture of SnO_2 nanopowder and graphene flakes dispersed in ethanol, the resulting solution was dried. The dried powder mixture was treated in the commercial microwave heater for heating process. Microwave-treated powder was again dispersed in ethanol and then the solution was spray-coated on SiO_2 substrate placed on a hot plate. Along with SnO_2/graphene nanocomposite, a small amount of secondary SnO_x ($x < 2$) nanoparticles were also deposited on the surface. Secondary SnO_x nanoparticles tend to increase as the microwave heating time is increased [52].

2.2.2 Synthesis of graphene/1-D metal oxides composites

One dimensional (1-D) nanostructures of the metal oxide like nanotubes, nanorods, nanofibers, nanowires, etc. are considered as most promising for the detection of analytes in gaseous phases [53, 54]. Owing to its large surface-to-volume ratio, large open porosity and most importantly one of its dimension is comparable to the Debye length which enhances the gas sensitivity significantly. Gas sensing performance is improved further in 1-D materials by using as a nanocomposite with graphene, GO and rGO. Here, we reviewed different methods used for the synthesis of 1-D metal oxide nanostructure and its composite with graphene and graphene derivatives.

ZnO nanowires (NW) and graphene hybrid architecture were reported by Yi et al. [55] where graphene sheets covered with thin metal layers were used as top electrodes for ZnO where graphene sheets coated with thin metal layers were employed as top electrodes for ZnO vertical-NW channels. The ZnO NWs-graphene/metal hybrid architectures maintained sufficient spaces between the NWs for easy and fast gas transport. However, ZnO nanorods (NRs) were synthesized by using hydrothermal reaction and graphene sheets were synthesized by the CVD method and transferred to the top of the ZnO NRs by PMMA treatment. The scanning electron micrograph of ZnO NRs-graphene/metal hybrid architectures are represented in **Figure 7(a)**. Single crystalline WO_3 nanorods on the surface of graphene were synthesized through a one-step hydrothermal method [56]. WO_3 nanorods with 3.5 wt.% graphene composites improved gas sensitivity of 25 times showing good selectivity towards NO_2. SEM image of WO_3 nanorods and graphene hybrids is shown in **Figure 7(b)**.

Large-scale sandwich-like heterostructures of ZnO nanorod arrays with reduced graphene oxide sheets were reported by Zou et al. [30] as shown in **Figure 7(c)**. Highly dense ZnO nanorods were grown by hydrothermal method and double sides coverage of reduced graphene sheets by ZnO NRs formed a sandwich like heterostructures of ZnO/graphene/ZnO for efficient ethanol detection.

Electrospinning is a potential and well-reported technique for the synthesis of nanofibers (NFs) network of metal oxides. N,N-dimethylformamide (DMF) and polyvinyl pyrrolidone (PVP) are mixed with target metal oxide precursor and the whole mixture is poured into a syringe having a suitable needle attached. A high voltage (a few kV, DC) is applied between the needle and the collector plate to get the NFs of the target metal oxides. In electrospinning method, composites of metal oxide NFs and graphene derivatives are synthesized by two different routes, that is, (i) GO or rGO solutions are added into the base mixture before electrospinning and (ii) synthesized NFs are decorated with GO or rGO solutions. rGO/Co_3O_4 NFs [57] and rGO/ZnONFs composites [54] were synthesized where rGO was added into the precursor before electrospinning and both the nanohybrids were tested towards different gases and vapors like NH_3, ethanol, etc. rGO/Co_3O_4 NFs [58], rGO/SnO_2 NFs [53], and rGO/WO_3 NFs composites [59] were synthesized where Co_3O_4, SnO_2, and WO_3 nanofibers were synthesized by electrospinning method first and then functionalized with rGO solution. All three composites were used to detect acetone in the selective route. **Figure 8(a)** and **(b)** represents the TEM and SEM images of

Figure 7.
SEM images of 1-D metal oxides and graphene nanocomposites synthesized by hydrothermal route (a) ZnO NRs-Gr/M hybrid architectures [55], (b) WO_3 nanorods/graphene composites [56], and (c) ZnO/G array structures [30].

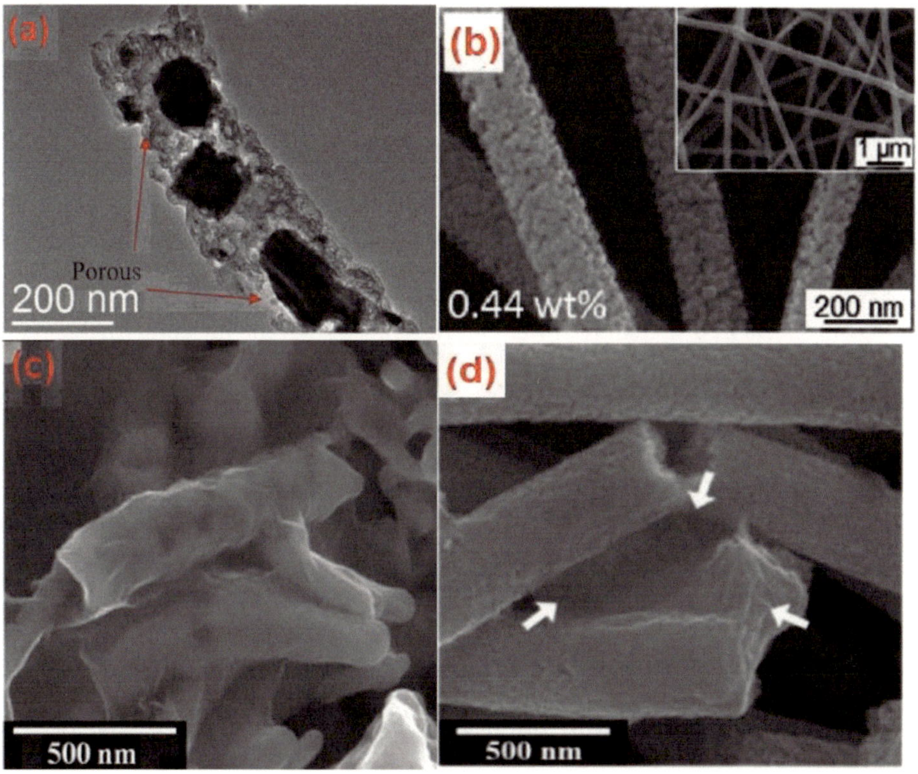

Figure 8.
(a) TEM images of rGO/Co_3O_4 NFs [57], (b) SEM image of rGO/ZnO NFs composite [54], SEM image of (c) rGO/Co_3O_4 NFs composites [58], and (d) rGO/SnO_2 NFs composites [53].

rGO/Co_3O_4 NFs [57] and rGO/ZnONFs composite [54], respectively. Here, rGO film was almost invisible due to the mixing of rGO in solution before electrospinning. rGO film was clearly visible in rGO/Co_3O_4 NFs [58] and rGO/SnO_2 NFs [53] composites in SEM images shown in **Figure 8(c)** and **(d)** as the rGO functionalization was carried out after electrospinning.

Synthesis of TiO_2 nanotubes array was done by electrochemical anodization of metallic titanium films [60] and rGO/TiO_2 nanotubes composite was synthesized by electrodeposition of rGO on TiO_2 nanotubes [61, 62]. Electrodeposition method was also used to synthesize ZnO nanorods and selective electrochemical etching of those nanorods to synthesized ZnO nanotubes. rGO/ZnO nanotubes hybrid structure was synthesized by dip-coating technique for efficient alcohol sensing application [63].

One step colloidal synthesis was employed for rGO/SnO_2 quantum wire nanocomposite for room temperature H_2S sensing [25]. Single crystal SnO_2 nanowire was directly grown on the platinum electrode by thermal evaporation and composites was formed by using CVD grown graphene layer for efficient NO_2 sensing [26]. Hydrolysis method was used in absence as well as in presence of GO to form ZnO nano-seed and GO supported ZnO nano-seed, respectively, and ultrathin ZnO nanorod/rGO mesoporous nanocomposites were synthesized for NO_2 sensing [29].

2.2.3 Synthesis of graphene/2-D metal oxides composites

Metal oxides nano-sheets and nameplates were functionalized with graphene and its derivatives for efficient gas sensing behavior. Ni-doped ZnO nanosheets were deposited on a p-Si substrate by using radio frequency (RF) sputtering techniques. GO was synthesized by Hummer's method and reduced thermally at high temperature (600°C). rGO flakes were then decorated on Ni-doped ZnO nanosheets by drop-casting method. rGO/Ni-doped ZnO nanosheets were used for low ppm hydrogen detection [64]. Highly wrinkled SnO_2/rGO composite was synthesized by one-time hydrothermal technique and used for the detection of ethanol at 250°C (**Figure 9(a)**) [65]. Nanocomposites of ZnO nanosheets and GO were synthesized for highly efficient acetone sensing. The nanocomposites sensor was flexible, high effective surface area and enhanced functional groups due to GO which were in favor of gas adsorption (**Figure 9(b)**) [66]. rGO/hexagonal WO_3nanosheets hybrid materials were fabricated through the hydrothermal method and post-annealing treatment. 2-D porous WO_3nanosheets were attached on rGO. The sensor based on 3.8 wt.% rGO/hexagonal-WO_3composites offered promising sensing performance to H_2S [67].

2.2.4 Synthesis of graphene/3-D metal oxides composites

3-D metal oxides like nanoflower and nanosphere were used to synthesize nanocomposites with graphene and its derivatives by the hydrothermal and sol-gel method [15, 68–70].

Hybrids with flower-like hierarchical ZnO and rGO were synthesized by the facile and mild solution-processed method. Compared with the pristine flower-like ZnO, NO_2 sensing was increased significantly in case of hierarchical rGO/ZnO hybrids [15]. A facile one-pot hydrothermal method was used to synthesize rGO/In_2O_3 composites. The flower-like hierarchical structure of In_2O_3 showed high effective surface area enhancing the active interaction sites. In the composite, rGO formed local p-n heterojunctions enhancing the gas sensing performance significantly. The rGO/In_2O_3 composite exhibited an excellent selectivity towards NO_2 in the wide concentration range from 10 ppb to 1 ppm [68]. α-Fe_2O_3/rGO nanocomposites with nanosphere-like α-Fe_2O_3 structure were synthesized by a hydrothermal route at 120°C. α-Fe_2O_3 nanosphere was 40–50 nm in diameter and constructed by a few nanometer-sized nanoparticles where rGO was intercalated single sheets. These nanocomposites showed excellent response and selectivity towards NO_2 at room temperature [69]. Graphene-WO_3 nanostructure with

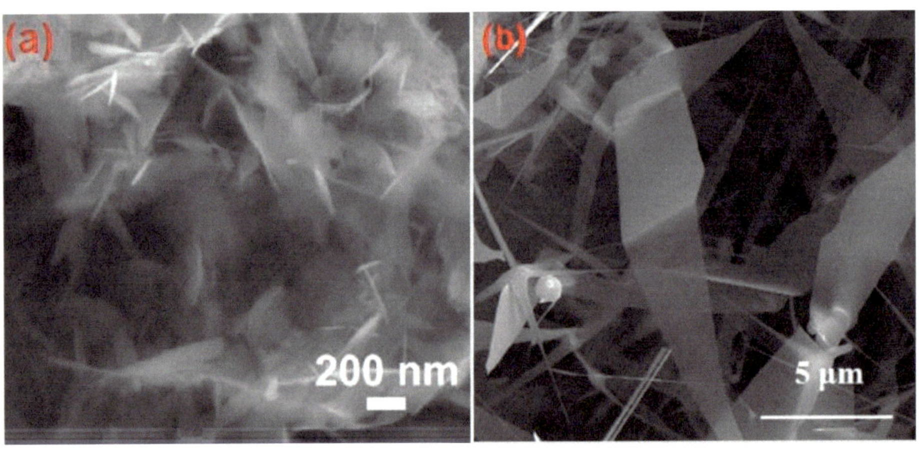

Figure 9.
SEM images of rGO/2-D metal oxide nanocomposites (a) wrinkled SnO_2/rGO composite [65] and (b) GO/ZnO nano-sheets [66].

Figure 10.
SEM images of (a) rGO/flower-like hierarchical ZnO composites [15], (b) rGO/In_2O_3 nanoflower composite [68], (c) rGO/α-Fe_2O_3 nanosphere [69], and (d) graphene-wrapped WO_3 nanosphere [70].

gauze-like graphene nanosheets wrapping up spherical WO_3 nanoparticles was synthesized by a facile sol-gel method. Graphene-wrapped WO_3 nanocomposites offered uniform nanospheres with 200–400 nm diameter. Graphene/WO_3 nanocomposites showed good sensitivity and selectivity to low concentrations of NO_2 gas at room temperature when pure WO_3 and graphene-based sensors did not show any response towards NO_2 at room temperature [70]. Scanning electron micrograph of 3-D metal oxides and graphene nanocomposites is shown in **Figure 10(a–d)**.

3. Assessments of gas/vapor sensing performances

3.1 Sensing performance of graphene/0-D metal oxides composites

Gas/vapor sensing performance of metal oxide nanoparticles (0-D) functionalized with graphene or graphene derivatives (GO, rGO) is represented in **Table 1** where 37 references are considered for performance assessment of graphene/0-D metal oxide oxides composites sensors.

Graphene/metal oxide NPs (0-D) composites and its gas sensing performance were explored extensively in case of SnO_2 where hydrothermal method was commonly used as the synthesis technique. The gas sensing performance of rGO/SnO_2 NPs was further improved by functionalization with Pd, Au, Pt, and Ag nanoparticles [31, 34, 41, 78]. ZnO nanoparticles occupied the second position to use as a nanocomposite with graphene and its derivatives other metal oxide

Composite material	Target gas/ vapor	Operating temperature (range) (°C)	Concentration and range (ppm)	Response magnitude	Response/ recovery time (s)	Ref.
Graphene-SnO_2 nanoparticles	NO_2	150 (30–190)	5 (1–5)	R_{gas}/R_{air} = 72.6	129/107	[52]
rGO/SnO_2 nanoparticles	Acetone	Room temperature	2000 (10–2000)	$R_{air} - R_{gas}/R_{air}$ = 9.72%	95/141	[33]
Graphene aerogel-SnO_2 nanoparticle	NO_2	Room temperature	200 (10–200)	$R_{gas} - R_{air}/R_{air}$ = −12%	190/224	[71]
Graphene-SnO_2 nanoparticles	H_2	50	100 (1–100)	I_{gas}/I_{air} = 6	1.1/1.1	[72]
rGO-SnO_2 quantum dots	H_2	200	500	$R_{air} - R_{gas}/R_{air}$ = 89.3%	~50/~155	[36]
	LPG	250	500	$R_{air} - R_{gas}/R_{air}$ = 92.4%	~80/~155	
rGO/SnO_2 nanoparticles	NH_3	200 (100–200)	1000 (25–2800)	—	—	[37]
rGO/SnO_2 nanoparticles	C_2H_2	180 (100–300)	50 (0.5–500)	R_{air}/R_{gas} = 12.4	54/23	[38]
rGO-SnO_2 nanoparticles	NO_2	30 (30–100)	1 (0.05–2)	R_{air}/R_{gas} = 3.8	14/190	[40]
rGO-SnO_2 nanoparticles	NO_2	Room temperature	5 (1–20)	$I_{air} - I_{gas}/I_{air}$ = 65.5%	12/17	[42]
Graphene/SnO_2 nanoparticles	NO_2	150 (25–350)	5	R_{gas}/R_{air} = 26342	13/—	[43]
rGO/SnO_2 nanoparticles	NO_2	50 (30–60)	5 (0.5–500)	R_{air}/R_{gas} = 3.31 (25% RH)	135/200	[35]
rGO/SnO_2 nanoparticles	SO_2	60 (22–220)	500 (10–500)	$R_{air} - R_{gas}/R_{gas}$ = ~22	144/210	[73]
Graphene/SnO_2 nanoparticles	NO_2	Room temperature	100 (0.3–100)	$G_g - G_o/G_o$ = ~11	—	[74]
Graphene/SnO_2 nanoparticles	Ethanol	350 (150–350)	1000 (50–1000)	965	1.8/~120	[76]
Graphene/SnO_2 nanoparticles	NO_2	60 (25–120)	4 (1–4)	$R_g - R_a/R_a$ = ~22	—	[75]
rGO/SnO_2 nanoparticles	CO	Room temperature	1600 (50–1600)	$R_g - R_a/R_a$ = 9.5%	~60/~60	[76]
Sulfonated graphene/SnO_2 nanoparticles	NO_2	Room temperature	5 (1–50)	$R_{air} - R_{gas}/R_{gas}$ = 1.203	40/357	[77]
Graphene-SnO_2 nanoparticle with doped Ni	Acetone	350 (150–350)	200 (1–50)	R_{air}/R_{gas} = 169.7	5.4/150	[32]
Graphene-Pd/SnO_2 nanoparticles	H_2 Ethanol	Room temperature	20,000 200 (25–200)	$R_o - R_{gas}/R_o$ = 11%(H_2), 14.8% (ethanol)	34/27	[31]
rGO/SnO_2 nanoparticles decorated Au NPs	NO_2	50 (30–60)	50 (5–100)	R_{gas}/R_{air} = 2.68	19/20	[34]

Composite material	Target gas/ vapor	Operating temperature (range) (°C)	Concentration and range (ppm)	Response magnitude	Response/ recovery time (s)	Ref.
rGO-SnO_2 nanoparticles (activated by Pt)	Methanol	110 (20–180)	500 (10–500)	R_{air}/R_{gas} = 203	6/21	[41]
rGO/SnO_2 nanoparticles (with Ag NPs)	NO_2	Room temperature	5 (0.5–500)	R_{air}/R_{gas} = 2.17 (25% RH)	49/339	[78]
Graphene-ZnO quantum dots	HCHO	Room temperature	100 (25–100)	$G_g - G_o/G_o$ = 1.1	30/40	[46]
Graphene/ZnO nanoparticles	C_2H_2	250 (25–350)	100 (30–1000)	R_{air}/R_{gas} = 143	100/24	[45]
rGO/ZnO nanoparticles	NO_2	50 (25–140)	50 (5–275)	$R_g - R_a/R_a$ = 32%	96/1552	[74]
rGO/ZnO nanoparticles	NO_2	Room temperature	5 (1–25)	$R_g - R_a/R_a$ = 25.6%	165/499	[80]
3-D rGO/ZnO nanoparticle	CO	200	1000 (1–1000)	$R_g - R_a/R_a$ = 85.2%	7/9	[81]
rGO-ZnO/Ag nanoparticles	C_2H_2	150 (25–250)	100 (1–1000)	R_{air}/R_{gas} =21.2	25/80	[44]
rGO/ZnO-Au nanoparticles	NO_2	80 (60–90)	100 (20–100)	$R_{air} - R_{gas}/R_{air}$ = 32.55	27/86	[47]
Graphene-NiO nanoparticles	H_2	250 (100–350)	2000 (400–2000)	$R_g - R_a/R_a$ = 52.4%	NA	[49]
rGO/NiO nanoparticles	CH_4	260 (20–400)	1000 (100–1000)	$R_{air} - R_{gas}/R_{gas}$ = 15.2	16/20	[39]
rGO/NiO NP with SnO_2 nanoplates	NO_2	Room temperature	60 (5–60)	$G_g - G_o/G_o$ = 62.28	220/835	[82]
rGO/CuO nanohybrid	NO_2	135 (25–225)	75 (1–75)	$I_{gas} - I_{air}/I_{air}$ = 51.7	50/105	[83]
rGO/$CuFe_2O_4$ nanoparticles	NH_3	110	200 (5–200)	$R_g - R_a/R_a$ = 24.41	3/6	[50]
rGO/Fe_3O_4 nanoparticles	Ethanol	Room temperature	1	$R_{air} - R_{gas}/R_{gas}$ = 1.86	—	[48]
	NO_2	200 (250–450)	2.5 (1–5)	$R_{air} - R_{gas}/R_{gas}$ = 4.68	—	
Graphene/WO_3 nanoparticle	NO_2	250 (200–300)	5 (1–20)	R_{gas}/R_{air} = 133	~25/—	[84]
Graphene-CeO_2 nanoparticles	NO_2	—	(10–200 ppm)	—	—	[85]

Table 1.
Summary of the performance of sensors fabricated by using the nanocomposites of metal oxide nanoparticles (0-D) and graphene or graphene derivatives (GO, rGO).

nanoparticles like NiO, CuO, WO_3, Fe_3O_4, CeO_2, etc. were reported as promising gas/vapor sensing composite materials with graphene and its derivatives. However, among all the target gases and vapors, NO_2 was the mostly explored gas and detected successfully by graphene/metal oxide nanocomposite sensors [79]. Other gases like H_2, NH_3, CO, C_2H_2, CH_4, SO_2, and organic vapors like acetone, ethanol, methanol, and formaldehyde were detected successfully by using graphene/NP metal oxide hybrids. Sensors were tested at different temperature range varying from room temperature ~25°C to 400°C. The average operating temperature of graphene/NP metal oxide nanocomposites was recorded below 150°C. Among the 37 research articles, room temperature sensing was reported by 13 groups of researchers mentioned in the literature. The detailed transient behavior of graphene/metal oxide NPs sensors is shown in **Figure 11(a–f)** where all the sensing was reported at room temperature. So, overall study confirms the lowering of gas sensing the temperature of metal oxide nanoparticles due to the formation of composites with graphene, GO, and rGO. Most of the article showed a lower detection limit of gases and vapors (<100 ppm). Depending on the surface morphology, sensing temperature and device structures the detection range varied from ppb to ppm level as shown in **Table 1**. Finally, the response magnitude and response/recovery time were fully dependent on the operating temperature and concentration range of the analyte. However, all the sensors showed adequate response magnitude towards the target gas/vapors. Response/recovery time increased and decreased with decrease and increase of operating temperature, respectively. Response time and recovery time varied from 1 s to 220 s and 1 s to 1552 s, respectively, as shown in **Table 1**. Wang et al. confirmed that the uniform distribution of SnO_2 nanoparticles on rGO sheets is an effective factor for enhanced NO_2 sensing performances [42]. The p-n junction existed in the interface of nanoparticle and rGO contributed to good room temperature NO_2 sensing properties which is associated with the valid electron flow from SnO_2 nanoparticle to rGO.

3.2 Sensing performance of graphene/1-D metal oxides composites

One dimensional (1-D) metal oxide like nanorods, nanotubes, nanowires, nanofibers, quantum nanowires, etc. functionalized by graphene, GO, and rGO

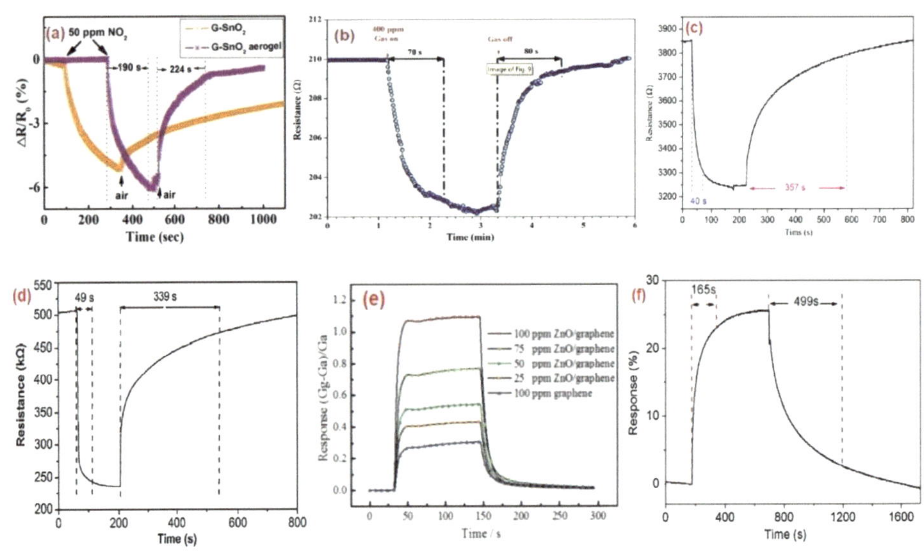

Figure 11.
Room temperature sensing of graphene (GO, rGO)/metal oxide nanoparticles composites. (a) NO_2 sensing by graphene aerogel/SnO_2 nanoparticle [71], (b) CO sensing by rGO/SnO_2 nanoparticles [76], (c) NO_2 sensing by sulfonated graphene/SnO_2 nanoparticles [77], (d) NO_2 sensing by rGO/SnO_2 nanoparticles with Ag NPs [78], (e) formaldehyde sensing by graphene-ZnO quantum dots [46], and (f) NO_2 sensing by rGO/ZnO nanoparticles [80].

Composite material	Target gas/vapor	Operating temperature (range) (°C)	Concentration and range (ppm)	Response magnitude	Response/recovery time (s)	Ref.
rGO/ZnO nanorods	NO_2	Room temperature	1 (0.120–1)	$R_{gas} - R_{air}/R_{air}$ = 50%	120/320	[86]
Functionalized graphene/ZnO nanorods	Ethanol	340 (200–370)	100	$R_{air} - R_{gas}/R_{air}$ = 93.5%	5/20	[28]
rGO/ZnO nanofibers	NO_2	400 (300–400)	5 (1–5)	R_{gas}/R_{air} = 119	143/259	[54]
	CO	400 (300–400)	5 (1–5)	R_{air}/R_{gas} = 22.6	—	
Non-oxidized graphene/ZnO nanofibers	Acetone	350 (250–450)	5 (1–5)	R_{air}/R_{gas} = 18.5	12.8/—	[59]
rGO/ZnO nanotubes	Ethanol	125 (27–150)	100 (1–800)	79.14%	41.1/98.32	[63]
rGO/ZnO nanorods	Ethanol	260	50 (5–50)	R_{air}/R_{gas} = ~27	<10/<10	[30]
rGO/ZnO nanorods	NO_2	Room temperature	1 (1–10)	$R_{air} - R_{gas}/R_{gas}$ = 1.19	75/132	[29]
Graphene/ZnO nanorod doped by Au/Ti	Ethanol	300	50 (10–50)	R_{air}/R_{gas} = ~90	—	[55]
Au/Pd functionalized rGO/ZnO nanofiber	CO	400 (300–450)	5 (1–5)	R_{air}/R_{gas} = 35.8	191.3/82.1	[87]
	C_6H_6	400 (300–450)	5 (1–5)	R_{air}/R_{gas} = 22.8	110.3/318.2	
rGO/SnO_2 nanofibers	H_2S	200	5 (1–5)	R_{air}/R_{gas} = 33.7	<198/<114	[53]
	Acetone	350 (150–400)	5 (1–5)	R_{air}/R_{gas} = 10.4	<198/<114	
Graphene/SnO_2 nanowires	NO_2	150 (100–250)	0.1 (0.01–0.1)	R_{air}/R_{gas} = 11	43/37	[26]
Graphene/SnO_2 nanorods	H_2S	260	50 (1–50)	R_{air}/R_{gas} = 130	5/10	[27]
rGO/SnO_2 quantum wires	H_2S	Room temperature	50 (10–100)	R_{air}/R_{gas} = 33	2/292	[25]
Nanoporous graphene hybrid-SnO_2 nanorods	CH_4	150 (100–200)	1000	$\|R_{gas} - R_{air}\|/R_{air}$ = 24.9 %	369/519.8217.9/242	[88]
rGO/TiO_2 nanotubes	H_2	200 (100–300)	480 (120–480)	$\|G_{gas} - G_o\|/G_o$ = 37.6	1110/<300	[60]
rGO/TiO_2 nanotubes	Methanol	Room temperature	800 (10–800)	$R_{air} - R_{gas}/R_{air}$ = 96.93%	18/61	[61]
GO/Co_3O_4 nanofibers	Acetone	300 (200–350)	5 (1–5)	R_{gas}/R_{air} = 2.29	—	[58]
rGO/Co_3O_4 nanowires	NH_3	Room temperature	50 (5–100)	$\Delta R/R_{air}$ = 53.6%	4/300	[57]

Composite material	Target gas/ vapor	Operating temperature (range) (°C)	Concentration and range (ppm)	Response magnitude	Response/ recovery time (s)	Ref.
rGO/Cu_2O nanowires	NO_2	Room temperature	2 (0.4–2)	$I_{gas} - I_{air}/I_{air}$ = 67.8%	—	[89]
Graphene/WO_3 nanorods	NO_2	300	20 (0.025–20)	R_{gas}/R_{air} = 202	—	[56]
rGO/αFe_2O_3 nanofibers	Acetone	375 (250–400)	100 (0.05–2)	R_{air}/R_{gas} = 8.9	3/9	[18]

Table 2.
Summary of the performance of the sensor fabricated by using the nanocomposites of 1-D metal oxides and graphene or graphene derivatives (GO, rGO).

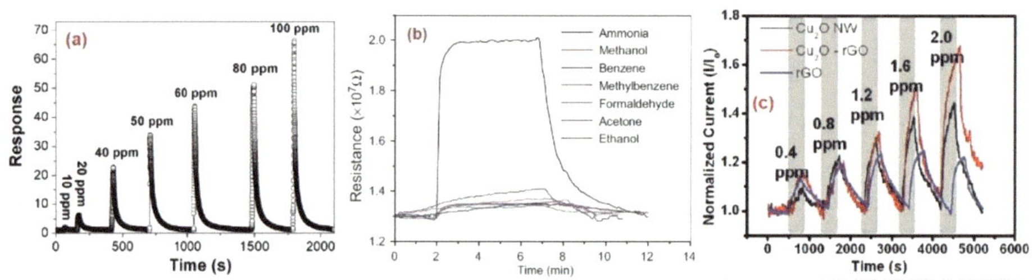

Figure 12.
Room temperature transient response of graphene/1-D metal oxide nanocomposites sensors. (a) H_2S detection by for rGO/SnO_2 quantum wires [25], (b) selective NH_3 sensing by rGO/Co_3O_4 nanowires [57], and (c) NO_2 sensing was observed for rGO/Cu_2O nanowires [89].

were reported as the promising gas sensing materials. A total of 21 reports on graphene/1-D metal oxide nanocomposites have been summarized in **Table 2** and compared in terms of target gas/vapors and its concentration, operating temperature, response magnitude, and response/recovery time. In the case of 1-D nanomaterials, ZnO was most explored metal oxide used to synthesize nanocomposites with rGO in the form of nanorods, nanotubes, and nanofibers. Other metal oxides are SnO_2 nanofibers/nanorods, TiO_2 nanotubes, Co_3O_4 nanofibers, WO_3nanorods, αFe_2O_3 nanofibers, etc. reported for nanocomposites with graphene for gas sensing application. NO_2 and ethanol were the mostly explored target gas and vapor in case of graphene/1-D metal oxide nanocomposites. Other selective gases and vapors are CO, H_2S, CH_4, H_2, NH_3, acetone, benzene, methanol, etc. Though the operating temperature range was in-between room temperature ~25°C and 450°C, average sensing temperature was slightly high (~200° C) in case of 1-D compared with 0-D metal oxide nanocomposites with graphene. However, the detection range of gases/vapors was quite small in case of graphene/1-D metal oxide nanocomposites where 1 ppm and below 1 ppm detection were reported frequently. Quite a high response magnitude and moderate response/recovery time were recorded in case of graphene/1-D metal oxide nanocomposites.

Figure 12(a) represents very efficient H_2S sensing for rGO/SnO_2 quantum wires sensor for the concentration range of 10–100 ppm at room temperature. Being a room temperature sensing, the sensor showed a very fast response of 2 s only [25]. Highly selective NH_3 sensing was reported for rGO/Co_3O_4 nanowires at room temperature as shown in **Figure 12(b)** where response time was only 4 s [57]. Improvised NO_2 sensing was observed for rGO/Cu_2O nanowires compared with the pure Cu_2O nanowires and pure rGO in the concentration range of 0.4–2 ppm at room temperature (**Figure 12(c)**) [89]. However, the overall study envisages the potential gas sensing of 1-D metal oxides functionalized by graphene, GO, and rGO.

The 1-dimensional structure of TiO_2 nanotubes (NTs) in its hybrid with rGO provided large amount of gas interaction sites which lead to high response magnitude of the sensor [61].

3.3 Sensing performance of graphene/2-D metal oxides composites

Table 3 shows gas sensing performance of 2-D metal oxides and GO (or rGO) nanocomposites where ZnO, SnO_2, and WO_3 nanosheets were used as 2-D materials. Relatively high operating temperature was reported for graphene/2-D metal oxide nano composites. Average sensing temperature was more than 200°C. Moderate response magnitude, response/recovery time were recorded for graphene/2-D metal oxide nano composites. Transient behavior of GO/ZnO nanosheets in the exposure of 100 ppm acetone at 240°C and rGO/hexagonal WO_3nanosheets in the exposure of 40 ppm H_2S at 350°C are shown in **Figure 13(a)** and **(b)** [66, 67].

3.4 Sensing performance of graphene/2-D metal oxides composites

The gas sensing performance of nanocomposites developed by 3-D metal oxide and graphene derivatives are shown in **Table 4**. Nanoflowers and nanosphere structures were reported here. Interestingly, all the nanocomposites showed their

Composite material	Target gas/ vapor	Operating temperature (range) (°C)	Concentration and range (ppm)	Response magnitude	Response/ recovery time (s)	Ref.
rGO/Ni-doped ZnO nanoplates	H_2	150	100 (1–100)	$\|R_{air} - R_{gas}\|/R_{air}$ = 63.8%	28/—	[64]
GO/ZnO nanosheets	Acetone	240	100 (50–500)	R_{air}/R_{gas} = 35.8%	13/7	[66]
rGO nanosheets/ wrinkled SnO_2	Ethanol	250 (150–300)	100 (5–5000)	R_{air}/R_{gas} = 80%	9/457	[65]
rGO/hexagonal WO_3 nanosheets	H_2S	350 (50–400)	40 (0.01–40)	$\|R_{gas} - R_{air}\|/R_{air}$ = 168.58%	7/55	[67]

Table 3.
Summary of the performance of a sensor fabricated by using the nanocomposites of 2-D metal oxides and graphene or graphene derivatives (GO, rGO).

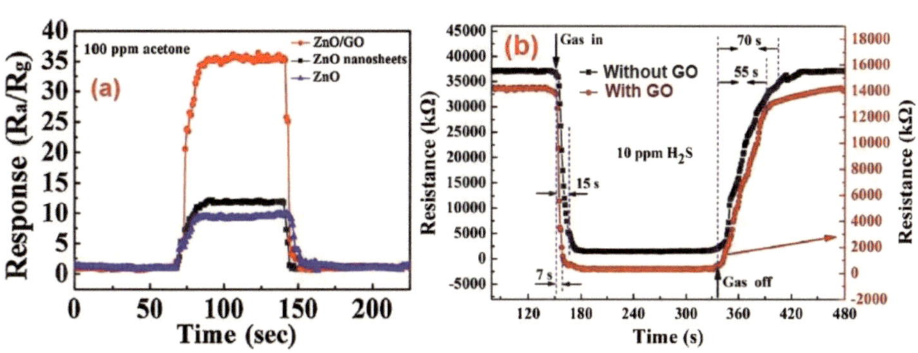

Figure 13.
Response behavior of (a) GO/ZnOnanosheets in the exposure of 100 ppm acetone at 240°C [66] and (b) rGO/hexagonal WO_3nanosheets in the exposure of 40 ppm H_2S at 350°C [67].

Composite material	Target gas/ vapor	Operating temperature (range) (°C)	Concentration and range (ppm)	Response magnitude	Response/ recovery time (s)	Ref.
rGO/ZnO nanoflower	NO_2	100 (50–150)	0.5 (0.005–0.5)	R_{gas}/R_{air} = 12	258/288	[15]
rGO/In_2O_3 nanoflower	NO_2	74 (25–110)	1 (0.01–1)	R_{gas}/R_{air} = 1337	208/39	[68]
rGO/Fe_2O_3 nanosphere	NO_2	Room temperature	90 (0.18–90)	$R_{air} - R_{gas}/R_{gas}$ = 150.63%	80/1648	[69]
rGO/CuO nanoflower	NO_2	Room temperature	1000 (0.25–1000)	$\|R_{gas} - R_{air}\|/R_{air}$ = 6.61 %	76/232	[90]
Graphene/ WO_3 nanosphere	NO_2	Room temperature	56 (7–56)	$I_{gas} - I_{air}/I_{air}$ = 40.8%	—	[70]
rGO/WO_3 porous nanoflakes	NO_2	90 (20–200)	10 (5–200)	R_{gas}/R_{air} = 5%	4.1/5.8	[91]

Table 4.
Summary of the performance of sensor fabricated by using the nanocomposites of 3-D metal oxides and graphene or graphene derivatives (GO, rGO).

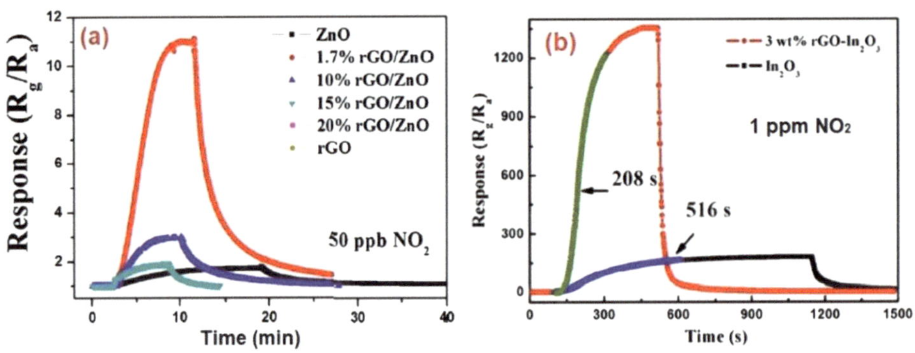

Figure 14.
NO_2 response of (a) rGO/ZnO nanoflower (1.7% rGO in ZnO) at 100°C [15] and (b) rGO/In_2O_3nanoflower at 74°C [68].

selective behavior towards NO_2. Operating temperature of the sensor was quite low and most of the cases room temperature sensing was reported. The detection range was quite high where lower and higher detection limit varied from a few ppb to 1000 ppm. Very high response magnitude was reported in case of graphene/3-D metal oxide nanocomposites. Reported response time and recovery time were quite high in the case of 3-D metal oxide composites compared with 0-D and 1-D metal oxide nanocomposites. Highly selective NO_2 sensing was reported for rGO/ZnO nanoflower (1.7% rGO in ZnO) as shown in **Figure 14(a)**. Promising NO_2 sensing was observed for rGO/In_2O_3 $_{nanoflower}$ where the response was poor for pure In_2O_3nanoflower as shown in **Figure 14(b)**. The nanoflower-shaped CuO nanostructure in its nanocomposite with rGO is effective to prevent the aggregation of rGO sheets and form porous structure with rGO, which greatly facilitate the adsorption and diffusion of gas molecules [92].

4. Conclusion

Hybrids of graphene/nanoscale metal oxides have been extensively discussed in this chapter where the major focused area was synthesis/fabrication of monohybrid and its performance assessment for gas/vapor sensing applications. Detailed literature survey confirmed that metal oxide nanoparticle (0-D) are the most reported nanostructure used for the synthesis of nanocomposites with graphene (and GO and rGO) for potential gas sensing application while 1-D metal oxides like nanorods, nanotubes, nanowires, etc. were in the second place. Use of 2-D and 3-D metal oxides were relatively less to form composites with graphene. In chemical synthesis, GO/rGO functionalization was carried out in two routes, that is, (i) mixing of GO/rGO in precursor before synthesizing a nanostructure and (ii) functionalization by GO/rGO after synthesis of nanostructures. Hydrothermal was the most popular method followed by solvothermal, sol-gel, spray pyrolysis, etc. reported to synthesize a composite of 0-D metal oxides. Hydrothermal, electrospinning, electrochemical anodization, etc. were used for the synthesis of graphene/1-D metal oxide composites. Most of the 2-D and 3-D nanocomposites were grown by low-cost chemical methods. Therefore, the graphene/nanoscale metal oxide composites can synthesize via a cost-effective way.

Among all the metal oxides, SnO_2 was mostly reported materials in 0-D structure used in composites with graphene. Other popular metal oxides are ZnO, WO_3 and TiO_2 mostly used for 1-D metal oxide hybrid. A large number of the report showed NO_2 selective behavior of rGO/metal oxide nanocomposite gas sensors especially for 3-D and 0-D metal oxide hybrids. Other reported gases/vapors are H_2, NH_3, CO, H_2S, C_2H_2, ethanol, methanol, acetone, benzene, etc. A significant variation was observed in case of operating temperature of the sensors in case of different nanoscale metal oxides. The average sensing temperature was highest in case of 2-D nanocomposites and decreased from 1-D, 0-D and 3-D. However, the overall study confirmed the relatively low-temperature detection of gases and vapors after the formation of composites of graphene and nanoscale metal oxides. The detection range was varying from lower ppb to higher ppm level but most of the report was confined near a low ppm range (1–100 ppm). A significant improvement was also observed in case of response magnitude and response time/recovery time.

Finally, we would like to conclude with the comment that gas/vapor sensing performance was improved significantly due to the formation of nanohybrid of nanoscale metal oxides and graphene derivatives like GO and rGO. Further study may be necessary with these nano thin-film sensors to encourage the performance in terms of high selectivity and long-term stability. Then these hybrid sensors to make these nanocomposite sensors more suitable for practical applications.

Acknowledgements

This work was supported in part by Department of Biotechnology grant (Letter No. BT/PR28727/NNT/28/1569/2018) and SPARC grant (SPARC/2018-2019/P1394/SL), Govt. of India.

Author details

Arnab Hazra[1*], Nagesh Samane[1] and Sukumar Basu[2]

1 Department of Electrical and Electronics Engineering, Birla Institute of Technology and Science (BITS)-Pilani, VidyaVihar, Rajasthan, India

2 Department of Physics and Materials Science, Jaypee University of Information Technology (JUIT), Solan, India

*Address all correspondence to: arnab.hazra@pilani.bits-pilani.ac.in

References

[1] Hosseini MS, Zeinali S, Sheikhi MH. Fabrication of capacitive sensor based on Cu-BTC (MOF-199) nanoporous film for detection of ethanol and methanol vapors. Sensors and Actuators B: Chemical. 2016;**230**:9-16. DOI: 10.1016/j.snb.2016.02.008

[2] Righettoni M, Amann A, Pratsinis SE. Breath analysis by nanostructured metal oxides as chemo-resistive gas sensors. Materials Today. 2015;**18**:163-171. DOI: 10.1016/j.mattod.2014.08.017

[3] Fratoddi I, Bearzotti A, Venditti I, Cametti C, Russo MV. Role of nanostructured polymers on the improvement of electrical response-based relative humidity sensors. Sensors and Actuators B: Chemical. 2016;**225**: 96-108. DOI: 10.1016/j.snb.2015.11.001

[4] Gupta CS, Chatterjee S, Ray AK, Chakraborty AK. Graphene-metal oxide nanohybrids for toxic gas sensor: A review. Sensors and Actuators B. 2015; **221**:1170-1181. DOI: 10.1016/j.snb.2015.07.070

[5] Bindra P, Hazra A. Impedance behavior of n-type TiO_2 nanotubes porous layer in reducing vapor ambient. Vacuum. 2018;**152**:78-83. DOI: 10.1016/j.vacuum.2018.03.008

[6] Tang X, Mager N, Vanhorenbeke B, Hermans S, Raskin J-P. Defect-free functionalized graphene sensor for formaldehyde detection. Nanotechnology. 2017;**28**:055501. DOI: 10.1088/1361-6528/28/5/055501

[7] Hazra A, Dutta K, Bhowmik B, Manjuladevi V, Gupta RK, Bhattacharyya P. Low temperature methanol sensing by p-type nano-titania: Correlation with defects states and Schottky barrier model. IEEE Transactions on Nanotechnology. 2015; **14**:187-195. DOI: 10.1109/TNANO.2014.2376633

[8] Hazra A, Dutta K, Bhowmik B, Bhattacharyya P. Highly repeatable low-ppm ethanol sensing characteristics of p-TiO_2-based resistive devices. IEEE Sensors Journal. 2014;**15**:408-416. DOI: 10.1109/JSEN.2014.2345575

[9] Bindra P, Hazra A. Capacitive gas and vapor sensors using nanomaterials. Journal of Materials Science: Materials in Electronics. 2018;**29**:6129-6148. DOI: 10.1007/s10854-018-8606-2

[10] Liu X, Chang S, Liu H, Hu S, Zhang D, Ning H. A survey on gas sensing technology. Sensors. 2012;**12**: 9635-9665. DOI: 10.3390/s120709635

[11] Jiménez-Cadena G, Riu J, Riusa FX. Gas sensors based on nanostructured materials. The Analyst. 2007;**132**:1083-1099. DOI: 10.1039/b704562j

[12] Singh E, Meyyappan M, Nalwa HS. Flexible graphene-based wearable gas and chemical sensors. ACS Applied Materials & Interfaces. 2017;**9**: 34544-34586. DOI: 10.1021/acsami.7b07063

[13] Wang Y, Zhang L, Hu N, Wang Y, Zhang Y, Zhou Z, et al. Ammonia gas sensors based on chemically reduced graphene oxide sheets self-assembled on Au electrodes. Nanoscale Research Letters. 2014;**9**:251. DOI: 10.1186/1556-276X-9-251

[14] Lu G, Ocola LE, Chen J. Reduced graphene oxide for room-temperature gas sensors. Nanotechnology. 2009;**20**: 445502. DOI: 10.1088/0957-4484/20/44/445502

[15] Liu J, Li S, Zhang B, Xiao Y, Gao Y, Yang Q, et al. Ultrasensitive and low detection limit of nitrogen dioxide gas sensor based on flower-like ZnO hierarchical nanostructure modified by reduced graphene oxide. Sensors and

Actuators B. 2017;**249**:715-724. DOI: 10.1016/j.snb.2017.04.190

[16] Venkatesan A, Rathi S, Lee I-Y, Park J, Lim D, Kim G-H, et al. Low temperature hydrogen sensing using reduced graphene oxide and tin oxide nanoflowers based hybrid structure. Semiconductor Science and Technology. 2016;**31**:125014. DOI: 10.1088/0268-1242/31/12/125014

[17] Cui S, Pu H, Lu G, Wen Z, Mattson EC, Hirschmugl C, et al. Fast and selective room-temperature ammonia sensors using silver nanocrystal-functionalized carbon nanotubes. ACS Applied Materials & Interfaces. 2012;**4**:4898-4904. DOI: 10.1021/am301229w

[18] Guo L, Kou X, Dinga M, Wanga C, Dong L, Zhang H, et al. Reduced graphene oxide/α-Fe_2O_3 composite nanofibers for application in gas sensors. Sensors and Actuators B. 2017; **244**:233-242. DOI: 10.1016/j.snb.2016.12.137

[19] Inyawilert K, Wisitsoraat A, Sriprachaubwong C, Tuantranont A, Phanichphant S, Liewhiran C. Rapid ethanol sensor based on electrolytically-exfoliated graphene-loaded flame-made In-doped SnO_2 composite film. Sensors and Actuators B. 2015; **209**:40-55. DOI: 10.1016/j.snb.2014.11.086

[20] Xiao Y, Yang Q, Wang Z, Zhang R, Gao Y, Sun P, et al. Improvement of NO_2 gas sensing performance based on discoid tin oxide modified by reduced graphene oxide. Sensors and Actuators B. 2016;**227**:419-426. DOI: 10.1016/j.snb.2015.11.051

[21] Rao CNR, Sood AK, Subrahmanyam KS, Govindaraj A. Graphene: The new two-dimensional nanomaterial. Angewandte Chemie, International Edition. 2009;**48**: 7752-7777. DOI: 10.1002/anie.200901678

[22] Randviir EP, Brownson DAC, Banks CE. A decade of graphene research: Production, applications and outlook. Materials Today. 2014;**17**: 426-432. DOI: 10.1016/j.mattod.2014.06.001

[23] Edwards RS, Coleman KS. Graphene synthesis: Relationship to applications. Nanoscale. 2013;**5**:38-51. DOI: 10.1039/C2NR32629A

[24] Mohan VB, Lau K-T, Hui D, Bhattacharyya D. Graphene-based materials and their composites: A review on production applications and product limitations. Composites Part B Engineering. 2018;**142**:200-220. DOI: 10.1016/j.compositesb.2018.01.013

[25] Song Z, Wei Z, Wang B, Luo Z, Xu S, Zhang W, et al. Sensitive room-temperature H_2S gas sensors employing SnO_2 quantum wire/reduced graphene oxide nanocomposites. Chemistry of Materials. 2016;**28**:1205-1212. DOI: 10.1021/acs.chemmater.5b04850

[26] Quang VV, Dung NV, Trong NS, Hoa ND, Duy NV, Hieu NV. Outstanding gas-sensing performance of graphene/SnO_2 nanowire Schottky junctions. Applied Physics Letters. 2014; **105**:1-5. DOI: 10.1063/1.4887486

[27] Zhang Z, Zou R, Song G, Yu L, Chen Z, Hu J. Highly aligned SnO_2 nanorods on graphene sheets for gas sensors. Journal of Materials Chemistry. 2011;**21**:17360-17365. DOI: 10.1039/c1jm12987b

[28] Song N, Fan H, Tian H. PVP assisted in situ synthesis of functionalized graphene/ZnO (FGZnO) nanohybrids with enhanced gas-sensing property. Journal of Materials Science. 2015;**50**: 2229-2238. DOI: 10.1007/s10853-014-8785-z

[29] Xia Y, Wang J, Xu JL, Li X, Xie D, Xiang L, et al. Confined formation of ultrathin ZnO nanorods/reduced graphene oxide mesoporous

nanocomposites for high-performance room-temperature NO_2 sensors. ACS Applied Materials & Interfaces. 2016;**8**: 35454-35463. DOI: 10.1021/acsami.6b12501

[30] Zou R, He G, Xu K, Liu Q, Zhang Z, Hu J. ZnO nanorods on reduced graphene sheets with excellent field emission, gas sensor and photocatalytic properties. Journal of Materials Chemistry A. 2013;**1**:8445-8452. DOI: 10.1039/c3ta11490b

[31] Dhall S, Kumar M, Bhatnagar M, Mehta BR. Dual gas sensing properties of graphene-Pd/SnO_2 composites for H_2 and ethanol: Role of nanoparticles-graphene interface. International Journal of Hydrogen Energy. 2018;**43**: 17921-17927. DOI: 10.1016/j.ijhydene.2018.07.066

[32] Singkammo S, Wisitsoraat A, Sriprachuabwong C, Tauntranout A, Phanichphat S, Liewhiran C. Electrolytically exfoliated graphene-loaded flame-made Ni-doped SnO_2 composite film for acetone sensing. ACS Applied Materials & Interfaces. 2015;**7**: 3077-3092. DOI: 10.1021/acsami.5b00161

[33] Zhang D, Liu A, Chang H, Xia B. Room-temperature high-performance acetone gas sensor based on hydrothermal synthesized SnO_2-reduced graphene oxide hybrid composite. RSC Advances. 2015;**5**: 3016-3022. DOI: 10.1039/C4RA10942B

[34] Zhang H, Wang L, Zhang T. Reduced graphite oxide/SnO_2/Au hybrid nanomaterials for NO_2 sensing performance at relatively low operating temperature. RSC Advances. 2014;**4**: 57436-57441. DOI: 10.1039/C4RA10474A

[35] Zhang H, Feng J, Fei T, Liu S, Zhang T. SnO_2 nanoparticles-reduced graphene oxide nanocomposites for NO_2 sensing at low operating temperature. Sensors and Actuators B: Chemical. 2014;**190**:472-478. DOI: 10.1016/j.snb.2013.08.067

[36] Mishra RK, Upadhyay SB, Khushwaha A, Kim TH, Murali G, Verma R, et al. SnO_2 quantum dots decorated on RGO: A superior sensitive, selective and reproducible performance for a H_2 and LPG sensor. Nanoscale. 2015;**7**:11971-11979. DOI: 10.1039/c5nr02837j

[37] Ghosh R, Nayak AK, Santra S, Pradhan D, Guha PK. Enhanced ammonia sensing at room temperature with reduced graphene oxide/tin oxide hybrid films. RSC Advances. 2015;**5**: 50165-50173. DOI: 10.1039/C5RA06696D

[38] Jin L, Chen W, Zhang H, Xiao G, Yu C, Zhou Q. Characterization of reduced graphene oxide (rGO)-loaded SnO_2 nanocomposite and applications in C_2H_2 gas detection. Applied Sciences. 2016;**7**:19. DOI: 10.3390/app7010019

[39] Zhang D, Chang H, Li P, Liu R. Characterization of nickel oxide decorated-reduced graphene oxide nanocomposite and its sensing properties toward methane gas detection. Journal of Materials Science: Materials in Electronics. 2016;**27**: 3723-3730. DOI: 10.1007/s10854-015-4214-6

[40] Wang Z, Zhang T, Han T, Fei T, Liu S, Lu G. Oxygen vacancy engineering for enhanced sensing performances: A case of SnO_2 nanoparticles-reduced graphene oxide hybrids for ultrasensitive ppb-level room-temperature NO_2 sensing. Sensors and Actuators B: Chemical. 2018;**266**: 812-822. DOI: 10.1016/j.snb.2018.03.169

[41] Peng R, Chen J, Nie X, Li D, Si P, Feng J, et al. Reduced graphene oxide decorated Pt activated SnO_2 nanoparticles for enhancing methanol sensing performance. Journal of Alloys and Compounds. 2018;**762**:8-15. DOI: 10.1016/j.jallcom.2018.05.177

[42] Wang Z, Han T, Fei T, Liu S, Zhang T. Investigation of microstructure effect on NO_2 sensors based on SnO_2 nanoparticles/reduced graphene oxide hybrids. ACS Applied Materials & Interfaces. 2018;**10**: 41773-41783. DOI: 10.1021/acsami.8b15284

[43] Tammanoon N, Wisitsoraat A, Spriprachuabwong C, Phokharatkul D, Tauntranout A, Phanichphant S, et al. Ultrasensitive NO_2 sensor based on Ohmic metal-semiconductor interfaces of electrolytically exfoliated graphene/flame-spray-made SnO_2 nanoparticles composite operating at low temperatures. ACS Applied Materials & Interfaces. 2015;**7**: 24338-24352. DOI: 10.1021/acsami.5b09067

[44] Iftekhar Uddin ASM, Phan DT, Chung GS. Low temperature acetylene gas sensor based on Ag nanoparticles-loaded ZnO-reduced graphene oxide hybrid. Sensors and Actuators B: Chemical. 2015;**207**:362-369. DOI: 10.1016/j.snb.2014.10.091

[45] Uddin ASMI, Chung GS. Synthesis of highly dispersed ZnO nanoparticles on graphene surface and their acetylene sensing properties. Sensors and Actuators B: Chemical. 2014;**205**: 338-344. DOI: 10.1016/j.snb.2014.09.005

[46] Huang Q, Zeng D, Li H, Xie C. Room temperature formaldehyde sensors with enhanced performance, fast response and recovery based on zinc oxide quantum dots/graphene nanocomposites. Nanoscale. 2012;**4**:5651-5658. DOI: 10.1039/c2nr31131c

[47] Liu S, Wang Z, Zhang Y, Dong Z, Zhang T. Preparation of zinc oxide nanoparticle-reduced graphene oxide-gold nanoparticle hybrids for detection of NO_2. RSC Advances. 2015; **5**:91760-91765. DOI: 10.1039/c5ra18680c

[48] Tung TT, Chien NV, Duy NV, Hieu NV, Nine MJ, Coghlan CJ, et al. Magnetic iron oxide nanoparticles decorated graphene for chemoresistive gas sensing: The particle size effects. Journal of Colloid and Interface Science. 2019;**539**:315-325. DOI: 10.1016/j.jcis.2018.12.077

[49] Kamal T. High performance NiO decorated graphene as a potential H_2 gas sensor. Journal of Alloys and Compounds. 2017;**729**:1058-1063. DOI: 10.1016/j.jallcom.2017.09.124

[50] Achary LSK, Kumar A, Barik B, Nayak PS, Tripathy N, Kar JP, et al. Reduced graphene oxide-$CuFe_2O_4$ nanocomposite: A highly sensitive room temperature NH_3 gas sensor. Sensors and Actuators B: Chemical. 2018;**272**: 100-109. DOI: 10.1016/j.snb.2018.05.093

[51] Neri G, Leonardi SG, Latino M, Donato N, Baek S, Conte DE, et al. Sensing behavior of SnO_2/reduced graphene oxide nanocomposites toward NO_2. Sensors and Actuators B: Chemical. 2013;**179**:61-68. DOI: 10.1016/j.snb.2012.10.031

[52] Kim HW, Na HG, Kwon YJ, Kang SY, Choi MS, Bang JH, et al. Microwave-assisted synthesis of graphene-SnO_2 nanocomposites and their applications in gas sensors. ACS Applied Nano Materials. 2017;**9**: 31667-31682. DOI: 10.1021/acsami.7b02533

[53] Choi SJ, Jong BH, Lee SJ, Min BK, Rothschild A, Kim ID. Selective detection of acetone and hydrogen sulfide for the diagnosis of diabetes and halitosis using SnO_2 nanofibers functionalized with reduced graphene oxide nanosheets. ACS Applied Materials & Interfaces. 2014;**6**: 2588-2597. DOI: 10.1021/am405088q

[54] Abideen ZU, Katoch A, Kim JH, Kwon YJ, Kim HW, Kim SS. Excellent gas detection of ZnO nanofibers by

loading with reduced graphene oxide nanosheets. Sensors and Actuators B: Chemical. 2015;**221**:1499-1507. DOI: 10.1016/j.snb.2015.07.120

[55] Yi J, Lee JM, Park WI. Vertically aligned ZnO nanorods and graphene hybrid architectures for high-sensitive flexible gas sensors. Sensors and Actuators B: Chemical. 2011;**155**: 264-269. DOI: 10.1016/j.snb.2010.12.033

[56] An X, Yu JC, Hu Y, Yu X, Zhang G. WO_3 nanorods/graphene nanocomposites for high-efficiency visible-light-driven photocatalysis and NO_2 gas sensing. Journal of Materials Chemistry. 2012;**22**:8525-8531. DOI: 10.1039/C2JM16709C

[57] Feng Q, Li X, Wang J, Gaskov AM. Reduced graphene oxide (rGO) encapsulated Co_3O_4 composite nanofibers for highly selective ammonia sensors. Sensors and Actuators B: Chemical. 2016;**222**:864-870. DOI: 10.1016/j.snb.2015.09.021

[58] Choi SJ, Ryu WH, Kim SJ, Cho HJ, Kim ID. Bi-functional co-sensitization of graphene oxide sheets and Ir nanoparticles on p-type Co_3O_4 nanofibers for selective acetone detection. Journal of Materials Chemistry B. 2014;**2**:7160-7167. DOI: 10.1039/C4TB00767K

[59] Choi SJ, Choi C, Kim SJ, Cho HJ, Jeon S, Kim ID. Facile synthesis of hierarchical porous WO_3 nanofibers having 1D nanoneedles and their functionalization with non-oxidized graphene flakes for selective detection of acetone molecules. RSC Advances. 2015;**5**:7584-7588. DOI: 10.1039/C4RA13791D

[60] Galstyan V, Ponzoni A, Kholmanov I, Natile MM, Comini E, Nematov S, et al. Reduced graphene oxide–TiO_2 nanotube composite: Comprehensive study for gas-sensing applications. ACS Applied Nano Materials. 2018;**1**:7098-7105. DOI: 10.1021/acsanm.8b01924

[61] Acharyya D, Bhattacharyya P. Highly efficient room-temperature gas sensor based on TiO_2 nanotube-reduced graphene-oxide hybrid device. IEEE Electron Device Letters. 2016;**37**: 656-659. DOI: 10.1109/LED.2016.2544954

[62] Gakhar T, Hazra A. Synthesis of GO loaded TiO_2 nanotubes array by anodic oxidation for efficient detection of organic vapor. Journal of Electronic Materials. 2019;**48**:5342-5347. DOI: 10.1007/s11664-019-07347-8

[63] Maity I, Acharyya D, Huang K, Chung P, Ho M, Bhattacharya P. A comparative study on performance improvement of ZnO nanotubes based alcohol sensor devices by Pd and rGO hybridization. IEEE Transactions on Electron Devices. 2018;**65**:3528-3534. DOI: 10.1109/TED.2018.2846784

[64] Bhati VS, Ranwa S, Rajamani S, Kumari K, Raliya R, Biswas P, et al. Improved sensitivity with low limit of detection of a hydrogen gas sensor based on rGO-loaded Ni-doped ZnO nanostructures. ACS Applied Materials & Interfaces. 2018;**10**:11116-11124. DOI: 10.1021/acsami.7b17877

[65] Zhao C, Gong H, Lan W, Ramachandran R, Xu H, Liu S, et al. Facile synthesis of SnO_2 hierarchical porous nanosheets from graphene oxide sacrificial scaffolds for high-performance gas sensors. Sensors and Actuators B: Chemical. 2018;**258**: 492-500. DOI: 10.1016/j.snb.2017.11.167

[66] Wang P, Wang D, Zhang M, Zhu Y, Xu Y, Ma X, et al. ZnO nanosheets/graphene oxide nanocomposites for highly effective acetone vapor detection. Sensors and Actuators B: Chemical. 2016;**230**:477-484. DOI: 10.1016/j.snb.2016.02.056

[67] Shi J, Cheng Z, Gao L, Zhang Y, Xu J, Zhao H. Facile synthesis of

reduced graphene oxide/hexagonal WO_3 nanosheets composites with enhanced H_2S sensing properties. Sensors and Actuators B: Chemical. 2016;**230**:736-745. DOI: 10.1016/j.snb.2016.02.134

[68] Liu J, Li S, Zhang B, Wang Y, Gao Y, Liang X, et al. Flower-like In_2O_3 modified by reduced graphene oxide sheets serving as a highly sensitive gas sensor for trace NO_2 detection. Journal of Colloid and Interface Science. 2017; **504**:206-213. DOI: 10.1016/j.jcis.2017.05.053

[69] Dong YL, Zhang XF, Cheng XL, Xu YM, Gao S, Zhao H, et al. Highly selective NO_2 sensor at room temperature based on nanocomposites of hierarchical nanosphere-like α-Fe_2O_3 and reduced graphene oxide. RSC Advances. 2014;**4**:57493-57500. DOI: 10.1039/C4RA10136G

[70] Jie X, Zeng D, Zhang J, Xu K, Wu J, Zhu B, et al. Graphene-wrapped WO_3 nanospheres with room-temperature NO_2 sensing induced by interface charge transfer. Sensors and Actuators B: Chemical. 2015;**220**:201-209. DOI: 10.1016/j.snb.2015.05.047

[71] Li L, He S, Liu M, Zhang C, Chen W. Three-dimensional mesoporous graphene aerogel-supported SnO_2 nanocrystals for high-performance NO_2 gas sensing at low temperature. Analytical Chemistry. 2015;**87**: 1638-1645

[72] Zhang Z, Zou X, Xu L, Liao L, Liu W, Ho J, et al. Hydrogen gas sensor based on metal oxide nanoparticles decorated graphene transistor. Nanoscale. 2015;**7**:10078-10084. DOI: 10.1039/C5NR01924A

[73] Tyagi P, Sharma A, Tomar M, Gupta V. A comparative study of RGO-SnO_2 and MWCNT-SnO_2 nanocomposites based SO_2 gas sensors. Sensors and Actuators B: Chemical. 2017;**248**:980-986. DOI: 10.1016/j.snb.2017.02.147

[74] Cui S, Wen Z, Mattson EC, Mao S, Chang J, Weinert M, et al. Indium-doped SnO_2 nanoparticle-graphene nanohybrids: Simple one-pot synthesis and their selective detection of NO_2. Journal of Materials Chemistry A. 2013; **1**:4462-4467. DOI: 10.1039/C3TA01673K

[75] Zhu X, Guo Y, Ren H, Gao C, Zhou Y. Enhancing the NO_2 gas sensing properties of rGO/SnO_2 nanocomposite films by using microporous substrates. Sensors and Actuators B: Chemical. 2017;**248**:560-570. DOI: 10.1016/j.snb.2017.04.030

[76] Shojaee M, Nasresfahani S, Sheikhi MH. Hydrothermally synthesized Pd-loaded SnO_2/partially reduced graphene oxide nanocomposite for effective detection of carbon monoxide at room temperature. Sensors and Actuators B: Chemical. 2018;**254**:457-467. DOI: 10.1016/j.snb.2017.07.083

[77] Liu S, Wang Z, Zhang Y, Li J, Zhang T. Sulfonated graphene anchored with tin oxide nanoparticles for detection of nitrogen dioxide at room temperature with enhanced sensing performances. Sensors and Actuators B: Chemical. 2016;**228**:134-143. DOI: 10.1016/j.snb.2016.01.023

[78] Wang Z, Zhang Y, Liu S, Zhang T. Preparation of Ag nanoparticles-SnO_2 nanoparticles-reduced graphene oxide hybrids and their application for detection of NO_2 at room temperature. Sensors and Actuators B: Chemical. 2016;**222**:893-903. DOI: 10.1016/j.snb.2015.09.027

[79] Kumar N, Srivastava AK, Patel HS, Gupta BK, Varma GD. Facile synthesis of ZnO-reduced graphene oxide nanocomposites for NO_2 Gas sensing applications. European Journal of Inorganic Chemistry. 2015; **2015**:1912-1923. DOI: 10.1002/ejic.201403172

[80] Liu S, Yu B, Zhang H, Fei T, Zhang T. Enhancing NO_2 gas sensing performances at room temperature based on reduced graphene oxide-ZnO nanoparticles hybrids. Sensors and Actuators B: Chemical. 2014;**202**: 272-278. DOI: 10.1016/j.snb.2014.05.086

[81] Ha NH, Thinh DD, Huong NT, Phuong NH, Thach PD, Hong HS. Fast response of carbon monoxide gas sensors using a highly porous network of ZnO nanoparticles decorated on 3D reduced graphene oxide. Applied Surface Science. 2018;**434**:1048-1054. DOI: 10.1016/j.apsusc.2017.11.047

[82] Zhang J, Wu J, Wang X, Zeng D, Xie C. Enhancing room-temperature NO_2 sensing properties via forming heterojunction for NiO-rGO composited with SnO_2 nanoplates. Sensors and Actuators B: Chemical. 2017;**243**: 1010-1019. DOI: 10.1016/j.snb.2016.12.062

[83] Li Z, Liu Y, Guo D, Guo J, Su Y. Room-temperature synthesis of CuO/reduced graphene oxide nanohybrids for high-performance NO_2 gas sensor. Sensors and Actuators B: Chemical. 2018;**271**:306-310. DOI: 10.1016/j.snb.2018.05.097

[84] Srivastava S, Jain K, Singh VN, Singh S, Vijayan N, Dilwar N, et al. Faster response of NO_2 sensing in graphene–WO_3 nanocomposites. Nanotechnology. 2012;**23**:205501. DOI: 10.1088/0957-4484/23/20/205501

[85] Zhang L, Fang Q, Huang Y, Xu K, Ma F, Chu PK. Facet-engineered CeO_2/graphene composites for enhanced NO_2 gas-sensing. Journal of Materials Chemistry C. 2017;**5**:6973-6981. DOI: 10.1039/C7TC01523B

[86] Triet NM, Duy LT, Hwang BU, Haneef A, Siddiqui S, Park KH, et al. High-performance Schottky diode gas sensor based on the heterojunction of three-dimensional nanohybrids of reduced graphene oxide-vertical ZnO nanorods on an AlGaN/GaN layer. ACS Applied Materials & Interfaces. 2017;**9**: 30722-30732. DOI: 10.1021/acsami.7b06461

[87] Abideen ZU, Kim JH, Mirzaei A, Kim HW, Kim SS. Sensing behavior to ppm-level gases and synergistic sensing mechanism in metal-functionalized rGO-loaded ZnO nanofibers. Sensors and Actuators B: Chemical. 2018;**255**: 1884-1896. DOI: 10.1016/j.snb.2017.08.210

[88] Kooti M, Keshtkar S, Askarieh M, Rashidi A. Progress toward a novel methane gas sensor based on SnO_2 nanorods-nanoporous graphene hybrid. Sensors and Actuators B: Chemical. 2019;**281**:96-106. DOI: 10.1016/j.snb.2018.10.032

[89] Deng S, Tjoa V, Fan HM, Tan HR, Sayle DC, Olvio M, et al. Reduced graphene oxide conjugated Cu_2O nanowire mesocrystals for high-performance NO_2 gas sensor. Journal of the American Chemical Society. 2012;**134**:4905-4917. DOI: 10.1021/ja211683m

[90] Yang Y, Tian C, Wang J, Sun L, Shi K, Zhou W, et al. Facile synthesis of novel 3D nanoflower-like Cu_xO/multilayer graphene composites for room temperature NO_x gas sensor application. Nanoscale. 2014;**6**: 7369-7378. DOI: 10.1039/c4nr00196f

[91] Hao Q, Liu T, Liu J, Liu Q, Jing X, Zhang H, et al. Controllable synthesis and enhanced gas sensing properties of a single-crystalline WO_3-rGO porous nanocomposite. RSC Advances. 2017;**7**: 14192-14199. DOI: 10.1039/C6RA28379A

[92] Zhang D, Jiang C, Liu J, Cao Y. Carbon monoxide gas sensing at room temperature using copper oxide decorated graphene hybrid nanocomposite prepared by layer-by-layer self-assembly. Sensors and Actuators B: Chemical. 2017;**247**: 875-882. DOI: 10.1016/j.snb.2017.03.108

Chapter 2

Crystalline Silicon Nitride Films on Si(111): Growth Mechanism, Surface Structure and Chemistry down to Atomic Scale

Subhashis Gangopadhyay

Abstract

A detailed investigation of the growth mechanism of ultra-thin silicon nitride (Si_3N_4) films on Si(111) substrates, their structure, morphology and surface chemistry down to atomic scale have been investigated using various surface analytical techniques such as low energy electron diffraction (LEED), scanning tunneling microscopy (STM) and ESCA microscopy. A radio frequency N_2 plasma source from Epi Uni-bulb has been used for the nitridation of atomically clean Si(111) surfaces. The substrate temperatures during the nitridation process were ranging from 600–1050°C and the plasma exposure times were varied from 5 s for initial nucleation up to 45 min for saturation thickness. The initial stage of N nucleation on Si(111), how the structure and morphology of the nitride films depend on thickness and temperature, surface atomic reconstructions and the nitride film chemical composition are discussed here. All findings are explained in terms of thermally activated inter-diffusion of Si and N atoms as well as the surface adatom diffusion/mobility.

Keywords: silicon nitride thin film, growth mechanism, surface reconstruction, STM, LEED

1. Introduction

In semiconductor technology, one of the most important parts is the formation of homogeneous insulating layers and are of high practical importance as it is an integral part all kinds of integrated circuits. Due to the recent miniaturization in nanotechnology, insulating layer thickness needs to be precisely controlled down to a few nanometer ranges. This nanometer scale insulating films not only reduce the devices size, but also provide a platform for novel device fabrication such as resonant tunneling diodes [1] or memory devices for magnetic tunnel junction [2].

Since last few decades, SiO_2 on silicon (Si) is found to be the most useful insulating material in VLSI device technology because of its high quality and superior homogeneity. However, continuous miniaturization of device size now demands a thickness of this insulating layer down to a few atomic layers. Here, the homogeneity of the insulating film in terms of morphology and chemical purity becomes more pronounced. Very small fluctuations in thickness of the oxide layer may lead

IntechOpen

to the break-down due to the enhanced electron tunneling as this insulating barriers is extremely thin. Hence, demand for new materials with a higher dielectric constant is of high priority which can replace the SiO_2 layer to overcome this issue. In this regard, crystalline silicon nitride (Si_3N_4) films received considerable attention to replace the existing SiO_2 gate dielectric materials, as it is compatible with existing Si processing technologies as well as larger dielectric constant and diffusion resistive materials properties [3, 4]. Plasma assisted amorphous silicon nitride layer has recently been used as high performance gate dielectric [5]. High thermal stability and refractive index of Si_3N_4 make it capable for high temperature structural ceramics and anti-reflective coating materials, respectively. Finally, crystalline Si_3N_4 on Si can also serve as a substrate for highly demanding GaN growth to integrate the opto-electronic devices to the well-established silicon based device technology [6, 7]. Therefore, better understanding of the initial Si_3N_4 films growth on Si, their structural and morphological evolution as well as chemical properties in a very local (atomic) scale are of high technological importance.

1.1 Structure and properties

Silicon nitride is a structural ceramics, which exhibits high mechanical strength at room as well as elevated temperature. It can also be useful because of its high fracture toughness. It shows a very high thermal stability, up to 1600°C in air and also has a much larger dielectric constant (ε = 7.5) as compared to the conventional SiO_2 (ε = 3.8). It is a semiconducting material with an energy band gap of about 4.7 eV, which is almost half of SiO_2 (9 eV). Usually the bulk Si_3N_4 has been produced by sintering method and structurally appeared as poly-crystalline ceramic. Crystalline silicon nitride appears with two different phases, such as α-phase and β-phase of Si_3N_4. Both the phases have space groups of $P31_c$ and $P6_3$ for α- and β-Si_3N_4, respectively and the structures appeared with a hexagonal symmetry. Although there were some contradiction about these two phases, finally it has been well accepted that the only stable phase is the β-Si_3N_4 phase whereas the α-Si_3N_4 phase is a meta-stable one and has a lattice constant along c-axis (0001) just double of the β-phase.

Various groups have already described the crystal structure of β-Si_3N_4 and for the bulk crystals. A lattice constant of 'a' = 7.595 Å in the hexagonal (X-Y) plane whereas 'c' = 2.902 Å in the vertical (Z) direction are obtained [9]. A schematic model of the β-Si_3N_4 structure (top view), containing 14 atoms in the lateral plane (normal to c-axis) can be seen in **Figure 1** where half of the atoms are placed below (at $Z = -c/4$ plane) and the other half above (at $Z = c/4$ plane) the X-Y plane. Local geometrical configuration indicates a combination of sp^3 and sp^2 hybridized orbital for Si and N atoms, respectively. This structure can also be considered as tetrahedrons of Si_3N_4 complex network, connected through their corners. Further structural information of β-phase of silicon nitride and how it is related to the α-phase has also be reported elsewhere [8, 9].

1.2 Different approaches

Formation of crystalline silicon nitride films has been performed using various growth methods. Chemical vapor deposition (CVD) growth process has been used to grow silicon nitride where a mixture of silicon-hydrogen (SiH_4, Si_2H_6, Si_3H_8) and nitro-hydro compounds (NH_3, N_2H_4, HN_3) are used. In general, resulting films are usually appeared in an amorphous phase, non-stoichiometric and significantly contaminated [10]. However, without supplying any Si -H_2 source compound, it also possible to obtain nitride layers by direct nitridation of a Si substrate at elevated

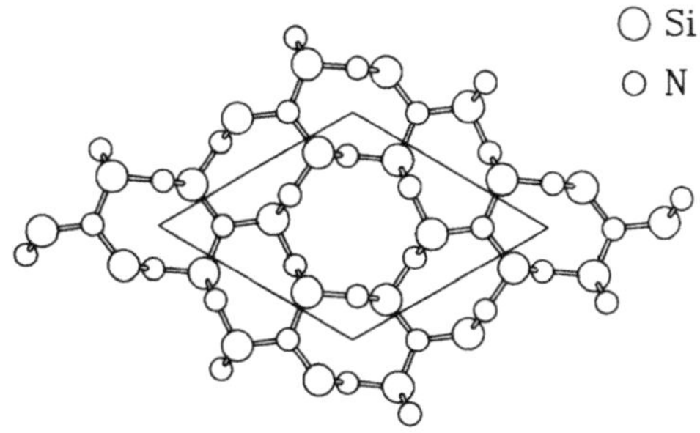

Figure 1.
Schematic top (0001) view of the β-Si_3N_4 structure model in a lateral (x-y) plane with 14 atoms. The bigger and smaller circles represent the Si and N atoms respectively, whereas the parallelogram indicates the unit cell [8].

temperature. Si(111) surface exhibits a threefold rotational symmetry. Moreover, its 2 × 2 unit cell is having a lattice mismatch of only about 1.1% with the 'a' axis of hexagonal Si_3N_4. These two properties make it a very compatible substrate for the growth of Si_3N_4 films in the (0001) lattice direction. It can be done by exposing the atomically clean Si substrate to various N_2 compound reactive gases such as NH_3 [11–17], NO [18–21] or other gaseous at high nitridation temperatures or by post exposure thermal annealing. In addition, ion bombardment methods [22–25] or sputter deposition technique have also been employed. However, use the pure nitrogen gas would be the simplest and easiest way for the nitridation of the Si surface. But a very high growth pressure is required due to the inertness of molecular nitrogen. This may lead to a huge out gassing and finally lead to a severe contamination within the nitride film. If we can provide an atomic nitrogen source, this problem can be solved. Therefore, nitrogen plasma sources can be used for the exposure of active nitrogen flux on Si surface. The atomic nitrogen exposure on Si(111) and Si(001) surfaces at relatively lower nitridation temperatures leads to the formation of amorphous nitride layer and appears with a highly disorder interface of nitride/Si. However, a crystalline interface and well-ordered films of hexagonal β-Si_3N_4 films have only been observed for nitridation temperature only above 700°C [13–17, 22–27]. Epitaxial β-Si_3N_4 formation on Si(111) by thermal annealing of N-irradiated Si(111) surface has been reported by Yamabe et al. [28]. Silicon nitride growth kinetics and surface thermodynamics at elevated temperature under ammonia exposure is recently been reported [29]. N_2-plasma assisted surface nitridation of Si(111) followed by vacuum annealing at high temperature (900°C) results in a better quality Si_3N_4 film [30].

Various analytical characterization methods have been employed to investigate the growth mechanism and structural properties of thermally grown nitride layers. In general, surface probing instruments collect the information on lateral averaging over the surface. As diffraction method, low energy electron diffraction (LEED), reflected high energy electron diffraction (RHEED) [28–29] are used for surface whereas low angle X-ray diffraction (XRD) is used for interface studies. Similarly, to investigate the chemical properties as well as bonding configuration, Auger electron spectroscopy (AES), photoelectron spectroscopy with X-ray (XPS) and ultraviolet light (UPS), electron energy loss spectroscopy (EELS), thermal desorption spectroscopy(TDS) etc. have been used. However, for atomic scale local information such as growth kinetics and surface reconstructions, direct microscopic imaging using low energy electron microscopy (LEEM), atomic force microscopy and scanning tunneling microscopy have been utilized. X-ray reflectivity (XRR) has also been carried out in some cases, to find the information about nitride layer thickness and subsurface interface.

2. Experimental details

Highly oriented p-type Si(111) wafers (boron doped, 0.02° miscut angle) were used as substrates and a clean Si(111) surface with well reconstructed 7 × 7 structure was prepared prior to the nitridation process. This preparation was routinely checked by both LEED and STM, and a clear 7 × 7 reconstruction was reproducibly observed. The sample temperature was measured using an infrared pyrometer with an uncertainty of 20°C. Si substrate nitridation process was performed by exposing the 7 × 7 reconstructed clean Si(111) surface to atomic nitrogen flux at different substrate temperatures. A commercial radio frequency (RF) plasma source from Epi Uni-bulb was used as atomic nitrogen sources. Plasma chamber pressure was maintained at about 10^{-5} mbar during plasma discharge. The pressure was controlled by a leak valve connected to the inlet line of N_2 gas. The plasma source was operated at with a RF power of 450 W. The exposure time of the active nitrogen flux on Si substrate was varied such a way that the coverage of the nitride films starts from sub-monolayer and ends up to the saturation thickness. Crystalline quality and surface reconstructions of the nitride layer were characterized by LEED. The surface morphology, growth kinetics, as well as nucleation process and the real space atomic structure were studied using an in-situ STM. Chemical composition and stoichiometry of the film are studied using ESCA microscopy and X-ray photoelectron spectroscopy, which finally provides the information about film thickness and its homogeneity.

3. Results

3.1 LEED results

An average information about the Si_3N_4 surface reconstruction grown on Si(111) and its crystalline quality after the nitridation process at different substrate temperatures ranging from 600–1100°C, have been investigated using low energy electron diffraction (LEED) method. Prior to any nitride formation, clean Si(111) surface appears with a sharp 7 × 7 LEED patterns. LEED patterns appear with a very faint "8 × 8" reconstruction for nitridation below 600°C. This finding indicates a pour crystallinity of the nitride films and mostly amorphous silicon nitrides later is formed. With increasing nitridation temperature, crystalline quality of the nitride films drastically improves and appears with a sharp "8 × 8" LEED patterns. After nitridation for about 15 min under the RF-plasma source at different temperatures, the evolution of the "8 × 8" structures has been shown in **Figure 2**. The LEED patterns of clean Si(111)-7 × 7 have been shown in **Figure 2(a)**. The yellow and cyan circles represent the (0,0) and (1,0) diffraction spots of the reconstructed Si(111) surface. After nitridation at 700°C, faint diffraction spots of Si_3N_4-"8 × 8" structure (1, 0) starts to appear. Increase the nitridation temperature results in a much brighter LEED spots along with an additional faint superstructures of "8/3 × 8/3" [6]. In case of a further increase in nitridation temperature to 920°C (**Figure 2(b)**), the LEED patterns get significantly sharpen. The "8/3 × 8/3" superstructures (red circles) appear in a much brighter contrast and dominate in lower beam energy. From the positions of the diffraction spots of the Si_3N_4-"8 × 8" structure and comparing those with the initial Si(111) diffraction spots, a periodicity of about 2.79 Å has been observed for the "8 × 8" structures. However for the "8/3 × 8/3" superstructures, the periodicity appears to be 10.2 Å.

For surface nitridation above 950°C, another type of surface reconstruction of Si_3N_4/Si(111) appears in LEED pattern, which is known as the 'quadruplet structure' or 3/4 × 3/4 reconstruction. This pattern appears in coexistence with

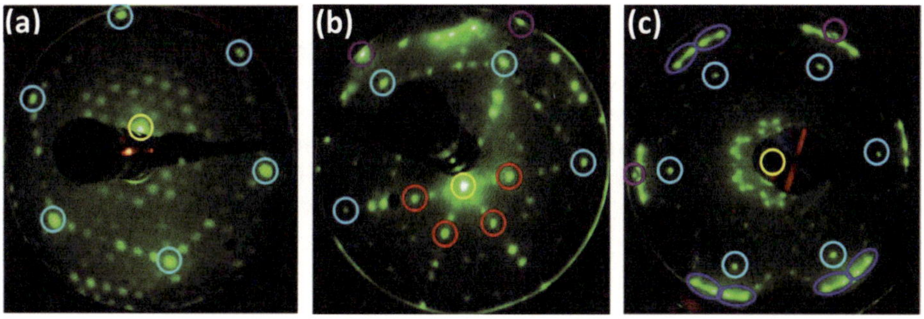

Figure 2.
LEED patterns show the evolution of the Si_3N_4-"structure with increasing nitridation temperature". (a) Clean Si(111)-7×7 structure: Yellow and cyan circles correspond to the (0,0) and the (1,0) diffraction spots. (b) Si_3N_4-"8 × 8" structures: Purple circles corresponds to the (1,0) diffraction spots of Si_3N_4-"8 × 8" structures, whereas the "8/3 × 8/3" superstructures are indicated by red circles (nitridation at 920°C, electron beam energy 43 eV). (c) Si_3N_4 'quadruplet structures': Intense quadruplet structures are indicated by blue ellipses, whereas the (1,0) diffraction spots of the Si_3N_4-"8 × 8" structure are represented by the purple circles (nitridation at 1050°C, electron beam energy 52 eV).

the existing "8 × 8" structure. With further increasing in nitridation temperature, the quadruplet structure starts to dominate over the "8 × 8" patterns. Clear LEED patterns of quadruplet structure have been observed only above 1000°C (**Figure 2(c)**). This is appeared in bright spots (blue ellipse) with the crystalline domains angular rotations of ±5° and ± 10°, with respect to the underlying Si lattice. The (1, 0) spots of the Si_3N_4 "8 × 8" structure has also been observed within the purple circles. By comparing the quadruplet structure LEED spots with the (1, 0) diffraction spot of Si(111), a surface lattice periodicity of about 2.88 Å, has been observed for the quadruplet structures. It has been observed that the quadruplet structures usually appear with a contaminated surface, sometimes even at lower nitridation temperatures [13, 31].

3.2 STM results

Exposure of atomic nitrogen on the Si(111) surface at elevated substrate temperatures results in silicon nitride formation. A clear understanding of the silicon nitride growth mechanism and its impact on the structural properties of the as grown film is of high fundamental interest. STM studies of various silicon nitride films after nitridation at various temperatures, as well as Si_3N_4 film thicknesses starting from sub-monolayer coverage up to the saturation thickness can be seen here. The initial nucleation stage of silicon nitride films, their various growth steps along with the evolution of surface morphology and finally surface atomic structures of are discussed with atomic precision.

Initial nucleation: In order to understand how the initial nucleation mechanism takes place and the absorption of N atoms on Si surface to form nitride layer, submonolayer coverage of Si_3N_4-"8 × 8" structures formation on Si(111) surface have been studied here. After a plasma exposure for 5 s, many small nucleation pits starts to appear within the terraces of the Si(111) surface (dark spots) as well as step edges exhibit many discrete nucleation patches within the upper terrace (also appeared in darker contrast) (**Figure 3**). For simplicity, henceforth, we will be mention those dotlike dark structures within the terrace area as *'nucleation pits'* and refer those step edge patches as the *'nucleation patches'* [32]. With increase in nitridation temperature, it has been observed that the density of the nucleation pits/patches both within terraces or at the step edges are significantly decreased, whereas the average size has drastically increased (**Figure 3**). Usually, the nucleation patches are larger in size as compared to the nucleation pits. Moreover, larger etch pits preferentially appear along the initial Si(111)-7 × 7 domain boundaries. For nitridation above 800°C, shape of the dark nucleation pits becomes regular (equilateral triangle). Triangular nucleation patches at the upper terrace of the step edges are also observed where the apex of the triangles point towards the [-1 -1 2] crystal direction, away from the steps. This direction of the nitride pit/patch growth is determined by comparing the faulted and unfaulted half unit cell of the Si(111)-7 × 7 (**Figure 3(d)**). A closer view STM image of this nitrified surface is shown in **Figure 4** (nitridation at 800°C). Nucleation etch pits of triangular

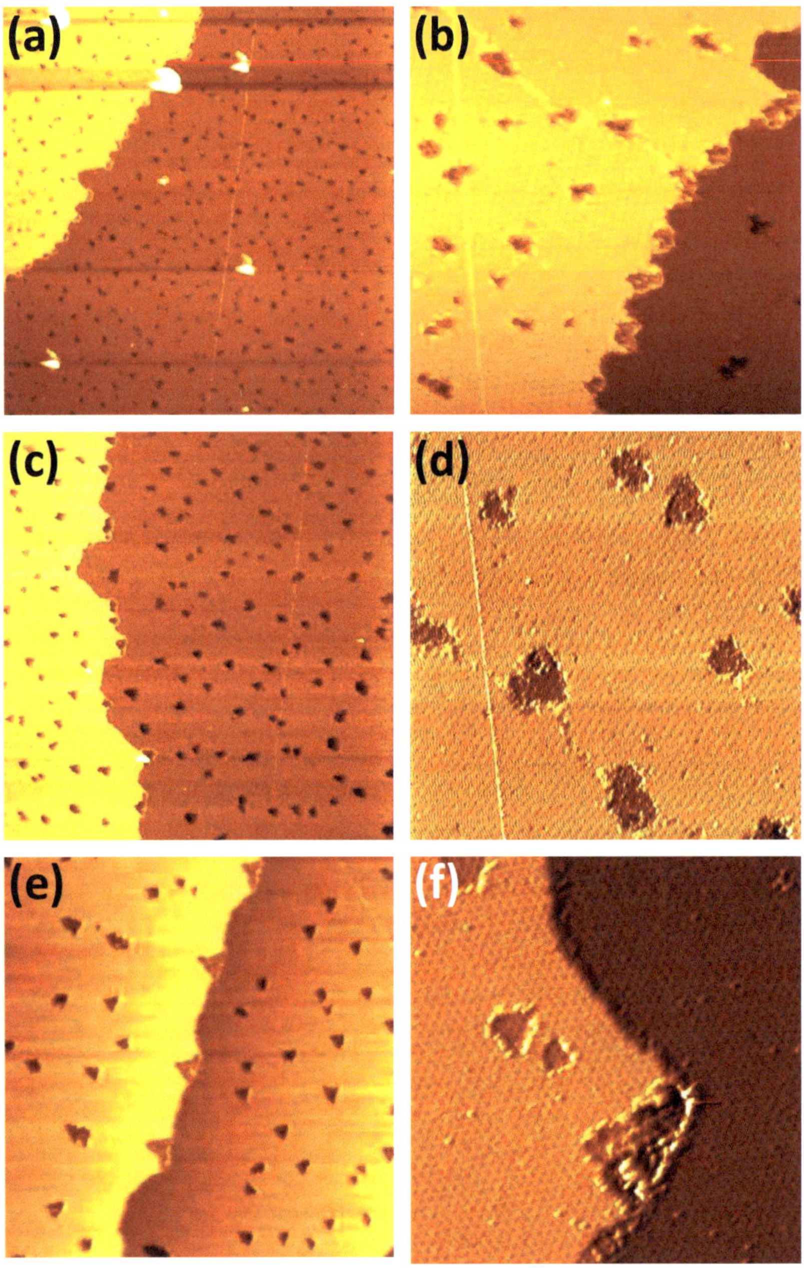

Figure 3.
STM images of submonolayer coverage silicon nitride nucleation with nitridation temperatures: [(a) and (b) 720°C, (c) and (d) 760°C, and (e) and (f)] 800°C, respectively, using rf-plasma exposure for 5 s. Scan areas: (a), (c) and (e) 400 × 400 nm² and (b), (d) and (f) 100 × 100 nm².

pattern can clearly be seen. Surprisingly, both nucleation pits appear with a clean Si(111)-7 × 7 structure with 1.5 Å lower than the Si terrace. Si and/or N adatoms are preferentially found around the outer periphery of the 7 × 7 nucleation pits and in STM image appear as bright dots (**Figure 4**). Sometimes, adatoms have also found within the larger nucleation pits. However, nucleation patches located at the step edges generally shows a disordered structure in STM imaging. **Figure 4** shows an interesting feature where a transition from Si(111)-7 × 7 nucleation patches to initial nucleation of "8 × 8" structures at the step edge has been observed. Inner side of the step shows an atomically resolved Si(111)-7 × 7 structure, whereas outer side of the step shows a Si_3N_4-"8 × 8" structure. Comparing the height difference it is confirmed that the patch appears with a similar height scale as the pits within terrace (1.5 Å, lower/higher) than the (upper/lower terrace). Furthermore, the Si(111)-7 × 7 structure exhibits the same rotational symmetry within nucleation pits and nucleation patches, having only translational shifts. Nitridation at 850°C results in atomically resolved honey-comb structures ("8 × 8") of Si_3N_4. In this case, Si(111)-7 × 7 structures are not found any more within the nucleation site.

Effect of annealing: The effect of the post-exposure thermal annealing at different temperatures for sub-monolayer coverage nitride films grown at 800°C is shown here in **Figure 5**. STM images depict the evolution of Si(111) surface after 800°C nitridation for 30 s, using RF-plasma source followed by subsequent post-annealing for one minute at 850°C and 900°C, respectively. As expected, nucleation pits (Si(111) terrace area) and nucleation patches (upper terrace of step edge) of triangular shapes are formed after 30 s nitridation at 800°C (**Figure 5(a)** and **(b)**), with an enhanced density and size as compared to earlier result presented in **Figure 4(e)** (5 s nitridation). After annealing at 850°C for 1 min, thermal energy increases the surface adatoms mobility, which significantly changes the surface morphology as shown in (**Figure 5(c)** and **(d)**). It has been observed that those triangular pits located away from the surface steps are not affected much, however, the patches/pits at the surface steps/close proximity of lower steps, are strongly affected. Some of the patches form free standing islands of silicon nitride getting detached from the upper terrace. Moreover, free standing nitride islands also appear in larger size, due to the coalescence between nucleation patches and pits. Comparing with earlier images (**Figure 5(a)**), it can be concluded that the surface steps are locally retracted towards the upper terrace, leading towards the formation of free standing nitride islands within the lower terrace

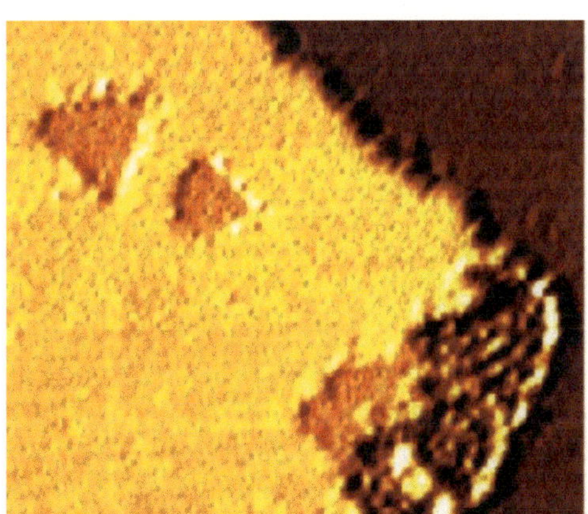

Figure 4.
Closer view STM image shows the formation of nucleation pits within Si(111)-7 × 7 surface terrace and at the step edge (nitridation at 800°C for 5 s). Scan areas 65 × 60 nm².

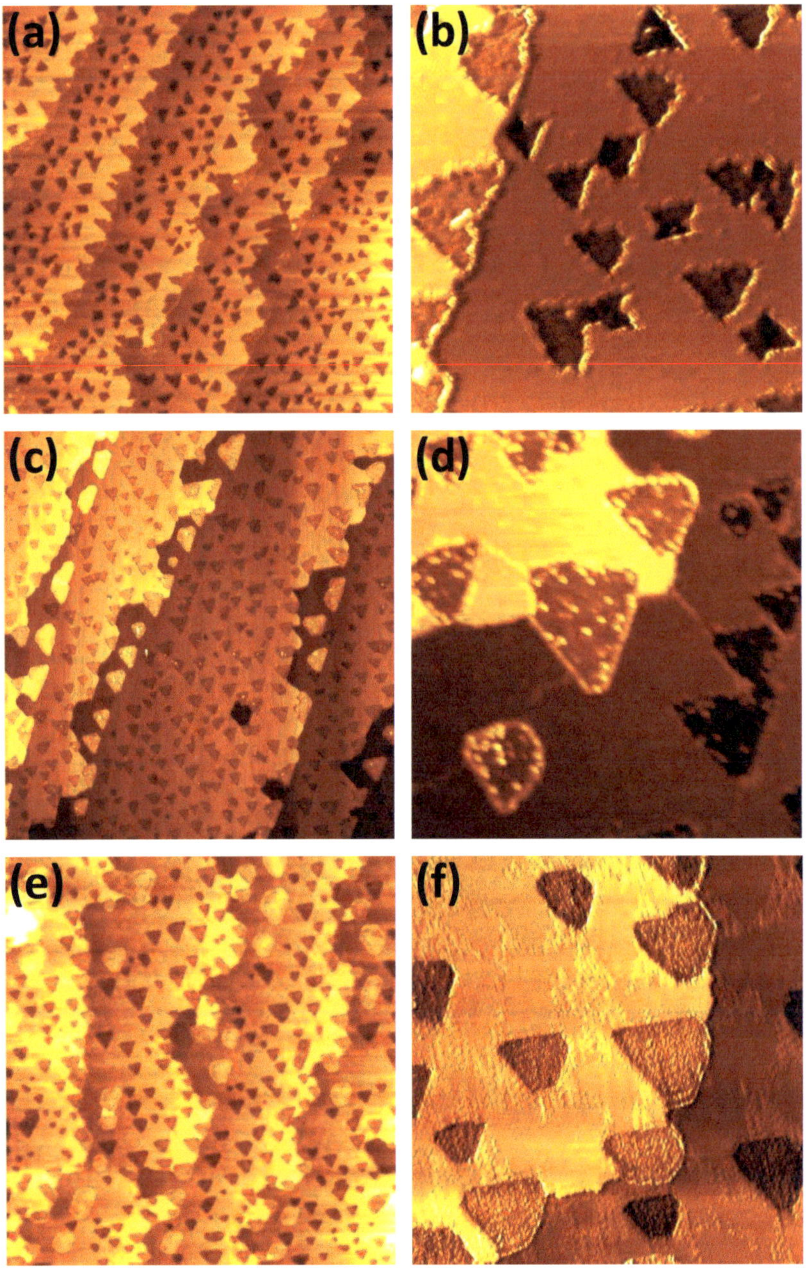

Figure 5.
STM images of evolution of Si(111) surfaces: (a) and (b) after nitridation at 800°C for 30 s, (c) and (d) post annealing at 850°C for 60 s, and (e) and (f) further post annealing at 900°C for 60 s, respectively. Scan areas: (a), (c) and (e) 1 × 1 μm² and (b), (d) and (f) 200 × 200 nm².

area. The shape of the free standing nitride islands also differs from the triangular pits/patches which have been discussed in later. Resulting changes in nitride surface morphology, after annealing at 900°C for another minute can be seen in **Figure 5(e)** and **(f)**. Further movement of the surface steps towards upper terrace and enlargement of the free standing nitride islands have been observed. Eventually, this effect may lead to a terrace breaking. In addition, nucleation of N adatoms at the initial (7 × 7) domain boundary regions has also been taken place, (appears as bright dots within the terrace), leading to the inter connection between two nitride pits/islands (**Figure 5(f)**).

The closer view STM images of free standing nitride island and nitridation at the initial domain boundary of Si(111)-7 × 7 surface are shown in **Figure 6**. The Si_3N_4 structures are now appeared with an ordered atomic arrangement of a honeycomb-like "8 × 8" (**Figure 6**). Within islands/pits, Si(111)-7 × 7 structures are not observed any more. A clearer appearance of the "8 × 8" structure also suggests a drastic improvement in nitride films crystalline quality. Apart from the structural improvements, the shape and size of nitride nucleation centers have significantly been altered. The "8 × 8" nucleation pits within the terrace area remain still triangular, however, the free standing islands become in truncated shape and start to become a hexagonal shape as shown in **Figure 6(a)** [32]. Direct nucleation of nitride layer (without forming pit/patch) has also started after second step of annealing at 900°C, at the domain boundary regions of initial the 7 × 7 terrace, as can be seen as bright dots in **Figure 6(b)**. This way, nitride layer connects between two different nucleation pits/islands and further proceeds to form a continuous layer on Si surface. The structural quality of the "8 × 8"- Si_3N_4 layer has also been improved with the nitridation/annealing temperature, which is also in agreement with our earlier LEED observations.

Thickness dependent surface morphology: STM images show the morphological evolutions of crystalline silicon nitride films grown on Si(111) surface at 850°C have been compared here with an increasing exposure durations of RF nitrogen-plasma. Structure and morphological evolution of Si(111) surface, exposed for nitridation for (a) 2 min, (b) 4 min, (c) 12 min and (d) 45 min, respectively, using RF-plasma are presented in **Figure 7**. Initial Si surface gets completely covered by "8 × 8" nitride layer after an exposure of RF plasma for 2 min (**Figure 7(a)**). Large Si_3N_4 islands with atomically flat top surfaces are observed throughout the surface. Initial terrace structure of Si(111)- 7 × 7 substrate gets completely broken and some of the nitride islands appear significantly higher (few bilayer steps of Si(111)) due to the effect of step bunching. Atomically resolved honeycomb-like structures of Si_3N_4 with "8 × 8" periodicity are obtained throughout the surface as shown within the inset of **Figure 7(a)**. For even longer exposure of RF-plasma, a thicker nitride layers is formed. The surface morphology of the nitride film grown after 4 min of nitridation is shown in **Figure 7(b)**. The surface appears in rough morphology with many grooves and holes and the initial terrace structures of Si(111) get partially broken. For even longer nitridation of 12 min a very different surface morphology is observed. Si(111) terrace structures are completely broken and large 2D islands with a flat top surfaces formed (**Figure 7(c)**). Step heights of these nitride islands

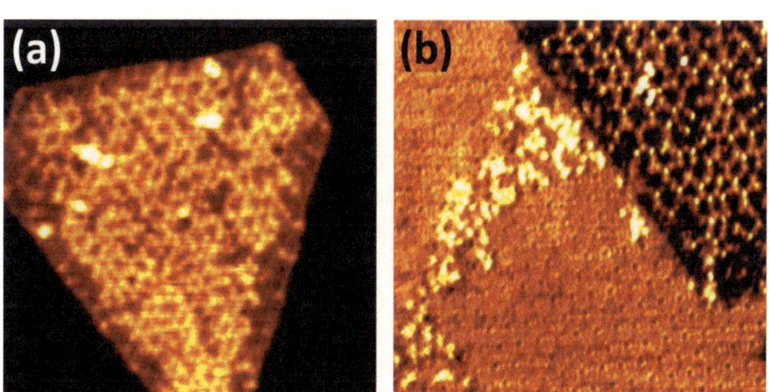

Figure 6.
Closer view STM images of (a) free standing nitride island detached from the step edge and (b) nitride formation within etch pit in the terrace area and in domain boundary.

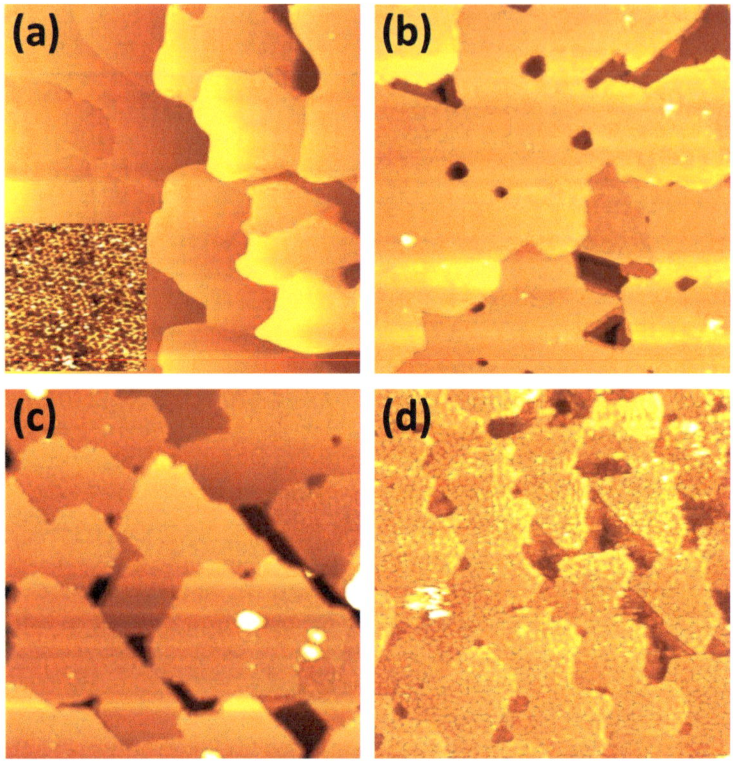

Figure 7.
STM images of silicon nitride film grown at 850°C with increasing duration of RF-plasma: (a) 2 min, (b) 4 min, (c) 12 min and (d) 45 min, respectively. Scan area: 1 × 1 μm². Inset of (a) shows atomically resolved honey comb structure of "8 × 8"- Si_3N_4.

are found to be an integer multiple of 2.9 Å. This finding suggests the formation of multilayer β-Si_3N_4. It also indicates the crystalline correlation of the as grown nitride film with the Si(111) substrate, such as: Si_3N_4 (0001) ‖ Si(111). Moreover, the holes and grooves appear even larger and deeper. In addition, the border of the nitride islands appears in a straight line manner. Very similar kind of structure is also observed within the holes and grooves. This observation confirms the formation of nitride layer within the holes/grooves and as a result finally a continuous Si_3N_4 layer is formed. After 45 min of nitridation, the Si_3N_4 film reaches to its saturation thickness as the surface morphology changes drastically (**Figure 7(d)**). Overall surface roughness is someway decreased. Holes/grooves are partially filled as they become smaller in size as well as shallower in depth. Along with, the top nitride surface becomes very granular and appears with many small clusters/blobs. A very similar kind of finding has also been reported in STM studies [29].

Surface reconstructions of β-Si_3N_4: Depending upon the nitridation parameters (temperature and duration), generally two types of silicon nitride surface reconstructions are observed in STM: (a) honeycomb-like "8 × 8" structures with a surface periodicity of 30.7 Å and(b) "8/3 × 8/3" superstructures of 10.2 Å surface periodicity. High temperature nitridation (> 1000°C) usually results in a so called 'quadruplet structures' (3/4 × 3/4) in LEED observation, however, atomically resolved structures of 3/4 × 3/4 reconstructions are not found in our STM study.

Honeycomb-like "8 × 8" reconstruction: A honeycomb-like "8 × 8" surface reconstruction of silicon nitride films are only observed in STM for a very thin Si_3N_4 films, having coverages below 2 ML. These honeycomb-like structures usually start to appear at relatively lower nitridation temperature (800°C) with high defect density with a poor order of crystal symmetry. Increase in nitridation temperature or post-nitridation thermal annealing (higher temperatures) significantly improves the ordering and structural symmetry. However, a thicker nitride film does not show any well resolved honeycomb-like surface in STM imaging.

Empty state STM images of Si_3N_4 surface grown at 850°C are shown in **Figure 8(a)** and **(b)**. Honeycomb-like "8 × 8" surface reconstruction of Si_3N_4 surface is formed after 30 s of RF-plasma nitridation. The structure appears as a defective and in quasi-periodic network of local disordering (**Figure 8(a)**). A closer view of this honeycomb-like structures nicely resolved with atomic resolution is depicted in **Figure 8(b)**. However, the short range lattice disorders can also be observed here along with the loop like structures. These disordered state can be attributed to the interfacial states of Si_3N_4/Si(111). An autocorrelation of this surface shows a periodicity of about 30.6 Å, indicating a "8 × 8"- Si_3N_4 structure [32].

"8/3 × 8/3" surface reconstruction: Apart from this "8 × 8" surface reconstructions, in case of relatively thicker Si_3N_4 films grown at a temperature between 900 and 980°C, another type of surface reconstructions known as "8/3 × 8/3" superstructure is also observed. A surface periodicity of 10.2 Å is found here and the empty state STM images of well reconstructed "8/3 × 8/3" structures appears with a long range symmetry, as shown in **Figure 8(c)** and **(d)**. The nitride film was formed by 20 min of RF-plasma exposure on clean Si(111) surface at 950°C. A long range atomic symmetry of "8/3 × 8/3" structure is nicely resolved here. This result also suggests that a highly crystalline Si_3N_4 films can also be formed by a systematically choosing the nitride growth parameters. However, few defect areas on this surface such as vacancies/missing atoms (dark areas) as well as adatoms (bright

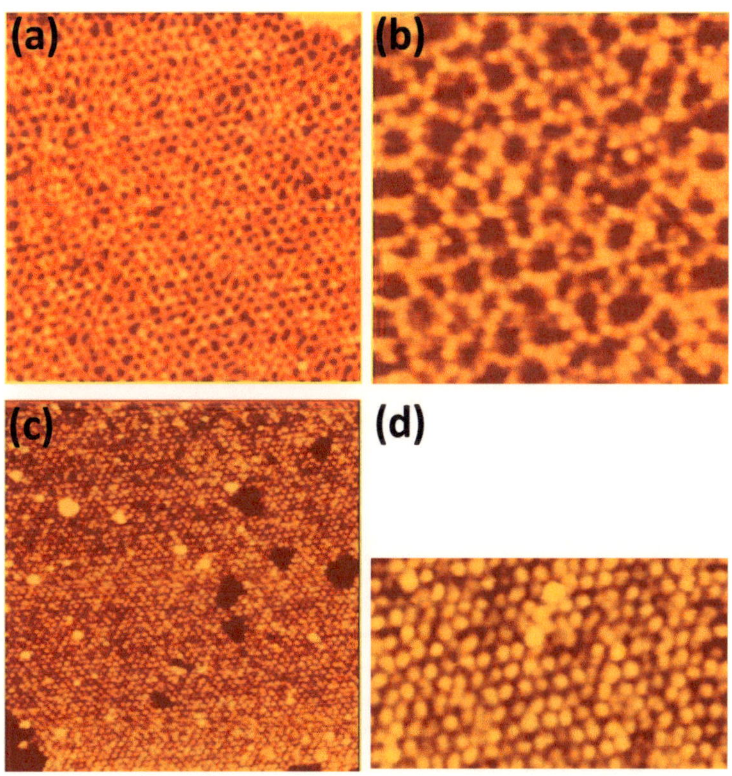

Figure 8.
Empty state STM images of Si_3N_4 surface showing atomically resolved surface of: (a) and (b) honey-comb like "8 × 8" surface reconstructions (3 V) and (c) and (d) 8/3 × 8/3 surface reconstructions (4 V).

areas) have also been observed. A closer view of this "8/3 × 8/3" surface can be seen in **Figure 8(d)**. Here, a weak modulation with 3-fold surface symmetry can also be observed by a careful inspection. This effect might be a related to the coexistence of the "8 × 8" LEED spots along with the "8/3 × 8/3" superstructure spots of nitride films. An autocorrelation measurement of this superstructure is also performed which shows a structural periodicity of about 10.2 Å, confirming the formation of "8/3 × 8/3" superstructure [6, 30].

3.3 ESCA results

Apart from STM imaging, ESCA microscopy technique is also employed to investigate the of surface chemical properties of the thin silicon nitride films grown on clean Si(111) substrates. Contrast observed in ESCA microscopy image usually occurs from two factors: (a) chemical inhomogeneity, i.e., any kind of compositional fluctuation and (b) film thickness inhomogeneity, i.e., surface roughening leading towards nitride layer thickness fluctuation. To perform a comparative study, two types of nitride samples are used here. In one hand, a flat and smooth nitride film grown at a relatively low nitridation temperature is used. In other hand, nitride film with a rough surface morphology grown by nitridation at higher temperature is tested.

In case of smooth nitride films, a very week modulation in ESCA image contrast has been observed for both scans using Si-2p bulk and nitride binding energies (images not shown here). ESCA microscopy image using N-1 s components also appears in a very similar manner. Both findings indicate towards the surface chemical homogeneity of the nitride film. To further confirm the chemical uniformity of this surface, high resolution XPS study has also been tested using of Si-2p photoemission line. A chemical shift of 2.9 eV has been found within the Si-2p spectra. This information can suggest the nitride film stoichiometry and in good agreement with the reported literature for Si_3N_4 compound formation [33, 34]. The thickness of the nitride film can also be estimated using this known Si_3N_4 stoichiometry. Total integral intensities of both nitride and bulk Si-2p components are used to calculate the film thickness. A detailed analysis of the nitride film thickness calculation has already been reported elsewhere [35, 36].

However, for nitride films of rough surface morphology, grown at higher nitridation temperatures appear with a strong contrast in ESCA microscopy images as shown in **Figure 9**. Photoelectron signals from: (a) N-1 s line and (b) Si-2p bulk line are used for the spectro-micrographs here. An inverted contrast in ESCA microscopy images is observed here appear with a strong contrast (**Figure 9(a)** and **(b)**). This finding can be explained in terms of nitride film thickness fluctuation. The existence of any bare Si(111) surface can be excluded from our earlier STM observations, where a continuous nitride layer with a thickness modulation was observed for this type of nitridation process.

In case of a smooth and uniform nitride films, the contrast in ESCA microscopy image for N-1 s line can be attributed to silicon nitride film stoichiometry, i.e., chemical inhomogeneity. In that case, Si-2p bulk component should be homogeneous and the ESCA microscopy is expected to be without any significant imaging contrast. In contrast, however, a strong contrast is observed also for the Si-2p bulk line. It is a clear indication of thickness fluctuation rather any chemical inhomogeneity within the nitride film. **Figure 9(c)** and **(d)** show a further investigation related to the film thickness variation of nitride films. **Figure 9(c)** shows a closer view of ESCA micrograph using the N-1 s line. Individual XPS spectra of the Si-2p binding energy measured at two different locations (marked bright and the dark areas within the ESCA micrographs) are recorded and the correlation between these

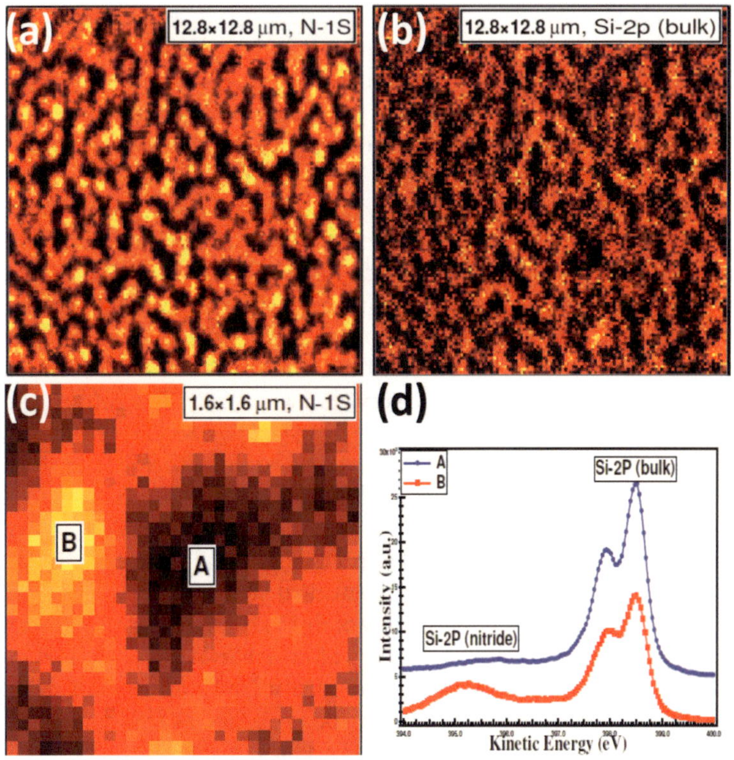

Figure 9.
ESCA micro-spectrograph of silicon nitride film grown at relatively higher nitridation temperature (950°C) (a) with N-1 s line, (b) with Si-2p bulk line, (c) closer view with N-1 s line and (d) Si-2p XPS spectra at the corresponding location mentioned within image (c).

two spectra is compared (**Figure 9(c)**). Both spectra clearly indicate the contrast within the ESCA micrograph is mostly due to the lateral fluctuation in the nitride film thickness rather not due to any chemical inhomogeneity.

4. Discussion

From our STM studies of silicon nitride formation it is quite clearly that the initial Si_3N_4 nucleation always starts at the surface steps of Si(111) substrate in two ways: (a) in one hand direct formation of nucleation patches at the step edges of initial Si(111) surface or (b) on other hand formation of additional steps by creating triangular nucleation etch pits within the 7 × 7-Si(111) terrace areas. Initial nucleation at the surface steps can be correlated to the defect induced nucleation of N atom. As more dangling bonds are available at the step edges, they act as chemically more active with respect to the terrace area of 7 × 7-Si(111) domains. As a result, it can facilitate the initial nucleation of nitride layer. In addition, surface steps give more degrees of freedom for the nitride layer growth leading towards the relaxation of nitride lattice strain induced from the mismatch between the deposited nitride material and the Si substrate lattice.

Larger size of the nitride nucleation centers with a lower number density for higher nitridation temperature are explained in terms of enhanced thermal diffusion of the N and Si adatoms. Thermal diffusion starts to dominate above 750°C, which results in a heterogeneous nucleation of pits/patches at substrate defects, i.e. at Si(111)-7 × 7 domain boundaries and surface steps. Si_3N_4 nucleation above 850°C occurs exclusively at the surface steps, which can concluded as follows. There are no such domain boundaries of 7 × 7 structure due to surface phase transition at about 830°C from the (7 × 7) → (1 × 1) [37]. Moreover, at higher nitridation temperature the surface diffusion of Si and N adatoms gets larger. The triangular shape of the etch pits/patches may be influenced by the three-fold crystal symmetry of the underlying Si(111)-7 × 7 substrate.

Morphological changes during post-annealing as well as improvement in structural quality can also be attributed to the temperature induced enhanced surface diffusion of Si and N adatoms. In one hand, high temperature promotes a better diffusion of surface Si atoms. On other hand, crystalline Si_3N_4 formation under atomic N exposure demands a proper supply of adatom species. Hence, an improvement in crystalline quality is expected for the higher nitridation temperature. LEED observations can also be explained in a very similar manner. The continuous supply of Si atoms can be linked to groove and hole formation on Si(111) surface by removal of the upper terrace. This will act as a Si source for further nitride formation. The enlargement of nitride nucleation patches or free standing islands can also be explained in terms of coalescence effect or by local nitride re-growth. The transition of free change standing nitride islands from a triangular to truncated /distorted-hexagonal shape can be explained by crystal rotational symmetry *i.e.,* three fold symmetry of Si(111) and the hexagonal symmetry of Si_3N_4. Nitride nucleation through etch pits formation around the domain boundary of initial Si(111)-7 × 7 is also related to defect induced nucleation, as found for nitridation along surface steps. 2D islands of Si_3N_4 with multiple steps/ step bunching can be explained as follows. Outer diffusion of Si-atoms can easily occur at the step-edges and it can act as a source for Si atoms for further nitride formation. In this way, steps are getting larger and deeper with nitridation temperature and provide the Si supply. As a result, it turns in to a deep crack formation. This type of surface morphology is generally found for 'quadruplet structures' where the growth temperature usually stays above 1000°C. However, for saturation thickness flat terrace area with little roughening is observed, which might be due to the ion induced damage, caused by the long exposure of plasma source.

The atomically resolved honey-comb like "8 × 8" surface reconstruction of the β-Si_3N_4 (0001) surface can only be observed for very thin nitride films (coverage below 2ML), which appears in a loop like way. In addition, no real long range symmetry has been found in STM imaging. This honeycomb-like appearance with many local disordering in STM images can be correlated in following ways. In one hand, the STM images can easily get influenced by the Si_3N_4-Si(111) interface states (very thin film). As there is a significant lattice mismatch between the substrate and epilayer, it may lead to a highly distorted bonding configuration. On other hand, STM imaging can also be modulated by the local electronic states of underlying Si(111) substrate (only valid for thin overlayer). In contrast, "8/3 × 8/3" super structures show a clear atomic reconstructions with long range symmetry. As it appears for a relatively thicker layer, background electronic influence such as sub-surface information can be ignored. However, a weak three fold surface modulation has been observed for thicker nitride films which is in good agreement with the proposed structural model of Bauer et al. [13]

STM findings of roughening of the silicon nitride surface at higher nitridation temperature are also complementary with our ESCA microscopy results. Nitride films appear as chemically homogeneous, however, contrast in the ESCA images are mostly due to the thickness fluctuation of the nitride film. The stoichiometry, i.e., chemical composition of the nitride films is found to be Si_3N_4 for both, smooth and rough surface morphology nitride films grown at lower and higher nitridation temperatures, respectively. The contrast in ESCA microscopy is only observed for higher nitridation temperatures as it results in formation of nitride films of inhomogeneous thickness.

5. Conclusions

In summary, high quality crystalline silicon nitride thin films have successfully been grown on clean Si(111) substrate by elevated temperature exposure of active nitrogen from a RF-plasma source. Initial nucleation process, nitridation temperature and atomic N exposure duration dependent films structure and morphology, surface atomic reconstructions and chemical properties of the β-Si_3N_4 /Si(111) have been investigated in details using different surface characterization techniques such as STM, LEED and ESCA microscopy. Initial Si_3N_4 nucleation strongly determine by the Si(111) surface defects. It always occurs at the step edges (upper terrace) and terraces by nucleation pit formation. Lower nitridation temperature generally results in nitride films of poor crystalline quality but appears with a smooth surface morphology. Whereas, a highly crystalline Si_3N_4 film can be achieved for nitridation at higher temperatures. Moreover, the surface morphology gets severely roughen by forming holes and grooves on Si(111) terrace. An atomically resolved honeycomb-like reconstruction of "8 × 8" surface periodicity is observed for very thin Si_3N_4 films. However, for thicker films grown at higher nitridation temperature shows an atomically resolved "8/3 × 8/3" superstructure in STM. Both the findings are complementary and in good agreement with LEED results. ESCA microscopy measurements confirm a Si_3N_4 stoichiometry of the nitride films. It also suggests a thickness fluctuation for a nitride growth at higher temperature. Finally, this type of crystalline Si_3N_4 films have a huge potential to successfully replace the existing SiO_2 dielectric layer on Si(111) for device technology. Furthermore, it can also provide a platform for crystalline growth of group III nitrides on Si(111), which can further integrate the optoelectronic devices to the existing well established Si based technology [38].

Acknowledgements

The author is very much grateful to Prof Jens Falta and his coworkers of University of Bremen, Germany, for all kind of experimental supports as well as all short of valuable scientific knowledge and discussions.

Author details

Subhashis Gangopadhyay
Department of Physics, Birla Institute of Technology and Science (BITS) Pilani, Rajasthan, India

*Address all correspondence to: subha@pilani.bits-pilani.ac.in

References

[1] Watanabe M, Suemasu T, Murakte S, Asada M. Negative differential resistance of metal ($CoSi_2$)/insulator (CaF_2) triple-barrier resonant tunneling diode. Applied Physics Letters. 1993;**62**:300

[2] Jiang X, van Dijken S, Parkin SSP. Giant magnetocurrent exceeding 3400% in magnetic tunnel transistors with spin-valve base layers. Applied Physics Letters. 2002;**80**:3364

[3] Garfunkel E, Gusev EP, Vul A, editors. Fundamental Aspects of Ultrathin Dielectrics on Si-Based Devices. Dordrecht: Kluwer Academic Publishers; 1998

[4] Gusev EP, Lu H-C, Garfunkel EL, Gustafsson T, Green ML. Growth and characterization of ultrathin nitrided silicon oxide films. IBM Journal of Research and Development. 1999;**43**:265

[5] Tsai S-J, Wang C-L, Lee H-C, Lin C-Y, Chen J-W, Shiu H-W, et al. Approaching defect-free amorphous silicon nitride by plasma-assisted atomic beam deposition for high performance gate dielectric. Scientific Reports. 2016;**6**:28326

[6] Gangopadhyay S, Schmidt T, Falta J. N-plasma assisted MBE grown GaN films on Si(111). Physica Status Solidi B. 2006;**243**:1416

[7] Nakada Y, Aksenov I, Okumura H. GaN heteroepitaxial growth on silicon nitride buffer layers formed on Si (111) surfaces by plasma-assisted molecular beam epitaxy. Applied Physics Letters. 1998;**73**:827

[8] Liu AY, Cohen ML. Structural properties and electronic structure of low-compressibility materials: β-Si_3N_4 and hypothetical β-C_3N_4. Physical Review B. 1990;**41**:10727

[9] Gruen R. The crystal structure of β-Si_3N_4: Structural and stability considerations between α- and β-Si_3N_4. Acta Crystallographica. Section B. 1979;**35**:800

[10] Hitchman MA, Jensen KF. Chemical vapour deposition. London: Academic Press; 1993

[11] van Bommel AJ, Meyer E. A low energy electron diffraction study of the PH_3 adsorption on the Si (111) surface. Surface Science. 1967;**8**:381

[12] Kubler L, Bischo JL, Bolmont D. General comparison of the surface processes involved in nitridation of Si(100)-2×1 by NH_3 and in SiN_x film deposition: A photoemission study. Physical Review B. 1988:13113

[13] Bauer E, Wei Y, Mueller T, Pavlovska A, Tsong IST. Reactive crystal growth in two dimensions: Silicon nitride on Si(111). Physical Review B. 1995;**51**:17891

[14] Yoshimura M, Takahashi E, Yao T. Initial stages of the nitridation of the Si(111) surface observed by scanning tunneling microscopy. Journal of Vacuum Science and Technology B. 1996;**14**:1048

[15] Wang X-S, Zhai G, Yang J, Cue N. Crystalline Si_3N_4 thin films on Si(111) and the 4×4 reconstruction on Si_3N_4(0001). Physical Review B. 1999;**60**:R2146

[16] Zhai G, Yang J, Cue N, Wang X-S. Surface structures of silicon nitride thin films on Si(111). Thin Solid Films. 2000;**366**:121

[17] Ahn H, Wu C-L, Gwo S, Wei CM, Chou YC. Structure determination of the Si_3N_4/Si(111)- (8×8) surface: A combined study of Kikuchi electron holography, scanning

tunneling microscopy, and ab initio calculations. Physical Review Letters. 2001;**86**:2818

[18] Wiggins MD, Baird RJ, Wynblatt P. Thermal nitridation of Si(111) by nitric oxide. Journal of Vacuum Science and Technology. 1981;**18**:965

[19] Nishijima M et al. Reactions of NO with the Si(111) (7 × 7) surface: EELS, LEED and AES studies. Surface Science. 1984;**137**:473

[20] Avouris P, Wolkow R. Atom-resolved surface chemistry studied by scanning tunneling microscopy and spectroscopy. Physical Review B. 1989;**39**:5091

[21] Roettger B, Kliese R, Neddermeyer H. Adsorption and reaction of NO on Si(111) studied by scanning tunneling microscopy. Journal of Vacuum Science and Technology B. 1996;**14**:1051

[22] Ha JS, Park K-H, Yun WS, Lee E-H, Park S-J. Evolution of surface morphology in the initial stage of nitridation of the Si(111)-7×7 surface by nitrogen ions. Journal of Vacuum Science and Technology B. 1997;**15**:1893

[23] Ha JS, Park K-H, Yun WS, Lee E-H, Park S-J. Interaction of low-energy nitrogen ions with an Si(111)-7×7 surface: STM and LEED investigations. Applied Physics A: Materials Science & Processing. 1998;**66**:S495

[24] Ha JS, Park K-H, Yun WS, Lee E-H. Nanometer scale selective etching of Si(111) surface using silicon nitride islands. Journal of Vacuum Science and Technology B. 1998;**16**:2806

[25] Ha JS, Park K-H, Yun WS, Ko Y-J, Kim SK. Interaction of nitrogen with Si(111)-7×7 surfaces at elevated temperatures. Surface Science. 1999;**426**:373

[26] Morita Y, Tokumoto H. Origin of the 8/3×8/3 superstructure in STM images of the Si(111)-8×8:N surface. Surface Science. 1999;**443**:L1037

[27] Morita Y, Tokumoto H. Bias-dependent atomic images of a quadruplet silicon-nitride monolayer on the Si(111) surface. Surface Science. 2000;**466**:L802

[28] Yamabe N, Yamamoto Y, Ohachi T. Epitaxial growth of b-Si_3N_4 by the nitridation of Si with adsorbed N atoms for interface reaction epitaxy of double buffer AlN(0001)/b-Si_3N_4/Si(111). Physica Status Solidi C: Current Topics in Solid State Physics. 2011;**8**:1552

[29] MansurovVG,MalinaTV,GalitsynYG, Shklyaev AA, Zhuravlev KS. Kinetics and thermodynamics of Si(111) surface nitridation in ammonia. Journal of Crystal Growth. 2016;**441**:12

[30] Wu C-L, Chen W-S, Su Y-H. N2-plasma nitridation on Si(111): Its effect on crystalline silicon nitride growth. N2-plasma nitridation on Si(111): Its effect on crystalline silicon nitride growth. Surface Science. 2012;**606**:L51

[31] Tabe M, Yamamoto T. Initial stages of nitridation of Si(111) surfaces: X-ray photoelectron spectroscopy and scanning tunneling microscopy studies. Surface Science. 1997;**376**:99

[32] Gangopadhyay S, Schmidt T, Falta J. Initial stage of silicon nitride nucleation on Si(111) by rf plasma-assisted growth. Surface Science and Nanotechnology. 2006;**4**:84

[33] Ermolieff A, Bernard P, Marthon S, Camargo da Costa J. Nitridation of Si (100) made by radio frequency plasma as studied by in situ angular resolved x-ray photoelectron spectroscopy. Journal of Applied Physics. 1986;**60**:3162

[34] Aballe L, Gregoratti L, Barinov A, Kiskinova M, Clausen T, Gangopadhyay S, et al. Interfacial interactions at Au/Si_3N_4/Si(111) and Ni/Si_3N_4/Si(111) structures with ultrathin nitride films. Applied Physics Letters. 2004;**84**:5031

[35] Schmidt T, Clausen T, Gangopadhyay S, Falta J, Heun S, Gregoratti L, et al. Spectro-microscopy of ultra-thin SiN films on Si (111). Nuclear Instruments and Methods in Physics Research Section B. 2003;**200**:79

[36] Falta J, Schmidt T, Gangopadhyay S, Clausen T, Brunke O, Flege JI, et al. Ultra-thin high-quality silicon nitride films on Si(111). Europhysics Letters. 2011;**94**:16003

[37] Fukaya Y, Shigeta Y. New phase and surface melting of Si(111) at high temperature above the (7×7)–(1×1) phase transition. Physical Review Letters. 2000;**85**:5150 and references therein

[38] Chen W-C, Yu T-Y, Lai F-I, Chen H-P, Lin Y-W, Kuo S-Y. Growth of catalyst-free hexagonal pyramid-like InN nanocolumns on nitrided Si(111) substrates via radio-frequency metal–organic molecular beam epitaxy. Crystals. 2019;**9**:291

Chapter 3

Nanostructured Silicon Sensors

Huseyn M. Mamedov

Abstract

Nanostructure porous silicon layers with systematically varied pore size of 8–70 nm were fabricated onto the p-type c-Si wafers using electrochemical anodizing method from HF + ethanol and HF + ethanol + $CdCl_2$ solutions (hereafter PS and PSCD, respectively). Gas and photo sensors based on c-Si/PS (or PSCD)/CdS and c-Si/PS/$Cd_{0.4}Zn_{0.6}O$ heterojunctions were synthesized by depositing CdS and $Cd_{0.4}Zn_{0.6}O$ films onto the c-Si/PS (or PSCD) substrates by electrochemical deposition (ED hereafter). The morphology of the PS and PSCD layers, CdS, and $Cd_{0.4}Zn_{0.6}O$ films was studied using scanning electron microscopy (SEM). Gas- and photosensitivity properties of heterojunctions were studied as a function of pore size. The optimal pore size is determined, which provides the maximum gas- and photosensitivity of heterojunctions in this study. It was established that the heterojunctions based on PSCD possess higher gas- and photosensitivity than heterojunctions based on PS.

Keywords: porous silicon, electrochemistry, nanostructure, thin film, photo sensors, gas sensors

1. Introduction

Currently, materials prepared by low-temperature electrochemical processing of semiconductors, in particular PS, formed by electrochemical etching of monocrystalline silicon (c-Si) are of particular interest. Anodic polarization of c-Si in hydrofluoric acid solution leads to formation of controlled network of pores of various morphology, size, and orientation. It is generally known that the surface modification of the silicon wafer plays a major role in the sensitivity enhancement of gas and photo sensors [1, 2, 7]. So, the porous surface of silicon layers participates in the processes of light absorption, gas adsorption, and desorption.

On the other hand, nanostructured PS has emerged as an attractive material in the field of photoelectronics due to its broadband gap, wide optical transmission range, favorable absorption spectrum, and surface texture. The surface roughness and low effective refractive index which can reduce reflection losses of sunlight radiation are the primary benefits offered by PS over c-Si [3–5]. A highly porous PS layer can enhance the efficiency of solar cells by increasing light trapping in the active region [6], solving the lattice mismatch problem, and surface reflection is also corrected due to the refractive index of silicon as reported by several other authors [7–10].

So, PS has disadvantages, also. The surface of the PS was covered with Si-H_x bond groups immediately after the deposition. In the process of storage in air, the Si-H_x bond groups are replaced by Si-O_x bond groups, and ultimately, silicon nanocrystallites are covered by an amorphous layer which is the main reason for the instability and degradation of all electrical, photovoltaic, photoluminescent, and sensor devices based on PS. There are various technological methods used for the passivating of the PS surface. For example, in [11], the passivating of PS was performed at the excretory hydrogen plasma. Passivating of PS can also be done by inserting different elements into the matrix. For example, in [12], the authors employed carbohydrate solutions in the PS matrix and then tried to carbide the matrix by thermal annealing. As the authors point out, carbohydrates decompose at a temperature of 2000°C and are converted to carbon and water vapor. Since carbohydrates are not volatile, almost all carbon in them remains in the pores. At this time, the size of the molecules of the selected carbohydrate should be smaller than the size of the pores. The authors have chosen sucrose as a conventional carbohydrate product. In some cases, the process of passage of hydrogen and oxygen in the PS was carried out directly with the deposition process. For this purpose, various salts (AuCl3, FeCl3, NaNO2, KIO3, CrO3, etc.) were added to the solution in the deposition process [13–18]. In [13, 14], gold and iron chloride salts were added at different concentrations during the dissolution of PS. The main purpose of the study was to replace the non-stable Si-H_x complexes by Si-Au or Si-Fe stable bonds. It has been established that anodizing in a metallic atmosphere not only stabilizes but also improves the electrical and optical parameters of PS.

Therefore, for the purpose of comparison, the results of investigations of the morphological, electrical, and photoelectrical parameters of heterojunctions p-Si/CdS and p-Si/$Cd_{0.4}Zn_{0.6}O$ based on PS [19, 20] and PSCD are considered in this chapter.

2. Preparation p-Si/PS (or PSCD)/CdS and p-Si/PS (or PSCD)/ $Cd_{0.4}Zn_{0.6}O$ heterojunctions

P-type single-crystal Si wafer with orientation of (100), resistivity of 0.01–2.5 Ohm cm, and thickness of 0.2–0.6 μm was etched through an electrochemical process to produce the porous structure. Before anodizing, the c-Si surface was cleaned from the SiO_2 oxide layer as well as contaminants, in an aqueous solution of hydrofluoric acid (HF), washed with deionized water at a temperature of 80°C and ethyl alcohol, and then dried in air. Anodizing of the c-Si substrate surface was carried out in a Teflon chamber with a platinum cathode [19, 20].

HF:ethanol (1,1) solutions with and without $CdCl_2$ (aqueous solution of $CdCl_2$ in 10:1 concentration was added to solution) were used for the formation of porous silicon. The anodizing current was 40–70 mA/cm^2. Depending on the anodizing time (30–1800 seconds) and potential in solutions, PS layers (prepared from solution without CdCl2) and PSCD layers (prepared from solution with CdCl2) with the pore sizes of 8–70 nm were prepared on the c-Si surface. For the comparison, electrical and photoelectrical properties of heterojunctions on the base of PS and PSCD layers, prepared at same anode potentials (20, 25, and 30 V), were investigated. After the formation of PS and PSCD layers, the samples were immersed in ethyl alcohol, dried in air, and placed in an electrochemical bath for deposition of nanostructured CdS and $Cd_{0.4}Zn_{0.6}O$ films.

Electrochemical deposition of CdS and $Cd_{0.4}Zn_{0.6}O$ films onto the c-Si/PS and c-Si/PSCD substrates was carried out at a temperature of 80°C in an aqueous solution of $CdCl_2$ + $Na_2S_2O_3$ and $Zn(NO_3)_2$ + $Cd(NO_3)_2$ (99.5% purities) salts, respectively (**Table 1**) [19–21]. Cyclic voltammetry based on obtaining and decoding the current–voltage dependences of the polarized electrode-electrolyte solution interface was used to control electrochemical reactions in each solution, and then in their combined solution with the same concentration and pH. Cyclic

Samples	Mole fraction of salts (mM)		Deposition current and potential	
CdS	$CdCl_2$	$Na_2S_2O_3$	J (mA/cm²)	U_c (V)
	200	50	20	-0.82
$Cd_{0.4}Zn_{0.6}O$	$Zn(NO_3)_2$	$Zn(NO_3)_2$	J (mA/cm²)	U_c (V)
	5.92	3.95	3.4	-1,2

Table 1.
Mole fraction of salts and deposition current and potential for the CdS and $Cd_{0.4}Zn_{0.6}O$ films.

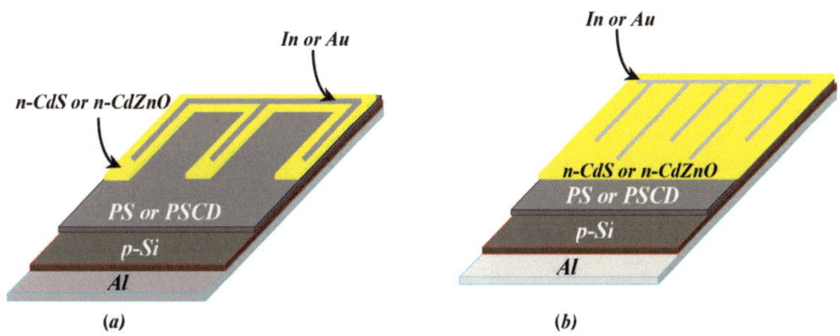

Figure 1.
Schematic structure of gas sensors (a) and solar cells (b) on the basis of PS and PSCD

voltammograms were scanned in the potential range from 1.2 to −1.2 V with the graphite (or Ag/AgCl) electrodes. Depending on the deposition time and pore size of the substrate, the 100–600-nm-thick CdS and $Cd_{0.4}Zn_{0.6}O$ films with various surface morphologies were deposited. All the films showed n-type conductivity.

In order to fabricate the heterojunctions, an ohmic In (or Au) electrode, in grid form, was evaporated on the CdS and $Cd_{0.4}Zn_{0.6}O$ films with an area of ~0.62 cm^2 (**Figure 1**). Aluminum (Al) was evaporated on the back side of the p-Si wafer as the ohmic electrode, followed by annealing at 500°C in a vacuum for 20 minutes.

3. Morphological, elemental, structural, and optical analyses

SEM images were obtained using a JEOL JSM-7600F and ZEISS Gemini 300 microscopes with energy dispersive spectroscopy (EDS) analysis.

Figure 2(a) illustrates the SEM images for PS surface formed at anodizing potential of 20 V, current of 10–40 mA/cm^2, and time of 1800 sec. As can be seen from the figure, only nonhomogeneously distributed cavities are formed on the surface of the p-Si surface, after the anodizing in solution without CdCl2. Only in some parts of the surface appear the pore cores.

EDS spectrum shows a very small amount of hydrogen in the surface of these samples (**Figure 2(b)**). This fact proves only electropolishing of the c-Si surface in HF + H2O + ethanol solution at low anode potentials.

Unlike that, SEM images for PSCD layers, formed at anodizing potential of 20 V and current of 40 mA/cm^2 in HF + $CdCl_2$ + H_2O + ethanol solution, show pores with very small dimension on the surface (**Figure 3(a)**). EDS spectrums confirm that Cd and hydrogen are on the surface of the layers (**Figure 3(b)**). The results show that Cd^{2+} ions together with Si^{2+} participate in charge exchange and accelerate the formation of initial growing piths on c-Si surface.

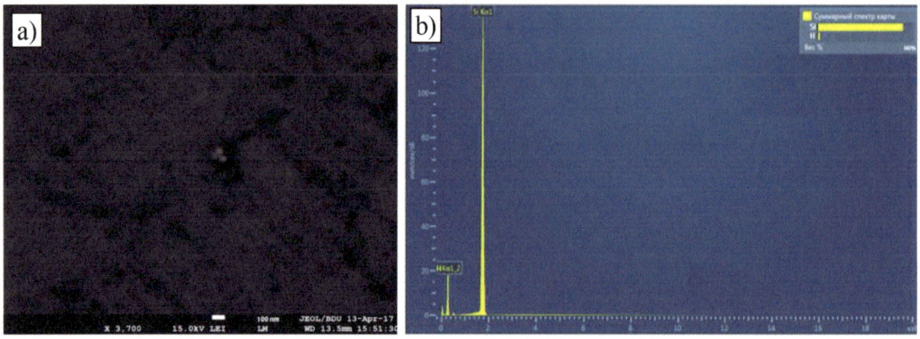

Figure 2.
SEM image (a) and EDS spectrum (b) of the PS samples prepared at anodizing voltage of 20 V and current of 40 mA/cm².

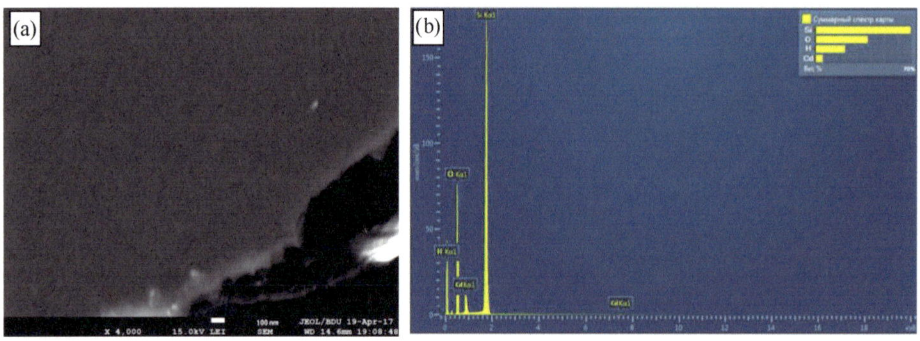

Figure 3.
SEM image (a) and EDS spectrum (b) of the PSCD samples prepared at anodizing voltage of 20 V and current of 40 mA/cm².

As a result, along with Si, cadmium atoms also deposit at the bottom and inside walls of the pores, causing the reduction of non-stable Si-H_x bond concentration and their replacement by Si-Cd bonds. Subsequently, the neutralizing Cd atoms in the pores as a result of the charge exchange determine the structure of the pores. EDS spectrum of PSCD layers indicates the presence of Cd atoms on the pores.

Increasing the anode potential up to 25–30 V (at current of 10–40 mA/cm^2) changes the nature of the anodizing—formation of pores is accelerating. SEM images show oval or spherical pores, which are distributed nonhomogeneously on the surface for PS samples (**Figure 4(a)**). The average dimensions of spherical-shaped pores were about 7–30 nm, and dimensions of oval-shaped pores were about 10–110 nm. Distribution of such pores on PS surface may be due to an irregular distribution of the anode current at Si-electrolyte boundary or relative weakening of the Si^{2+} ion neutralization process because of "charge deficiency" in certain parts of the surface.

EDS spectrums show the oxygen in PS layers after keeping in the open air (**Figure 4(b)**). True, the PS samples also have Si-H_x bonds, but the unstable bonds lead to degradation of parameters of the devices (gas sensors and solar cells). Si-O_x bond leads also to an increase of the PS resistivity.

SEM investigations show that, unlike PS layers, pores in PSCD layers, prepared at anodizing voltage of 30 V, are distributed homogeneously and almost show spherical form cavities (**Figure 5(a)**). The subsequent increase of anode voltage (up to 40 V) almost does not change the size (**Table 2**) and shape of the pores. This fact shows that the Cd^{2+} ions promote a uniform distribution of charges and anode voltage across the entire surface at the silicon-electrolyte interface. Uniform

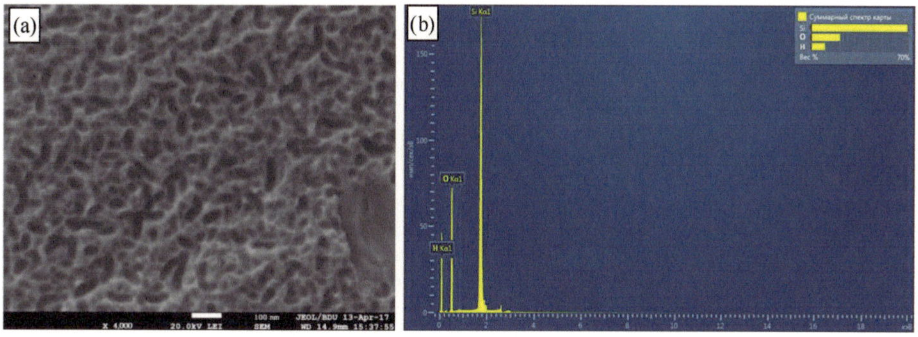

Figure 4.
SEM image (a) and EDS spectrum (b) of the PS samples prepared at anodizing voltage of 30 V and current of 40 mA/cm^2.

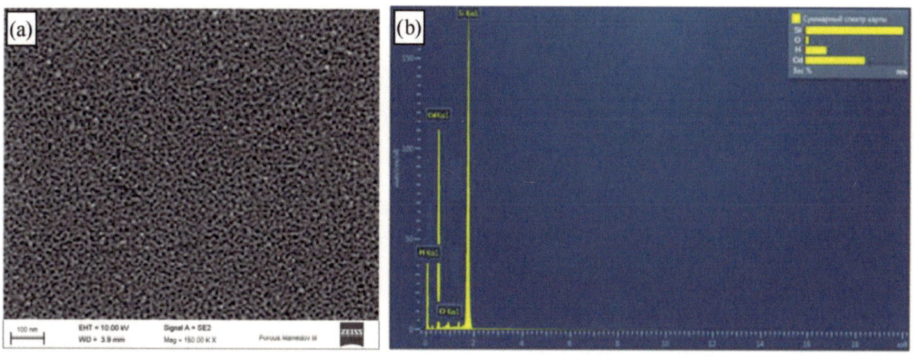

Figure 5.
SEM image (a) and EDS spectrum (b) of the PSCD samples prepared at anodizing voltage of 30 V and current of 40 mA/cm^2.

Samples	Anodizing voltage (V)	Anodizing current density (mA/cm^2)	Anodizing time (s)	Pore sizes (nm)
PSCD1	30	40	1800	8÷11
PSCD2	30	55	1800	10÷16
PSCD3	30	70	1800	30÷70
PSCD2	34	55	1800	10÷15
PSCD2	36	55	1800	11÷17
PSCD2	40	55	1800	10÷19
PSCD2	30	55	1200	9÷16
PSCD2	30	55	400	10÷17

Table 2.
Anodizing parameters of PSCD layers.

distribution of voltage on the silicon surface results in formation of pores with spherical shape. It is established that the size of the pores can be regulated only by anodizing current (**Table 2**). EDS spectrums testify an increase of Cd and a decrease of oxygen concentrations in the pores (**Figure 5(b)**).

It was possible to increase the size of the pores to 70–80 nm by increasing the anode current up to 70 mA/cm^2 (**Figure 6(a)** and **(b)**). It should also be noted that with the increase of the anode current, the location degree of Cd atoms in the pores

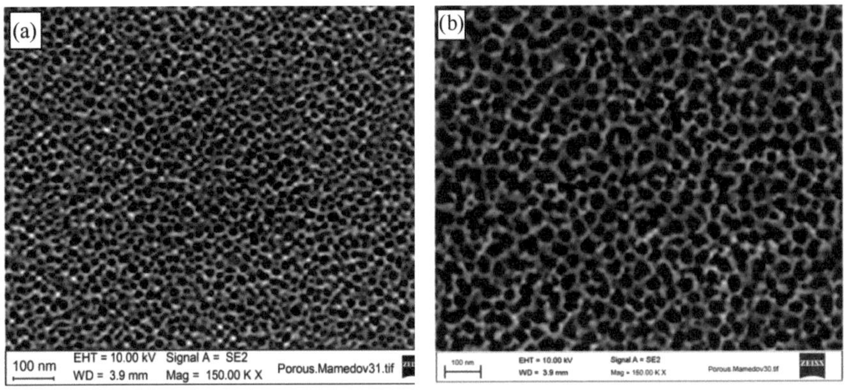

Figure 6.
SEM images of PSCD samples prepared at anodizing voltage of 30 V and currents of 55 mA/cm² (a) and 70 mA/cm² (b).

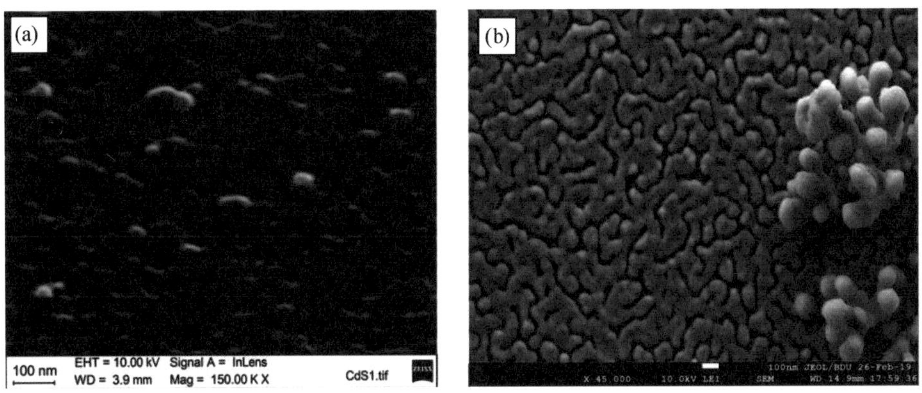

Figure 7.
SEM images and AFM images of CdS (a) and $Cd_{0.4}Zn_{0.6}O$ (b) films deposited on c-Si/PSCD1 substrates.

also varies. Thus, EDS spectrums testify an increase of Cd and a decrease of oxygen concentrations in the pores, with increasing the anode current up to 55–60 mA/cm². This fact shows that the low concentration of Si-O_x bonds is formed in the pores, i.e., the most of Si-H_x bonds are replaced by Si-Cd bonds (**Figure 5(b)**). Note that those EDS spectrums were drawn for samples immediately after depositing and after saving 1 month. In both cases, the same result is obtained—oxygen concentration in the pores has not changed. This also indicates that the PSCD layers prepared in metal ionic environment (in HF + $CdCl_2$+ ethanol solution) are stable.

But, further increase in current value up to 70 mA/cm² leads to the decrease of concentration of the Cd atoms and increase of oxygen concentration, as confirmed by EDS studies. However, Si-O_x bonds are not stable and show variable valence, which allows the use of PSCD layers as sensors of various types of gases.

SEM and AFM images confirmed that the morphological properties of CdS and $Cd_{0.4}Zn_{0.6}O$ films are governed by the pore size of PS. SEM images of thin films CdS (deposited at potential −0.82 V [19, 20]) and $Cd_{0.4}Zn_{0.6}O$ (deposited at potential −1.2 V [21]) onto the c-Si/PSCD (PSCD1, PSCD2, and PSCD 3) are shown in **Figures 7–9**, respectively. As seen from figures, the size of crystallites of the CdS and $Cd_{0.4}Zn_{0.6}O$ films can be controlled by selecting the pore size of silicon. The films deposited onto the substrate PSCD1 shows micro-texture structure. The grain sizes were ~20–70 μm for CdS films, and films of $Cd_{0.4}Zn_{0.6}O$ show mosaic microcrystalline (~0.12–150 μm) structure with some defects.

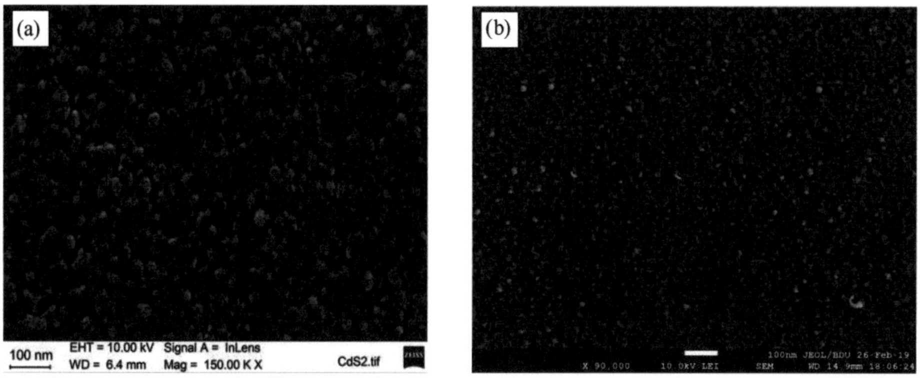

Figure 8.
SEM images and AFM images of CdS (a) and $Cd_{0.4}Zn_{0.6}O$ (b) films deposited on c-Si/PSCD2 substrates.

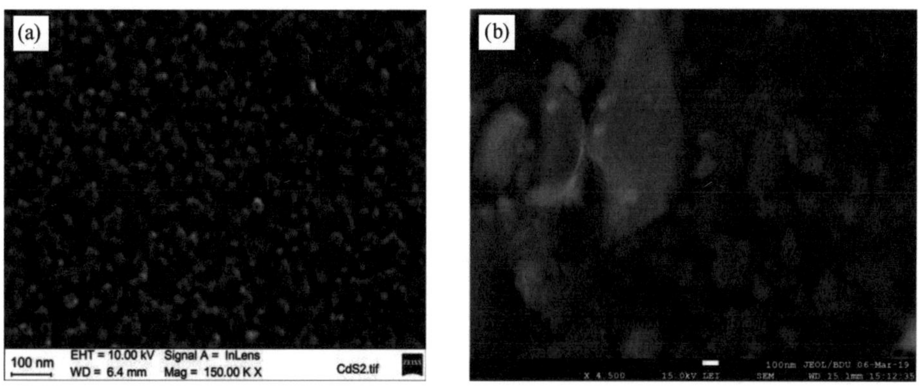

Figure 9.
SEM images and AFM images of CdS (a) and $Cd_{0.4}Zn_{0.6}O$ (b) films deposited on c-Si/PSCD3 substrates.

As seen from **Figure 8**, when PSCD2 is used as substrate, the size of the nanograins gets reduced due to the controlled process of nucleation, and nano-size grains are uniformly distributed at the surface. The size of particles, as determined by SEM, is found as 80–190 nm for CdS and 10–40 nm for $Cd_{0.4}Zn_{0.6}O$ films.

The morphology of the films changes dramatically, when replacing substrates of PSCD2 with PSCD3 (with a pore size of 30 nm), since the crystallite sizes in both films increase and their uneven distribution is observed from SEM images (**Figure 9**). The size of the grains was different from each other, indicating irregular growth rate of the grains.

The crystal phase structure of films CdS and $Cd_{0.4}Zn_{0.6}O$ was analyzed with a Shimadzu XRD-6100 diffractometer. X-ray diffraction patterns of $Cd_{0.4}Zn_{0.6}O$ and CdS films deposited on glass/SnO_2 substrates confirm that they possess a hexagonal wurtzite and cubic structures, respectively. $Cd_{0.4}Zn_{0.6}O$ films have a polycrystalline structure with a strongly preferred (002) c-axis orientation (as evidenced by the high intensity of the (002) diffraction peak). X-ray diffraction pattern of the CdS films shows peaks at 27, 44, and 52° corresponding to the planes (111), (220), and (311), respectively. The d-spacing values obtained match those of the corresponding reference material (JCPDS card No. 36-1451 for $Cd_{0.4}Zn_{0.6}O$ and No. 10-0454 for CdS films) well.

The optical transmission spectrums were recorded with an Evolution 300 spectrophotometer. The band gap of $Cd_{0.4}Zn_{0.6}O$ and CdS films was determined by extrapolating the straight-line section of the (αhν)2 versus hν curves. The band gap of the $Cd_{0.4}Zn_{0.6}O$ and CdS films was estimated to be 2.95 and 2.41 eV, respectively.

4. Electrical characterization and gas sensitivity of p-Si/PS/CdS and p-Si/PS/$Cd_{0.4}Zn_{0.6}O$ heterojunctions

Electrical properties of heterojunctions were studied depending on the crystallite and pore size. Current-voltage (*J-V*) characteristics show that all heterojunctions possess rectifying properties. The rectification factor and current passage mechanism through the junction region depends strongly on the pore and crystallite size. **Figure 10** represents the *J-V* characteristics for as-deposited heterojunctions of p-Si/PSCD/CdS and p-Si/PSCD/$Cd_{0.4}Zn_{0.6}O$ based on PSCD1, PSCD2, and PSCD3 substrates. The pass direction corresponds to positive polarity of the external bias on the *c*-Si layers.

The rectification at U = 1 V increases from 2340 to 4670 for c-Si/PSCD/CdS and from 780 to 1650 for c-Si/PSCD/$Cd_{0.4}Zn_{0.6}O$ heterojunctions, when replacing the PSCD1 substrates with PSCD2. As seen from **Table 3**, rectification in both types of heterojunctions is decreased sharply, when PSCD3 is used as substrate. It is assumed that the change in rectification factor value depending on the pore size is due to the oxygen or nitrogen molecules, because in order to remove excess water from pores and films, heterojunctions were dried in air, just after the deposition (as-deposited heterojunctions). In this case, adsorbed oxygen or nitrogen molecules to the silicon pores (mainly for PSCD3 substrates) create the acceptor states in junction region and thus increase recombination acts. It is established that the degree of adsorption depends on pore size. To confirm this fact, heterojunctions were heat-treated in a vacuum at 50–70°C, and *J-V* characteristics were taken in a vacuum. It is established that the rectification in heterojunctions deposited onto the PSCD2 substrates remained almost unchanged (**Table 4**). But the rectification in both types of heterojunctions on the base of PSCD3 substrates increases sharply, which, by our opinion, is due to desorption of oxygen or nitrogen molecules (**Table 4**).

To prove this, we carried out additional experiments on these heterojunctions. Direct current at U = 1 V was measured in air with different concentrations. Direct current in heterojunctions with PSCD2 has not changed and in heterojunctions with PSCD3 decreased with increasing air concentration (**Figure 11**).

Dependences of direct current at U = 1 V for heterojunctions with PSCD1, PSCD2, and PSCD3, on the concentration of various gases (oxygen, nitrogen, methane, and ethanol vapor), have been investigated. As seen from **Figure 12**, the direct current decreases at an insignificant change of the gas concentration. It is suggested that heterojunctions with PSCD3 can be used as gas sensors.

Experiments carried out in various atmospheres on heterojunctions with PSCD1 showed insignificant changes of current, and junctions with PSCD2 do not show

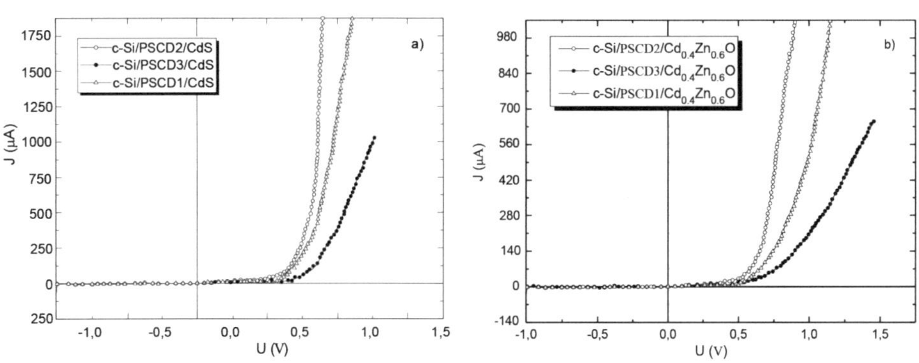

Figure 10.
Dark J-V curves for as-deposited c-Si/PSCD/CdS and c-Si/PSCD/$Cd_{0.4}Zn_{0.6}O$ heterojunctions prepared onto the PSCD1, PSCD2, and PSCD3 substrates.

Samples	Substrates	Ideality factor	Rectification factor
c-Si/PSCD/CdS	PSCD1	1.5	2340
	PSCD2	1.24	4670
	PSCD3	2.67	1076
c-Si/PSCD/$Cd_{0.4}Zn_{0.6}O$	PSCD1	1.66	780
	PSCD2	1.35	1650
	PSCD3	2.84	240

Table 3.
Electrical parameters of c-Si/PSCD/CdS and c-Si/PSCD/$Cd_{0.4}Zn_{0.6}O$ as-deposited heterojunctions depending on the pore size in Si.

Samples	Substrates	Ideality factor	Rectification factor
c-Si/PSCD/CdS	PSCD1	1.5	2346
	PSCD2	1.24	4680
	PSCD3	1.63	1954
c-Si/PSCD/$Cd_{0.4}Zn_{0.6}O$	PSCD1	1.66	785
	PSCD2	1.35	1670
	PSCD3	1.75	1687

Table 4.
Electrical parameters of heterojunctions c-Si/PSCD/CdS and c-Si/PSCD/ $Cd_{0.4}Zn_{0.6}O$ heat-treated in a vacuum at 50–70°C, depending on the pore size in Si.

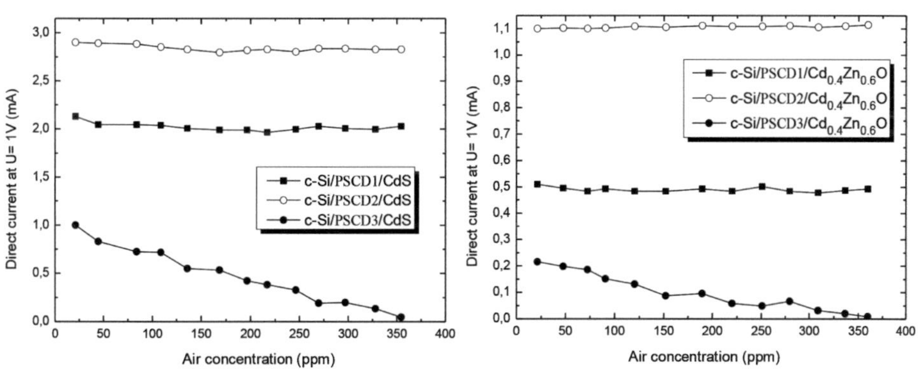

Figure 11.
Dependence of direct current at U = 1 V on the air concentration for as-deposited heterojunctions of c-Si/PSCD/CdS and c-Si/PSCD/$Cd_{0.4}Zn_{0.6}O$ with different pore size.

any changes of direct current, i.e., they have stable parameters. This fact testifies that heterojunctions with PSCD1 and PSCD2 can be used for manufacturing solar cells.

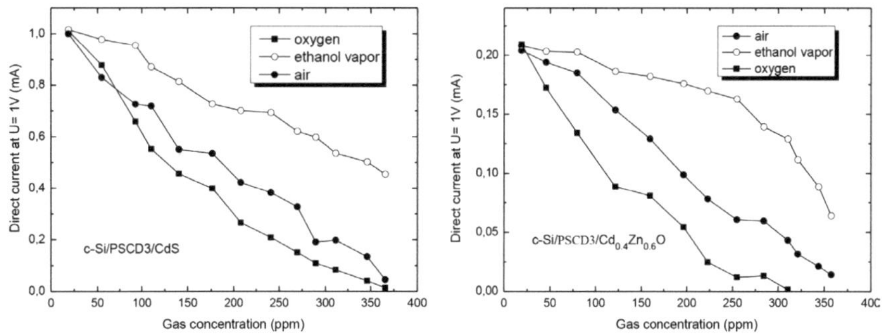

Figure 12.
Dependence of direct current at U = 1 V on the concentration of different gases for heterojunctions c-Si/PSCD3/CdS and c-Si/PSCD3/$Cd_{0.4}Zn_{0.6}O$.

The *J-V* characteristics of heterojunctions with PSCD1 and PSCD2, in natural log scale and temperature dependences of diode saturation current, testify that there is a tunnel-recombination mechanism of current through the junction region at direct voltages, and *J-V* characteristics of these heterojunctions can be presented by expression:

$$J = J_{0J}\left[exp\left(\frac{eU}{nkT}\right) - 1\right] \tag{1}$$

Here, J_0 is the saturation current density, U is the applied voltage, e is the electron charge, n is the ideality factor, k is the Boltzmann constant, and T is the temperature.

At room temperature the ideality factor values for heterojunctions with PSCD1 were approximately 1.5 and 1.66, respectively, for the heterojunctions c-Si/PSCD1/CdS and c-Si/PSCD1/$Cd_{0.4}Zn_{0.6}O$. The decrease of ideality factor has been observed when PSCD2 substrates is used (**Table 4**).

Figure 13 shows the room temperature capacitance-voltage (*C-V*) characteristics of heterojunctions c-Si/PSCD/CdS and c-Si/PSCD/$Cd_{0.4}Zn_{0.6}O$, with PSCD1 and PSCD2 substrates. The linearity of the *C-V* characteristics in $C^{-2}(V)$ coordinates indicates a sharp distribution of uncompensated acceptor impurities within the space charge region for both types of heterojunctions with PSCD2. Unlike it, *C-V* characteristics of heterojunctions c-Si/PSCD/CdS and c-Si/PSCD/$Cd_{0.4}Zn_{0.6}O$, with PSCD1 substrates, have peculiarities, typical for heterojunctions with the presence of defects at the junction region. Since the *C-V* characteristics of these junctions are poorly linearized in $C^{-2}(V)$ coordinates, its slope changes by frequency of the alternating signal.

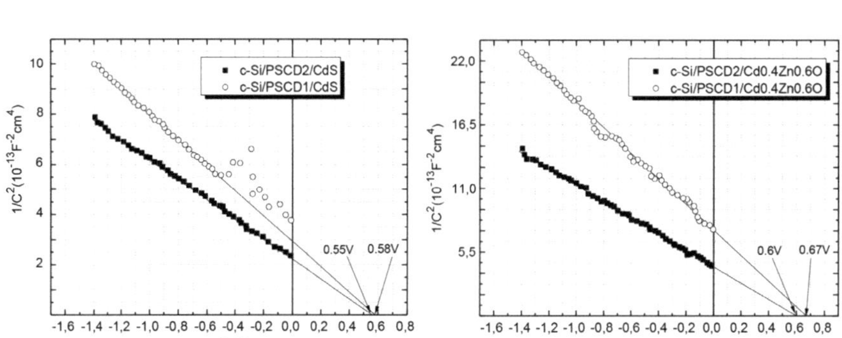

Figure 13.
Capacitance-voltage characteristics of heterojunctions c-Si/PSCD/CdS and c-Si/PSCD/$Cd_{0.4}Zn_{0.6}O$ based on PSCD1 and PSCD2 substrates

As seen from the figures, the value of cutoff voltage (V_c) is greater than heterojunctions based on PSCD2 substrates. It can be explained by the participation of oxygen (as Si-O_x bonds) on the surface of PSCD1 substrates, which can form the oxide layer (as CdO or $Cd_{1-x}Zn_xO$) between the porous silicon and CdS ($Cd_{1-x}Zn_xO$) films. This is confirmed by the frequency dependence of straight sections of the $C^{-2}(V)$ slopes, which indicates an increase of heterojunction capacity with increasing frequency.

5. Photosensitivity of p-Si/PS/CdS and p-Si/PS/$Cd_{0.4}Zn_{0.6}O$ heterojunctions

The simulated *J-V* characteristics under AM1.5 sunlight exposure show the higher photovoltaic performance of the heterojunctions c-Si/PSCD/CdS and c-Si/PSCD/$Cd_{0.4}Zn_{0.6}O$ based on PSCD substrates. As shown from these investigations, the sign of open circuit photovoltage (U_{oc}) does not change in all regions of photosensitivity. However, the values of U_{oc} and J_{sc} are non-monotonically dependent on the type of PSCD—maximum photosensitivity shows the heterojunctions based on PSCD2 substrates (**Figure 14**).

Spectral distribution of photocurrent (J_{ph}) in heterojunctions c-Si/PSCD/CdS and c-Si/PSCD/$Cd_{0.4}Zn_{0.6}O$ with different types of PSCD was investigated in a wavelength range of 300–1300 nm (**Figure 15**). It is established that the profile of J_{ph} spectrum depends on the pore size and morphology of CdS and $Cd_{0.4}Zn_{0.6}O$ films. As seen from the figure, short-wavelength peak for heterojunctions c-Si/PSCD/CdS and c-Si/PSCD/$Cd_{0.4}Zn_{0.6}O$ based on PSCD1 is observed, respectively, at

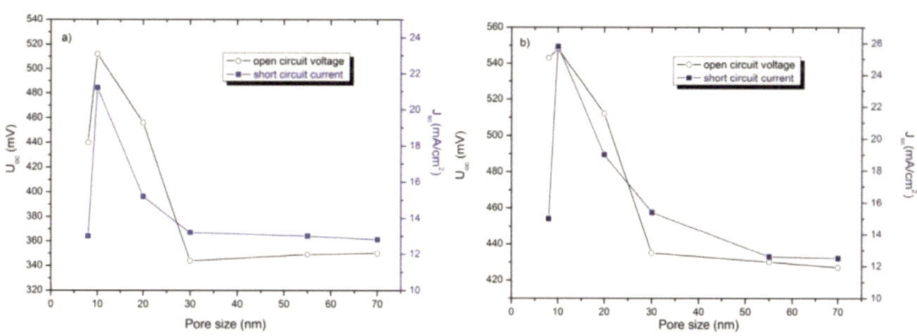

Figure 14.
Dependences of U_{oc} and J_{sc} on the pore size in silicon for heterojunctions c-Si/PSCD/$Cd_{0.4}Zn_{0.6}O$ (a) and c-Si/PSCD/CdS (b).

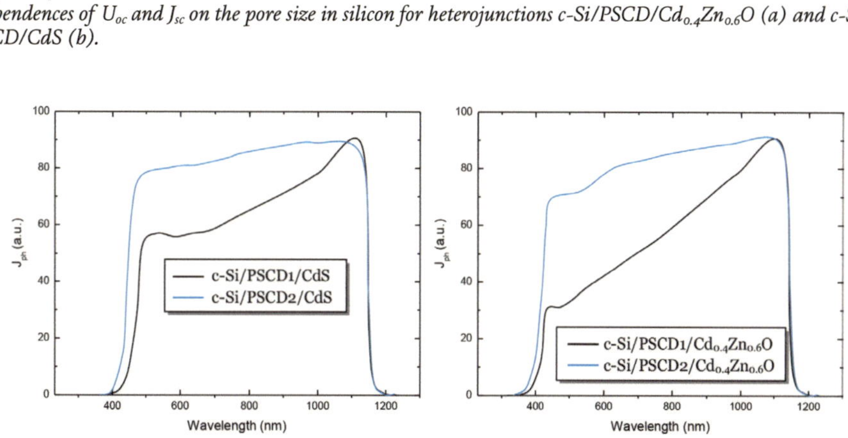

Figure 15.
Spectral distribution of photocurrent for heterojunctions c-Si/PSCD/CdS and c-Si/PSCD/$Cd_{0.4}Zn_{0.6}O$ based on PSCD1 and PDCS2 substrates [19–22].

Samples	Substrates	U_{oc}, mV	J_{sc}, mA/cm²	FF	η, %
c-Si/PSCD/CdS	PSCD1	543	15	0.62	5.05
	PSCD2	548	25.8	0.68	9.61
	PSCD3	435	15.4	0.6	4.01
c-Si/PSCD/$Cd_{0.4}Zn_{0.6}O$	PSCD1	440	13	0.55	3.15
	PSCD2	512	21.2	0.61	6.62
	PSCD3	344	13.2	0.52	2.36

Table 5.
Photoelectrical parameters of heterojunctions c-Si/PSCD/CdS and c-Si/PSCD/$Cd_{0.4}Zn_{0.6}O$ heat-treated in a vacuum at 50–70°C, depending on the pore size in Si.

510 and 420 nm, which corresponds to the band gap of CdS and $Cd_{0.4}Zn_{0.6}O$ films. However, long-wavelength peak of spectrum at 1125 nm is due to the direct inter-band transitions in *c*-Si. It can be seen from **Figure 15** that heterojunctions of c-Si/PSCD/CdS and c-Si/PSCD/$Cd_{0.4}Zn_{0.6}O$ demonstrate a good photo response in the wavelength range of 512–1125 nm and 425–1125 nm, which is not observed by us in heterojunctions of c-Si/A_2B_6 [23].

It is assumed that this is due to light absorption in porous silicon. The slight shifting of a short-wavelength peak in both type of heterojunctions to the more short-wavelength region of the spectrum, with increasing pores size (for PSCD2), is associated by us with the nano-structural properties of CdS and $Cd_{0.4}Zn_{0.6}O$ films [19–22]. However, an increase of the optical path of light in nanostructured films leads to an increase of degree of the light absorption; therefore, heterojunctions based on PSCD2 show greater efficiency compared to heterojunctions based on PSCD1 (**Table 5**).

6. Conclusions

Gas and photo sensors based on c-Si/PS (or PSCD)/CdS and c-Si/PS/$Cd_{0.4}Zn_{0.6}O$ heterojunctions were fabricated by electrodepositing CdS and $Cd_{0.4}Zn_{0.6}O$ films onto the c-Si/PS (or PSCD) substrates. SEM images confirmed that the morphological properties of CdS and $Cd_{0.4}Zn_{0.6}O$ films are governed by the pore size of PS and PSCD. The higher gas and photo sensitivities in heterojunctions based on PSCD were achieved when the pore sizes were 30 and 10 nm, respectively. The pronounced PS pore diameter and anodizing regime dependence of the photo- and gas sensitivity are the main novelty of the present work. We believe that this phenomenon could be exploited in other similar heterojunction types and even in hybrid solar cells containing, e.g., a perovskite charge separation layer.

Author details

Huseyn M. Mamedov
Faculty of Physics, Department of Physical Electronics, Baku State University, Baku, Azerbaijan

*Address all correspondence to: mhhuseyng@bsu.edu.az

References

[1] Nalin HM, Jibinlal A, Divesh NS. Application of porous silicon in solar cell. In: Proceedings of International Conference on Nanomaterials for Energy Conversion and Storage Applications (NECSA 2018). Gujarat, India: AIP Publishing; 29-31 January 2018;**1961**:030019-030022. DOI: 10.1063/1.5035221

[2] Giuseppe B. Porous silicon gas sensing. In: Canham L, editor. Handbook of Porous Silicon. 2nd ed. Switzerland: Springer; 2014. pp. 1251-1262. Available from: https://doi.org/10.1007/978-3-319-71381-6

[3] Santinacci L, Gonçalves A, Simon N, Etcheberry A. Electrochemical and optical characterizations of anodic porous n-InP(1 0 0) layers. Electrochimica Acta. 2010;**56**:878-888. DOI: 10.1016/j.electacta.2010.09.031

[4] Raid AI, Alwan MA, Ahmed SA. Preparation and characteristics study of nano-porous silicon UV photodetector. Applied Nanoscience. 2016;**7**:9-15. DOI: 10.1007/s13204-016-0544-9

[5] Naser MA, Al-Dourib Y, Alwan MA, Allaa AJ, Ghassan EA. Characteristics of nanostructure silicon photodiode using laser assisted etching. Procedia Engineering. 2013;**53**:393-399. DOI: 10.1016/j.proeng.2013.02.051

[6] Oh D, Kim TW, Cho W, Kwack KD. Effects of a H2SO4 treatment on the optical properties in porous Si layers and electrical properties of diode devices fabricated with a H_2SO_4 treated porous Si layer. Journal of Ceramic Processing Research. 2008;**9**:57-60

[7] Granitzer P, Rumpf K. Porous silicon—A versatile host material. Materials. 2010;**3**:943-998. DOI: 10.3390/ma3020943

[8] Korotcenkov G. Porous Silicon: From Formation to Application. Vol. 1: Formation and Properties. Taylor and Francis Group: CRC Press; 2015. p. 432. DOI: 10.1201/b19342

[9] Korotcenkov G. Porous Silicon: From Formation to Application. Vol. 2: Biomedical and Sensor Applications. Taylor and Francis Group: CRC Press; 2016. p. 424. DOI: 10.1201/b19342

[10] Korotcenkov G, editor. Porous Silicon: From Formation to Application. Vol. 3: Microelectronics, Optoelectronics and Energy Technology Applications. Taylor and Francis Group: CRC Press; 2016. p. 430. DOI: 10.1201/b19342

[11] Venger EF, Holiney RY, Matveeva LA, Vasin AV. The influence of hydrogen plasma on the spectrum of electromotive reflections and the spectrum of electronic states of porous silicon. Semiconductors. 2003;**37**:103-107

[12] Sreseli OM, Goryachev DN, Osipov VY, Belyakov LV, Vul SP, Serenkov IT, et al. Preparation and study of carbidized porous silicon. Semiconductors. 2002;**36**:574-580

[13] Primachenko VE, Kononets JF, Bulakh BM, Venger EF, Kaganovich ÉB, Kizyak IM, et al. The electronic and emissive properties of Au-doped porous silicon. Semiconductors. 2005;**39**:565-571

[14] Shevchenko OY, Goryachev DN, Belyakov LV, Sresli OM. Optical properties of iron-passivated nanoporous silicon. Semiconductors. 2010;**44**:642-646

[15] Xu YK, Adachi S. Properties of light-emitting porous silicon photoetched in

aqueous HF/FeCl3solution. Journal of Applied Physics. 2007;**101**(10):103509. DOI: 10.1063/1.2733752

[16] Xu YK, Adachi S. Properties of light-emitting porous silicon formed by stain etching in HF/KIO3 solution under light illumination. Journal of Applied Physics. 2008;**103**(10):103512. DOI: 10.1063/1.2924423

[17] Fathauer RW, George T, Ksendzov A, Vasquez RP. Visible luminescence from silicon wafers subjected to stain etches. Applied Physics Letters. 1992;**60**(8):995-997. DOI: 10.1063/1.106485

[18] Karbassian F. Porous Silicon. In: Taher G, editor. Porosity - Process, Technologies and Applications. UK: IntechOpen; 2018. p. 1-36. DOI: 10.5772/intechopen.68404

[19] Mamedov HM, Muradov MB, Konya Z, Kukovecz A, Kordas K, Shah SI, et al. Fabrication and characterization of c-Si/porous-Si/CdS/$Zn_xCd_{1-x}O$ heterojunctions for applications in nanostructured solar cells. Photonics Letters of Poland. 2018;**10**:73-75. DOI: 10.4302/plp.v10i3.813

[20] Mamedov HM, Kukevecz A, Konya Z, Kordas K, Shah SI, Mamedov VU, et al. Characteristics of c-Si/porous-Si/CdS heterojunctions. News of higher educational institutions. Physics (Russian Journal of Applied Physics). 2018;**61**:96-101. DOI: 10.1007/s11182-018-1584-2

[21] Mamedov HM, Mamedov VU, Mamedova VJ, Ahmedova KM, Tagiyev EB, Agazade LE. Nano-structure solar cells on the base of p-Si/$Cd_{1-x}Zn_xO$ thin film heterojunctions. Journal of Optoelectronics and Advanced Materials. 2018;**20**:468-473

[22] Mamedov HM, Shah SI, Chirakadze A, Mamedov VU, Mamedova VJ, Ahmedova KM. Photovoltaic performance of p-Si/$Cd_{1-x}Zn_xO$ heterojunctions. Photonics Letters of Poland. 2018;**10**:26-28. DOI: 10.4302/plp.v10i1.797

[23] Abdinov AS, Mamedov HM, Hasanov HA, Amirova SI. Photosensitivity of p, n-Si/n-$Cd_{1-x}Zn_xS$ heterojunctions manufactured by a method of electrochemical deposition. Thin Solid Films. 2005;**480-481**:388-391

Chapter 4

Environmental Gas Sensors Based on Nanostructured Thin Films

Nithya Sureshkumar and Atanu Dutta

Abstract

Since the discovery of electron microscopes, nanomaterials and nanotechnology have been influencing academic and industrial research greatly for bringing out newer and better products with improved materials' properties. In the field of environmental gas sensors too, the demonstration of nanomaterials for sensing various gases has become a common practice. Environmental gas pollution has turned out to be a huge concern in the society due to the progress of civilization. The awareness of health hazard for different toxic/polluting gases and rectification measure by imposing stricter norms has prompted extensive research to develop efficient gas sensors to detect trace level of pollution from various sources. Thin film, ultrathin film, and nanostructure materials of metal oxide semiconductor, polymer, metal, carbon nanotube, graphene, etc. with or without sensitizers have been investigated for sensing various toxic gases. New device structures have been fabricated to achieve high sensitivity, selectivity, fast response, etc. The microstructure and thickness of film are found to influence the performance greatly. Various methods of preparations and mechanism of sensing are being explored. All these aspects and the challenges were discussed in this chapter.

Keywords: gas sensor, nanostructure, thin film, carbon nanotube, metal oxide semiconductor

1. Introduction

Sensor as given in the *Handbook of Modern Sensors* by Fraden [1] is defined as "A device that receives a stimulus or signal such as chemical, physical or biological signal and responds or convert into an electrical signal." Sensor is an essential tool for everyday life which is used to identify a specific task assigned to it. The best sensors are already there in human body, namely, eye to see, ear to hear, nose to smell, tongue to taste, and skin to feel the surrounding. However, there are several threats which need to be identified well in advance, and appropriate measures can be taken to prevent any accident or health hazard. Due to urbanization, development of industries, and requirements of society, we have invited several intimidations to our society. Environmental gas pollution is one of them. The society is now concerned about how due to various reasons air pollution is harming our planet. It is impossible to find solutions for all these problems very quickly as civilization and its progress cannot be stalled. Therefore, efforts are being made to avoid and minimize environmental pollution and put safeguards for human health. Specifically, gases such as hydrocarbons (HCs), nitrogen oxides (NO_x), carbon monoxide (CO), carbon dioxide (CO_2), sulfur dioxide (SO_2), ammonia (NH_3), hydrogen sulfide (H_2S), volatile organic compounds (VOCs), etc., which are released or emitted in the environment, cause either directly or indirectly health problems and accidental situations. The sources of such accidental release vary with pollutant gas, and mostly industries and human amenities are responsible. Since the presence of such hazard gases in low concentrations cannot be smelled by human nose, improvised sensors or electronic noses are required to detect these gases efficiently. The detection of gas results in changing properties of the sensor which in turn activates an actuator circuit for alarm alert. Several governmental norms are published for health safety on exposure concentration and time of exposure [2]. Likewise, stringent norms have been imposed on each of the possible sources of all these pollutant gases.

The research of environmental gas sensors started since the 1960s with metal oxide films. In 1962, Seiyama et al. [3] reported thermally evaporated ZnO thin film sensor of thickness 1–100 nm for the detection of several volatile organics, hydrocarbon, and carbon dioxide at temperature above 400°C. Taguchi demonstrated SnO_2-based semiconductor sensor. He patented his work [4] and formed the famous company named Figaro Engineering Inc. Due to the strict environmental pollution norms imposed for emission of various toxic gases, research on gas sensors and materials expanded exponentially in the last 50 years. Several 10,000 research papers were published to claim suitable gas sensors. In 1980, Advani et al. proposed a SnO_2 thin film for H_2 gas sensor [5]. The device based on SnO_2 thick film gas sensor was reported by Heiland et al. [6] and thin film by Sberveglieri [7]. More efforts were given to develop sensors based on thin films to detect toxic gases mostly at room temperature. Several review articles on the progress of thin film sensors are available in the literature.

1.1 Importance of sensors in day-to-day life

The rapid growth of industrializations in modern life has enhanced the risk of pollution deteriorating the atmospheric environment around us. Thus it is necessary to control and monitor such pollutants to avoid health hazards.

Table 1 depicts potential gas sensor applications for various reasons. Several commercial sensors are available for gas detection in different environments. However, the demand of compact, reliable, cheap, selective, and low power consuming sensors is still not yet met. Therefore, extensive research is going on to satisfy the specific need of the situation.

1.2 Thin film

There is no accurate definition of thin film but it is limited by thickness. The film below 1 μm is considered as thin film, and film above it is considered as thick film. But, normally when a solid material is coated on solid support, called substrate, by either physical or chemical process, it is called as a thin film. It should be highlighted here that it is not only a small thickness (<1 μm) that endows the great and special distinguished characteristics but also the microstructure which is equally important. Thin films characteristically have different properties from the thick films [8]. Many researchers are investigating thick and thin film gas sensors using different materials like metal oxides, ceramic and polymer materials, etc. But, the problems associated with these sensors are limited by the measurement accuracy and long-term stability [9]. To solve this problem, nanotechnology plays a major role to rectify and provide huge opportunity to establish the next-generation gas sensor devices with enhanced sensing performance in terms of fast response and recovery, good reversibility, low power consumption, and excellent sensitivity at extremely low gas concentration by using nanostructured sensing materials [10–12].

Applications
Environmental control: [NO_X, SO_X, HCl, CO_2, volatile organic compounds (VOCs)] • Weather stations • Pollution monitoring
Safety: [hydrogen sulfide (H_2S), ammonia (NH_3), nitrogen dioxide (NO_2), oxygen (O_2) deficiency, carbon monoxide (CO), etc.] • Leak detection • Fire detection • Boiler control • Toxic/explosive/flammable gas detections
Automobiles: [ammonia (NH_3), carbon dioxide (CO_2), nitrogen oxide (NO_x)] • Gasoline vapor detection • Filter control • Car ventilation control
Medicine: [inorganic gases (e.g., NO, N_2O, and CO), VOCs (e.g., isoprene, acetone, pentane, ethane)] • Disease detection • Breath analysis
Industrial production: [hydrogen sulfide (H_2S), methane, chlorine, ammonia (NH_3), carbon dioxide (CO_2), oxygen (O_2)] • Process control • Fermentation control

Table 1.
Examples for gas sensor applications.

2. Nanostructure film materials for gas sensor application

From the literature it is seen that nanostructured materials have attracted keen attention for several applications including gas sensing. These include semiconductor metal oxide nanowires, nanoribbons, nanorods, carbon and other nanotubes, diamonds, graphene, and combination of these which make more interesting nanostructures. A major attraction of the nanomaterials in the gas sensor application is because of high surface area to volume ratio. This possibly deliberates high sensitivity due to the excellent adsorption of gas species on the available molecular binding site and thus often increases sensing capability [13]. Based on the materials, the following gas sensors can be considered for research and developments.

2.1 Semiconducting metal oxide nanostructure gas sensor

To fabricate the nanostructured gas sensors, semiconducting metal oxides are the most widely used material. The schematic diagram of semiconducting metal oxide thin film gas sensor is shown in **Figure 1**. The electrical conductivity of semiconductor gets modified once the surface is exposed to sensing gas. Nanostructure metal oxide thin film produces very high sensitivity due to high surface activity and unique microstructure. High adsorption of gas species on the surface and subsequent catalytic activity with the adsorbed species enhance sensing performance and thus reducing response time compared to the conventional microstructure gas sensor.

Semiconducting metal oxides in the form of nanowire, nanobelt, and nanoparticles are also widely used for detecting various gases. The majority of the sensor work reported in the literature so far are based on various metal oxide semiconductors. These materials can be doped easily or sensitized by appropriate metal to tailor-made sensitivity, selectivity, and also operating temperature. Very recently, a comprehensive review article was published by Li et al. to document all different types of such gas sensors for detecting several toxic gases [14]. The metal oxide gas sensors have an advantage of working in the lower-temperature range with high sensitivity. However, metal oxide nanosensors are more expensive than the thick film or bulk material gas sensors because of the deposition of nanothin film using highly sophisticated techniques and also the use of advanced microscopic analysis for characterizing the materials to improve its sensing characteristics [15].

2.2 Polymer-based nanosensors

In recent year, organic polymer-based nanosensors have been specially developed for detecting aromatic compounds, medicine, ionic species, amines, organic vapor, and gases [16]. Generally, the conducting polymers are coated on a substrate in the form of thin film. The sensor structures can either be with film with two electrodes or field-effect transistor with controlled electric field to monitor sensing current. Many such structures are also available in the literature. **Figure 2** shows such structures. The sensitivity of the polymer-based gas sensor depends on the sensing area of the device and thickness of the film. Addition of metal oxides, carbon nanoparticles, or fibers in the polymer-based sensors leads to the increase of sensitivity. The deposition techniques of thin film are very simple and the process is cheap also. The main problems with polymer-based sensor are lower stability with temperature and time and their lack of selectivity. The studies have been carried out to improve the stability and selectivity of polymer-based sensors by adding the carbon nanoparticles and fibers, but still some problems are yet to be solved.

2.3 Metal and metal complex nanosensor

Nanosensor with metal or normally metal nanoparticles dispersed on the substrate surface is also proposed [18]. In this case, very high surface exposure with

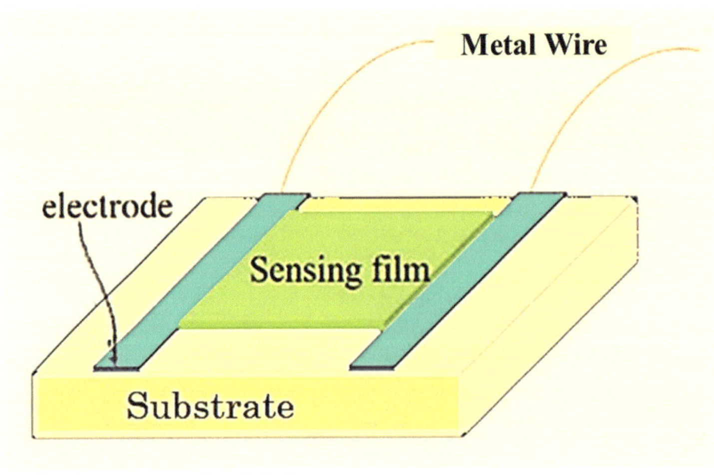

Figure 1.
Schematic of metal oxide thin film gas sensor.

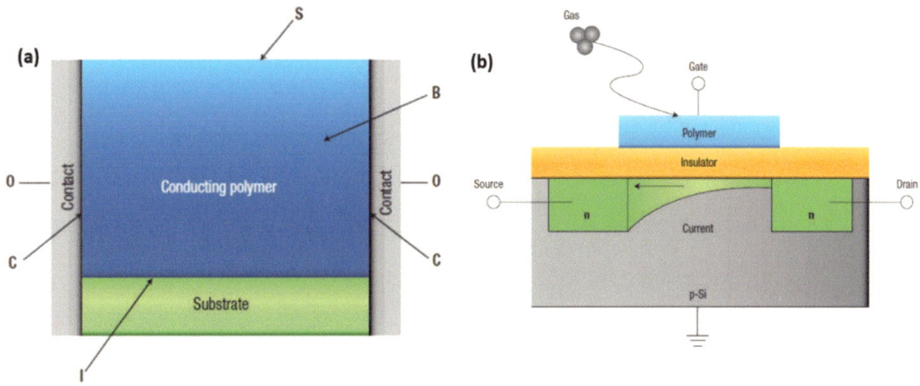

Figure 2.
(a) Chemiresistor device structure; S, surface; I, interface with insulating substrate; C, contact. (b) Conducting polymer active layer in field-effect transistor (FET) (reproduced from Ref. [17]).

respect to volume results more adsorption of gases. When these nanoparticles interact with gas molecules, change of electronic properties including that of substrate takes place rapidly. Vaporizing metal precursor deposition method produces metal film or metal nanoparticles, and subsequent annealing treatment produces these kinds of nanosensor. Maximum studies reported so far use metal oxide as a substrate with metal nanoparticle dispersed for such nanosensor device surface. Some of these sensors work at low temperature. Metal complex nanosensors have their unique property, i.e., ability to create reversible interaction with other atoms. Many studies, reported as metal complex, mostly act as a receptor for various types of molecules. This activity can increase the sensitivity and selectivity of different sensors [19]. The advantage of using metal complex as a sensing layer creates reversible interaction between the gas molecules and the device. The changing chemical environment or increasing the temperature affects the bonding formation during the gas sensing process. This type of sensors, based on the receptors, is designed to interact with specific gases to increase selectivity. However, synthesizing metal complex sensing layers for nanosensors is generally expensive.

2.4 Carbon-based nanosensor

Among the nanostructured gas sensor, carbon-based materials are proven to be very promising gas sensors due to high surface area, excellent intrinsic electrical properties, and good thermal and chemical stability [20–22]. A typical carbon-based nanostructured (CNT-carbon nanotube) gas sensor is shown in **Figure 3**. In the twenty-first century, the carbon allotrope, graphene, has been found to be an extraordinary material for many applications due to its excellent electronic, optical, chemical, thermal, and mechanical properties [23]. Graphene is defined as one-atom-thick planar with sp^2 hybridized 2D honeycomb crystal lattice of carbon. The stack of few atomic layers of graphene considered as a graphene sheets exhibits the semimetallic electronic behavior with zero bandgap energy state [24]. Graphene is differentiated as pristine graphene (PG), graphene oxide (GO), and reduced graphene oxide (RGO).

In gas sensing application, graphene materials have unique sensing capability based on two-dimensional structural property of PG, GO, and RGO due to superior electron transport phenomena [25]. In 2007, the charge density of graphene was found to increase by the adsorbed gas molecules. Schedin developed the first graphene-based gas sensor [26]. The adsorption of gas molecules leads to the change

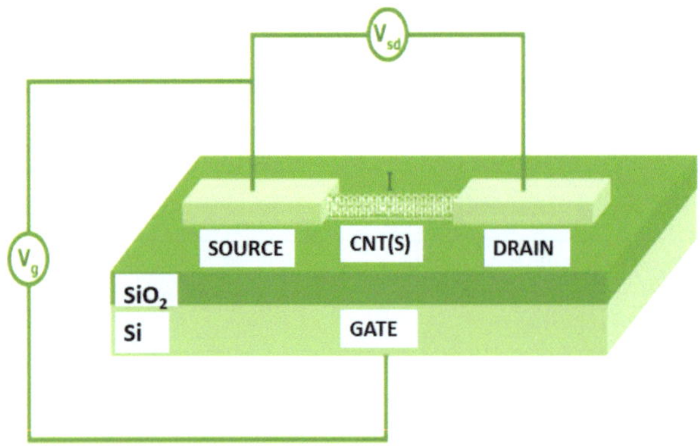

Figure 3.
Schematic diagram of a FET device based on CNT(s) (adapted from Ref. [9]).

of the electrical conductivity, i.e., local carrier concentration of graphene that attributed to the electron acting as donor/acceptor. All the varieties of graphene have different surface functional group and unique electrical properties, which play vital role in the sensing mechanism. In pristine graphene (PG), the adsorption of gas molecules can induce even very few charge carriers, resulting in a notable change in electrical conductivity.

Graphene oxide (GO) is different from pristine graphene due to its physical and chemical properties. GO is considered as promising material for gas sensing application because of highly rich oxygen functional group [27]. Prezioso et al. [28] using GO sensor device studied NO_2 gas sensing performance by varying the gas concentration at different operating temperatures to optimize the sensing efficiency. The oxygen functional group on the surface is mainly attributed to the sensing activity, and also when the device tested in both oxidizing and reducing environment, a typical p-type sensing response is observed. GO-based gas sensors have been used to detect the low concentration of nitrogen dioxide, hydrogen, and other gases also. GO-based sensor device reveals the high and fast sensing performance as compared to the RGO-based device. However, reduced graphene oxide (RGO) is easily synthesized from graphene oxide. By removing oxygen-containing groups and without changing the conjugated structure from GO, reduced graphene oxide (RGO) sheet can be obtained. In gas sensing application, RGO has more advantages than pristine graphene (PG) like inexpensive to prepare, possible to modify structure, fine tuning of structure, etc. [29, 30]. Many reported as available on the use of RGO for detecting gas molecules like NH_3, NO_2, H_2, etc. [31–33]. These sensors are tested in lower gas concentrations, and they exhibit high sensing response and reversibility at ambient temperature with low power consumptions. The limitation of RGO in practical application of low selectivity in gas sensor device is the same as the pristine graphene (PG). More detail of literature can be referred in the recent sensor review article based on carbon nanotube and graphene by Mao et al. [21].

3. Techniques used for thin film fabrications

The first gas sensor reported in the literature is based on thin film semiconductor. Since sensitivity, selectivity, and long-term stability are the main parameters for sensor for mass applications, various methods have been used to prepare sensor devices. Developing ultrathin film sensor with enhanced sensitivity at low concentration is challenging. In the field of gas sensor applications based on semiconductor thin film, the morphology and thickness of film play a crucial role for sensing properties of the materials. Several standard techniques are available for the fabrication of thin film. Thermal evaporation, electron beam evaporation, laser-assisted evaporation and sputtering, molecular beam epitaxy, etc. are well-known physical deposition techniques.

On the other hand, many useful techniques such as hydrothermal, sol-gel, chemical vapor deposition (CVD), spray CVD, microwave, dip coating, spin coating, etc. are used for depositing ultrathin films. The steps of making thin films using the above techniques are either standardized or as per the requirements; some modifications are done to achieve good quality films. Standard methods can be referred from various handbooks of thin film preparation [34–36]. The surface of thin film should be very active for gas sensing. Therefore, each of the techniques has some advantages and limitations. Grain size, film thickness, surface roughness (surface area) porosity, etc. significantly transform the film for effective detection of gas molecules down to ppm or sub-ppm level at room temperature.

In most of the films, surface adsorption-desorption of gases modifies the electrical conductivity of the material. Very recently, using slow evaporation rate of 0.02–0.3 nm/s in thermal evaporation technique, copperphthalocyanine (CuPc) organic semiconductor was deposited with thickness 10–40 nm to fabricate organic thin film transistor for ppm level NH_3 detection in room temperature [37]. The structure is shown in **Figure 4**. Using RF sputtering method, a 30-nm NiO film was deposited over sapphire and subsequently 1-nm platinum film was deposited over it by thermal evaporation for the detection of ammonia efficiently at low concentration at elevated temperature [38]. Several such works with ultrathin film have been reported in the literature. Since the demonstration of graphene-based sensor for detection of individual gas molecule almost a decade ago, several graphene-based sensors have been reported in the literature following newer techniques to improve the performance of the sensors [39, 40].

Generally, CVD, exfoliation, and other methods are used to prepare graphene for sensing. Wu et al. have recently developed graphene-based thin film using microwave plasma-enhanced chemical vapor deposition (MPCVD) followed by deposition of the fragmented 3D structure on substrate to develop nanoporous graphene (Gr) thin film [41]. Both NH_3 and CO_2 were detected efficiently using this

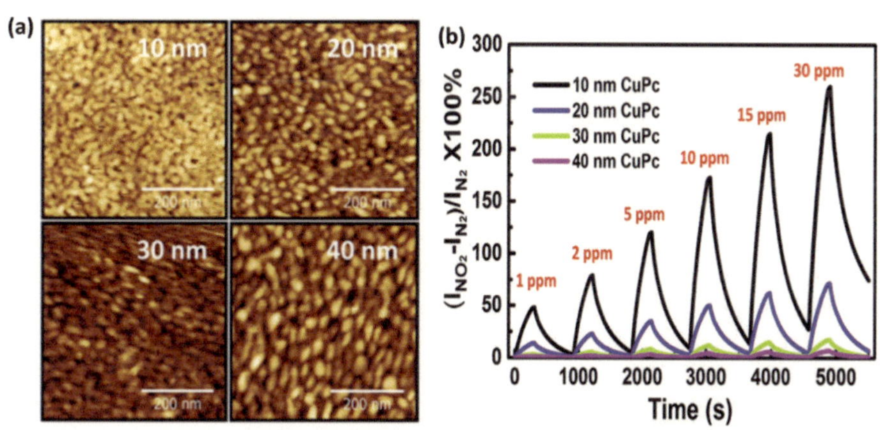

Figure 4.
(a) Topography image of AFM of CuPc films with different thicknesses on PMMA dielectric. (b) Response curves for the four types of devices to NO_2 pulses (reproduced with Ref. [37]).

film as conductance channel. Tian et al. have reviewed graphene-based gas sensor material preparation and characteristics [42].

4. Basic properties and gas sensing principle for nanostructured thin film devices

The popular environmental gas sensors operate on the principle of chemisorption. The adsorption and desorption processes of gas molecules on the active surface play a crucial role for sensing. Change in resistance or electrical current under an applied voltage in the presence of sensing gas is considered as the sensing signal response. Therefore, stability of the surface for long-term operation and also for repeatability of performance, the microstructure, and film thickness play a major role.

For instance, in the case of CuPc ultrathin film [37] and SnO_2 [43], the variation of microstructures of materials greatly improves the sensing performance for NO_2 and formaldehyde (VOC), respectively. These microstructures and performance are shown in **Figures 4** and 5.

4.1 Gas sensing principle for sensor device

The metal oxide gas sensor works on chemi-resistance principle. When the gas molecule interacts with metal oxide surface, it acts as either an acceptor or donor. This changes the resistivity or electrical conductivity of thin film. The resistivity of the metal oxide semiconducting thin film depends on the majority carrier in the film and also gas molecule nature, i.e., whether it is oxidized or reduced in the ambient temperature [44]. Surface adsorption sites ensure appropriate interaction of gas molecules with the material. In the case of n-type (electron being majority carrier), the surface is generally get depleted with electrons by the appearance of oxygen ion species (O^-, O^{2-}, etc.), and upon exposure to sensing gas, these species react with gas molecules to revert back electron on the surface, thereby increasing conductivity. The creation of these oxygen species on the surface is material specific and broadly dependent of temperature. In the case of p-type (hole being majority carrier), similar situation arises. The sensing behavior of metal oxide thin film sensor is shown in **Figure 6**.

Among the various nano gas sensor materials particularly, carbon-based materials like carbon nanotube (CNT), graphene, graphene oxide, and reduced graphene oxide of nanosensors play an imperative role for sensing applications. A two-dimensional sp^2 bonded allotropy of carbon called as graphene. A characteristic of high surface area to volume of carbon-based materials leads to a great potential for gas sensor applications. Especially, graphene acts as good device material due to its intrinsic properties. The graphene-based thin film sensor is shown in **Figure 7**. It has very large surface to volume ratio that can enhance adsorption of gas molecules, and hence response becomes fast and high. The working principle of graphene-based gas sensor depends on the catalytic and electronic properties of the active layers. For example, in the case of pristine graphene layer, it is electronically active but not sensing to detecting gas. In that case, surface modification or functionalization techniques are used to attain both electronic and sensing properties of graphene layer. But, this kind of surface modification technique should enhance the surface activity to adsorb the detecting gas molecules and quickly transfer the generated charges to the electrode.

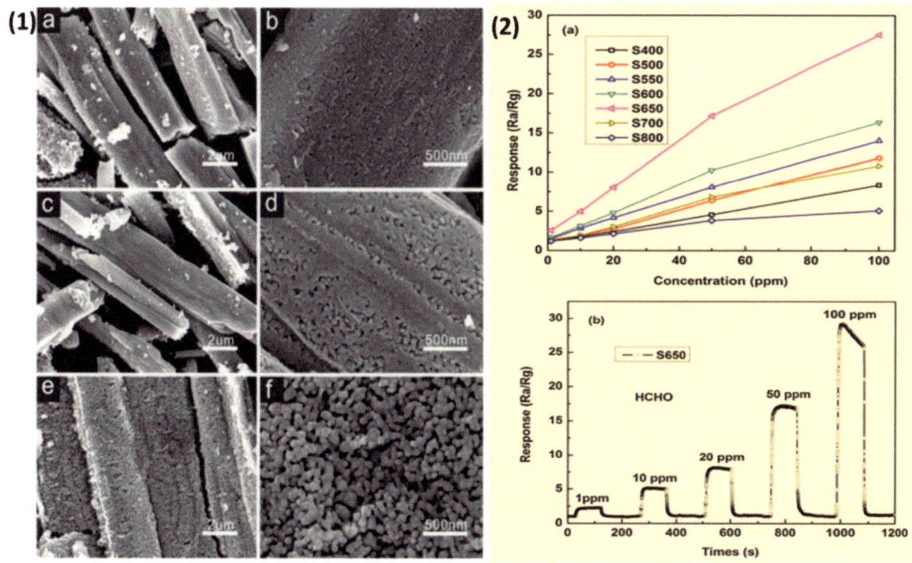

Figure 5.
(1) SEM images of tin oxide calcined at different temperatures (a, b) S400; (c, d) S650; (e, f) S800. **(2)** *(a) S400–S800 based sensor response as a function of formaldehyde concentration; (b) dynamic response-recovery curves of S650 to formaldehyde gas with different concentrations (Reproduced with Ref. [43]).*

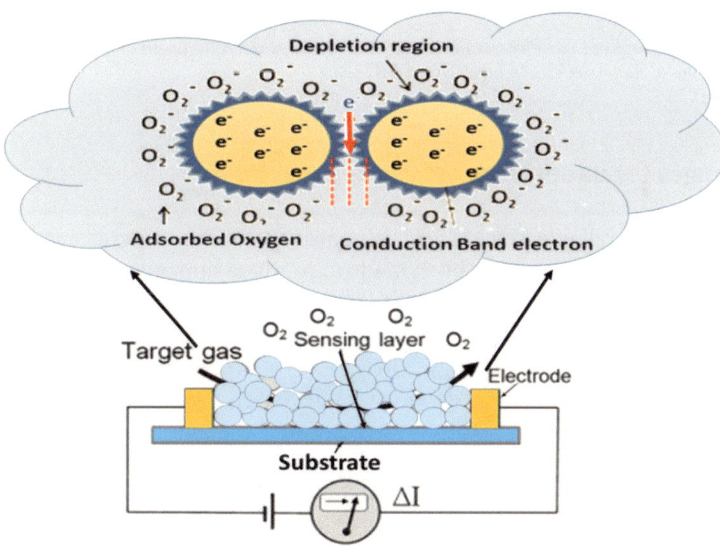

Figure 6.
Schematic of metal oxide thin film gas sensor mechanism.

Furthermore, the electronic properties of graphene changed by metal nanoparticles influence the increase of selectivity and sensitivity in gas sensing characteristics. **Figure 8** shows graphene decorated with metal nanoparticle-based sensor devices with interdigitated electrodes. Graphene with suitable sensitizer can improve the sensing performance and rectify the selectivity issues. Graphene-based functional nanohybrids can also facilitate fabrication of the nanosensor device [46].

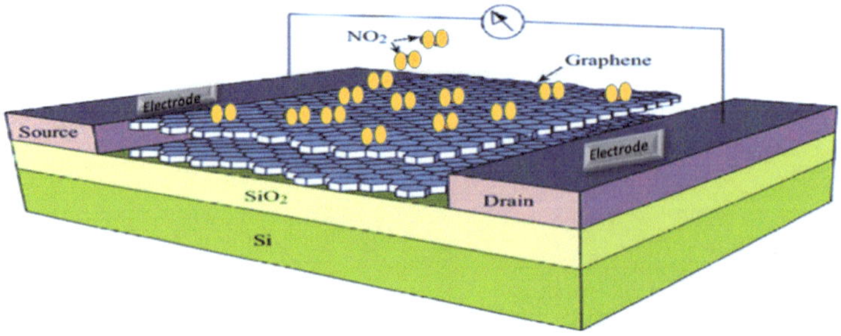

Figure 7.
Schematic of graphene-based thin film gas sensor (adapted from Ref. [45]).

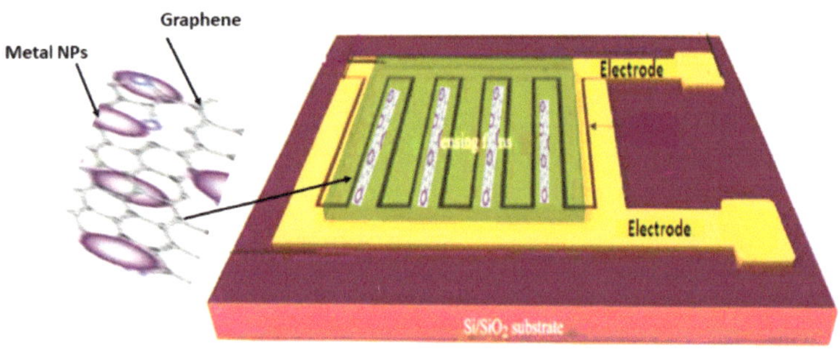

Figure 8.
Schematic image of graphene-based sensor devices with interdigitated electrodes and a graphene channel decorated with metal nanoparticles (adapted from Ref. [45]).

5. Challenges for nanothin film-based gas sensors

The demand of reliable sensors for toxic gases and VOCs is increasing rapidly for the safe life. A lot of progress has been made to produce sensor especially in the form of thin film or nanostructure materials to exploit effectively sensing properties. The challenging parts of the devices are uniform fabrication of nanostructure in terms of thickness, shape, and morphology. The process is expensive also. Sensitivity, effect of temperature, stability of detection, selectivity, response and recovery time, repeatability, and durability are the concern for all these sensors for gas detection. Sensors during operation suffer from multiple limitations which require several optimization procedures meticulously. Commercial production of these ultrathin film or nanostructure sensors in large scale will depend on how all these properties of sensor materials are achieved uniformly. However, on the positive note, because of so much sophistication in technologies and materials engineering it, it will be possible to achieve high-quality and cheap sensor devices to be available in the market very soon.

6. Conclusion

The importance of environmental sensors has been increasing exponentially due to the increase of toxic gas emission and at the same time imposition of several emission norms in the industries and permissible health hazard limits announced by health organizations. Looking at the historical aspects of the sensor development works, the chapter starts with the scope of various gas sensors to make environment risk free. Defining thin film, ultrathin film, and nanomaterials, efforts were made

DOI: http://dx.doi.org/10.5772/intechopen.89745

to categorize sensors and structures based on materials although there is always an overlapping research prospect. Various standard methods of preparation were mentioned. However, every sensor research based on nanotechnology is bringing some improvised method to fabricate sensor structures modifying available techniques. Sensing mechanisms were elaborated. Since to prepare nanostructure is tricky and expensive, more efforts should be given to achieve sensor fabrication viable, sustainable, and useful for mass production with acceptable quality.

Conflict of interest

The authors declare no conflict of interest.

Author details

Nithya Sureshkumar and Atanu Dutta*
Division of Physics, School of Advanced Sciences, Vellore Institute of Technology, Chennai, India

*Address all correspondence to: atanudutta@vit.ac.in

References

[1] Fraden J. Handbook of Modern Sensors: Physics, Design and Applications. Springer Science, Business Media; 2004

[2] United State Departmental of Labor, Occupational safety and Health Administration. Washington, DC; 20210. Available at : https://www.osha.gov/dsg/annotated-pels/

[3] Seiyama T, Kato A, Fujiishi K, Nagatani M. A new detector for gaseous components using semiconductive thin films. Analytical Chemistry. 1962;**34**(11):1502-1503

[4] Taguchi N. Gas detecting element and making of it. US Patent Specification. 1970. 3644795

[5] Advani GN, Jordan AG. Thin films of SnO2 as solid state gas sensors. Journal of Electronic Materials. 1980 Jan 1;**9**(1):29-49

[6] Heiland G. Physical and chemical aspects of oxidic semiconductor gas sensors. Chemical Sensor Technology. 1988;**1**:15-38

[7] Sberveglieri G. Recent developments in semiconducting thin-film gas sensors. Sensors and Actuators B: Chemical. 1995 Feb 1;**23**(2-3):103-109

[8] Ho SM, Vanalakar SA, Ahmed G, Vidya NS. A review of nanostructured thin films for gas sensing and corrosion protection. Mediterranean Journal of Chemistry. 2018 Jun 1;7(6)

[9] Jimenez-Cadena G, Riu J, Rius FX. Gas sensors based on nanostructured materials. Analyst. 2007;**132**(11):1083-1099

[10] Yang RD, Gredig T, Colesniuc CN, Park J, Schuller IK, Trogler WC, et al. Ultrathin organic transistors for chemical sensing. Applied Physics Letters. 2007 Jun 25;**90**(26):263506

[11] Patra MK, Manzoor K, Manoth M, Negi SC, Vadera SR, Kumar N. Nanotechnology applications for chemical and biological sensors. Defence Science Journal. 2008 Sep 1;**58**(5):636-649

[12] Sharma S, Madou M. A new approach to gas sensing with nanotechnology. Philosophical Transactions of the Royal Society A: Mathematical, Physical and Engineering Sciences. 2012 May 28;**370**(1967):2448-2473

[13] Bogue R. Nanomaterials for gas sensing: A review of recent research. Sensor Review. 2014 Jan 14;**34**(1):1-8

[14] Li Z, Li H, Wu Z, Wang M, Luo J, Torun H, et al. Advances in designs and mechanisms of semiconducting metal oxide nanostructures for high-precision gas sensors operated at room temperature. Materials Horizons. 2019;**6**(3):470-506

[15] Arafat MM, Dinan B, Akbar SA, Haseeb AS. Gas sensors based on one dimensional nanostructured metal-oxides: A review. Sensors. 2012;**12**(6):7207-7258

[16] Harris PD, Arnold WM, Andrews MK, Partridge AC. Resistance characteristics of conducting polymer films used in gas sensors. Sensors and Actuators B: Chemical. 1997 Aug 1;**42**(3):177-184

[17] Ziadan KM. Conducting polymers application. In: Gomes ADS, editor. New Polymers for Special Applications. IntechOpen; September 12th 2012. DOI: 10.5772/48316. Available from: https://www.intechopen.com/books/new-polymers-for-special-applications/conducting-polymers-

[18] Kimura T, Goto T. Ir–YSZ nano-composite electrodes for oxygen sensors.

Surface and Coatings Technology. 2005 Aug 1;**198**(1-3):36-39

[19] Brousseau LC, Aurentz DJ, Benesi AJ, Mallouk TE. Molecular design of intercalation-based sensors. 2. Sensing of carbon dioxide in functionalized thin films of copper octanediylbis (phosphonate). Analytical Chemistry. 1997 Feb 15;**69**(4):688-694

[20] Hanna Varghese S, Nair R, G Nair B, Hanajiri T, Maekawa T, Yoshida Y, et al. Sensors based on carbon nanotubes and their applications: A review. Current Nanoscience. 2010 Aug 1;**6**(4):331-346

[21] Mao S, Lu G, Chen J. Nanocarbon-based gas sensors: Progress and challenges. Journal of Materials Chemistry A. 2014 Mar 25;**2**(16):5573-5579

[22] Llobet E. Gas sensors using carbon nanomaterials: A review. Sensors and Actuators B: Chemical. 2013 Mar 31;**179**:32-45

[23] Geim AK, Novoselov KS. The rise of graphene. Nanoscience and Technology: A Collection of Reviews from Nature Journals. 2010:11-19

[24] Novoselov KS, Geim AK, Morozov SV, Jiang D, Zhang Y, Dubonos SV, et al. Electric field effect in atomically thin carbon films. Science. 2004 Oct 22;**306**(5696):666-669

[25] Lu G, Ocola LE, Chen J. Reduced graphene oxide for room-temperature gas sensors. Nanotechnology. 2009 Oct 7;**20**(44):445502

[26] Schedin F, Geim AK, Morozov SV, Hill EW, Blake P, Katsnelson MI, et al. Detection of individual gas molecules adsorbed on graphene. Nature Materials. 2007 Sep;**6**(9):652

[27] Chen D, Feng H, Li J. Graphene oxide: Preparation, functionalization, and electrochemical applications. Chemical Reviews. 2012 Aug 14;**112**(11):6027-6053

[28] Prezioso S, Perrozzi F, Giancaterini L, Cantalini C, Treossi E, Palermo V, et al. Graphene oxide as a practical solution to high sensitivity gas sensing. The Journal of Physical Chemistry C. 2013 May 8;**117**(20):10683-10690

[29] Toda K, Furue R, Hayami S. Recent progress in applications of graphene oxide for gas sensing: A review. Analytica Chimica Acta. 2015 Jun 9;**878**:43-53

[30] Chua CK, Pumera M. Chemical reduction of graphene oxide: A synthetic chemistry viewpoint. Chemical Society Reviews. 2014;**43**(1):291-312

[31] Lee HK, Lee J, Choi NJ, Moon SE, Lee H, Yang WS. Efficient reducing method of graphene oxide for gas sensor applications. Procedia Engineering. 2011 Jan 1;**25**:892-895

[32] Lu G, Yu K, Ocola LE, Chen J. Ultrafast room temperature NH_3 sensing with positively gated reduced graphene oxide field-effect transistors. Chemical Communications. 2011;**47**(27):7761-7763

[33] Wang J, Singh B, Maeng S, Joh HI, Kim GH. Assembly of thermally reduced graphene oxide nanostructures by alternating current dielectrophoresis as hydrogen-gas sensors. Applied Physics Letters. 2013 Aug 19;**103**(8):083112

[34] Grosso D, Boissière C, Faustini M. Chapter 9: Thin film deposition techniques. In: Levy D, Zayat M, editors. The Sol-Gel Handbook. Wiley Online Library

[35] Gould RD, Kasap S, Ray AK. Chapter 28: Thin films. In: Kasap S, Capper P, editors. Handbook

of Electronic and Photonic Materials. Springer; 2017 Oct 4

[36] Mattox DM. Handbook of Physical Vapor Deposition (PVD) Processing, 2nd ed. 2010

[37] Jiang Y, Huang W, Zhuang X, Tang Y, Yu J. Thickness modulation on semiconductor towards high performance gas sensors based on organic thin film transistors. Materials Science and Engineering: B. 2017 Dec 1;**226**:107-113

[38] Chen HI, Hsiao CY, Chen WC, Chang CH, Chou TC, Liu IP, et al. Characteristics of a Pt/NiO thin film-based ammonia gas sensor. Sensors and Actuators B: Chemical. 2018 Mar 1;**256**:962-967

[39] Schedin F, Geim AK, Morozov SV, Hill EW, Blake P, Katsnelson MI, et al. Detection of individual gas molecules adsorbed on graphene. Nature Materials. 2007;**6**:652-655

[40] Yuan W, Liu A, Huang L, Li C, Shi G. High-performance NO_2 sensors based on chemically modified graphene. Advanced Materials. 2013;**25**:766-771

[41] Wu J, Feng S, Li Z, Tao K, Chu J, Miao J, et al. Boosted sensitivity of graphene gas sensor via nanoporous thin film structures. Sensors and Actuators B: Chemical. 2018 Feb 1;**255**:1805-1813

[42] Tian W, Liu X, Yu W. Research progress of gas sensor based on graphene and its derivatives: A review. Applied Sciences. 2018 Jul;**8**(7):1118

[43] Xu K, Zeng D, Wu J, Mao Q, Tian S, Zhang S, et al. Correlation between microstructure and gas sensing properties of hierarchical porous tin oxide topologically synthesized on coplanar sensors' surface. Sensors and Actuators B: Chemical. 2014 Dec 15;**205**:416-425

[44] Shankar P, Rayappan JB. Gas sensing mechanism of metal oxides: The role of ambient atmosphere, type of semiconductor and gases—A review. Science Letters Journal. 2015;**4**(4):126

[45] Wang T, Huang D, Yang Z, Xu S, He G, Li X, et al. A review on graphene-based gas/vapor sensors with unique properties and potential applications. Nano-Micro Letters. 2016 Apr 1;**8**(2):95-119

[46] Varghese SS, Lonkar S, Singh KK, Swaminathan S, Abdala A. Recent advances in graphene based gas sensors. Sensors and Actuators B: Chemical. 2015 Oct 31;**218**:160-183

Chapter 5

Synthesis and Characterization of CoO-ZnO-Based Nanocomposites for Gas-Sensing Applications

Parthasarathy Panchatcharam

Abstract

CoO-ZnO composite nanofibers were synthesized through electrospinning technique. CoO-ZnO composite nanofibers were fabricated by doping zinc (zinc acetate dihydrate) and varied concentrations of cobalt (cobalt oxide) in the ratio of 1, 3, and 5 wt%, respectively. By modifying the solvent and the electrospinning parameters, different tests were carried out to optimize the morphological properties of the synthesized composite nanofibers. The morphological characterization was performed by scanning electron microscopy (SEM) with a field emission gun. The atomic composition of the nanofibers was analyzed by energy-dispersive X-ray (EDX) spectroscopy using a solid-state detector. Gas-sensing performances are done at different temperatures like at room temp, 50°C, and 100°C to find out the optimum operating temperature for detecting acetone gas. The sensitivity studies of CoO-ZnO composite nanofiber were carried out over different concentrations of acetone gas from 50 to 250 ppm. The sensitivity of this sensor developed is found to be increasing with increase in temperature and also increases if dopant concentration increases when compared with pure nanofibers. The sensitivity analysis proved a fact that uncalcinated CoO-ZnO composite nanofibers can be helpful in the detection of diabetics at the early stage with acetone concentration in the breaths.

Keywords: gas sensor, nanocomposites, metal oxide semiconductor, electrospinning, acetone sensing

1. Introduction

Semiconducting metal oxides have brought incredible attention as chemical sensors due to its characteristic resistivity and sensitivity changing features in an ambient environment [1, 2]. Being an N-type semiconductor, ZnO has been extensively used as a gas-sensing material owing to its wide band gap of 3.37 eV and high exciton binding energy of 60 meV at ambient temperature [3–9]. It has drawn tremendous attention in the last few decades due to its specific electrical, catalytic, and photochemical optoelectronic properties. It is the most widely studied semiconductor in gas-sensing applications because of its significant response to different reducing or oxidizing gases such as carbon monoxide, hydrogen, ammonia, ethanol, and acetone [10–12]. But some intrinsic disadvantages including high operating temperature, slow response and recovery time, and less sensitivity hinder the further advance in the development of ZnO-based gas sensors. So, boundless

IntechOpen

efforts have been made in the last few years to overcome these limitations employing different methods, including noble metal doping, structure optimization, and heterostructure fabrication [13–15].

One such method to enhance gas-sensing properties is doping impurity elements [16]. Among the available methods for synthesizing nanostructures, electrospinning is a simple, adaptable, and low-cost technique for fabricating organic and inorganic nanofibers with significant lengths, uniform diameters, and various compositions [17]. During the electrospinning process, high voltage is applied to a polymer droplet being suspended at the tip of a syringe needle [18]. In this we present electrospun CoO-ZnO composite nanofibers with different concentrations like 1, 3, and 5 wt% of cobalt oxide. Calcination of cobalt oxide and zinc oxides leads to the formation of CoO-ZnO composite nanofibers having tunable diameters and significant porous structures. We studied the properties of the CoO-ZnO composite nanofibers fabricated, and specific dopants are used to improve the selectivity. The sensitivity of the nanofibers over acetone gas at lower concentrations has been examined, and the outcome of the fabricated nanofiber for sensing the acetone levels in breath of diabetic patients is sensed in low concentrations.

2. Related works

Alali et al. successfully synthesized $ZnO/CoNiO_2$ hollow nanofibers by an electrospinning method and postcalcination treatment. $ZnO/CoNio_2$ hollow nanofibers gave an excellent response to ammonia solution, and further a great enhancement was achieved in ammonia sensing with these hollow fiber structures. $ZnO/CoNio_2$ offered the best choice of ammonia-sensing materials by providing a response of 40 at 80°C [19]. Sun et al. synthesized pure and Er-doped ZnO nanofibers by electrospinning for high sensitivity detection of ethanol. They demonstrated that the diameter of the ZnO nanofibers decreases from 200 to 70 nm with the increase of Er content and the Er doping significantly increased the ethanol-sensing sensitivity of ZnO nanofibers at an optimum operating temperature of 240°C. This research also found that at the optimal Er content of 1.0 at%, the sensitivity of the nanofibers is 3.7 times larger than that of pure ones [20]. Wang et al. synthesized pure and Cu-doped ZnO fibers through electrospinning technology. The results of this research revealed that H_2S gas-sensing properties of ZnO nanofibers were effectively improved by Cu doping: 6 at% Cu-doped ZnO nanofibers showed a maximum sensitivity to H_2S gas, and the response to 10 ppm H_2S is one order of magnitude higher than the one of pure ZnO nanofibers [21]. ZnO nanofibers were fabricated by an electrospinning method using a solution containing sol-gel precursors and solvent. It had been observed that the crystallinity of ZnO nanofibers improved with the increase in annealing temperature. The diameters of ZnO nanofibers after annealing above 600°C ranged from 35 to 100 nm. This work also confirmed the fact that the activation energy of ZnO nanofibers for electrical conduction was inversely proportional to the annealing temperature, and they also showed CO gas-sensing capacity at concentration as low as 1.9 ppm [22]. One-dimensional $ZnO\text{-}SnO_2$ hollow nanofibers were synthesized by Shouli bhai et al. by one-step electrospinning method and annealing treatment. The research depicted that the 20 atm% Zn composite not only exhibits the highest response that is 9 and 5.2 times higher than that of pristine ZnO and SnO_2, respectively. It also exhibited rapid response, excellent selectivity, and stability at low operating temperature of 90°C, thereby promising to be a sensing material for the detection of NO_2 [23]. Pure and Ag-doped $ZnO\text{-}SnO_2$ hollow nanofibers had been synthesized through electrospinning method by Ma et al. The sensor based on the Ag-doped $ZnO\text{-}SnO_2$ hollow nanofibers exhibited excellent gas-sensing performance at low operating temperature at 200°C and the fast response and recovery characteristics at low concentration of 1 ppm. This demonstrated that Ag-doped ZnO-SnO_2 could be used as a significant material for selective detection of low-concentration ethanol gas [24]. Moghaddam et al. synthesized polythiophene nanocomposite nanofibers containing ZnO nanoparticles through a self-assembly process in the presence of CTAB as the surfactant. Gas-sensing tests showed that the chemiresistor based on the as-prepared hybrid has high sensitivity, excellent repeatability, long-term stability, and short response time to ammonia gas at room temperature. ZnO/PT hybrid-sensing mechanism to ammonia gas was presumed to be the effect of p-n heterojunction between ZnO nanoparticles and PT [25]. Nano-crystalline ZnO nanofiber mats were synthesized through combined sol-gel electro-spinning techniques followed by calcination in which poly(styrene-co-acrylonitrile) and zinc acetate were used as the binder and precursor, respectively. The average diameter of the ZnO nanofibers decreased from 400 to 60 nm, while their grain size and crystallinity were enhanced by increasing the calcination temperature. Due to their high surface area and superhydrophilicity, these ZnO nanofiber mats were highly sensitive in sensing gaseous ammonia, and the sensitivity of these mats increased as a function of their calcination temperatures [26].

3. Experimental

3.1 Preparation of Co-doped ZnO composite nanofibers

The fabrication of CoO-ZnO composite nanofibers was achieved by electrospinning followed by calcination. In this 0.595 g of zinc acetate dihydrate and varied concentrations of cobalt oxide in the ratio of 1, 3, and 5 wt% were added to 10 ml of dimethylformamide solvent. All the reagents were purchased from Sigma-Aldrich Corporation. The solution was magnetically stirred for 5–6 hours at room temperature; subsequently 1 g of polyvinylpyrrolidone (PVP) and 8 ml of were added to the mixture and kept for further stirring of about 2–3 hours at room temperature. The resulting solutions were electrospun using commercial electrospinning apparatus (EC-DIG Electrospinning, IME Technologies). The composites were ejected from the needle of a syringe applying an electrical field, as high voltage of about 18 kV and at the needle with collector distance of 20 cm. Composite nanofibers were collected in the stationary mode on the surface of silicon substrates clamped on top of a conductive circular collector.

By modifying the solvent and the electrospinning parameters, different tests were carried out to optimize the morphological properties of the synthesized composite nanofibers. While decomposition, we employed a negative and positive voltage between the needle and the substrates was 14–20 cm, and the flow of the solutions were set to 5 μl/min. The deposition times were kept constant at 3 min. The morphological characterization was performed by scanning electron microscopy (SEM) with a field emission gun. The atomic composition of the nanofibers was analyzed by energy-dispersive X-ray (EDX) spectroscopy using a solid-state detector. Gas-sensing performances are done at different temperatures like at room temp, 50°C, and 100°C to find out the optimum operating temperature for detecting acetone gas. The sensitivity studies of CoO-ZnO composite nanofiber were carried out over different concentrations of acetone gas from 50 to 250 ppm.

3.2 Gas-sensing mechanism

When Zn and O combine, Zn loses two valence electrons to O; thus, due to loss of an outer shell, the Zn atom shrinks in size, while O atom increases in size due to addition of an outer shell. The wide difference in size between zinc and oxygen atom will allow large space while providing foreign atoms to incorporate. Cobalt oxide-zinc oxide composite in the presence of ambient oxygen undergoes chemical absorptive changes by capturing electrons in conduction band which is known as recombination where oxygen molecule is adsorbed on its surface. This will decrease the barrier height for electrons to transport there by reducing the resistance value.

Activation energy is required for adsorption and desorption of oxygen and detection of gas. This in turn will control the recovery time and response as well. Temperature plays a major role in controlling this mechanism. The basic sensing mechanism in this thin-film sensors includes that at temperature which is less than or equal to 100°C where the adsorption mechanism takes place, that is, acetone gas passes over the sensing film oxygen molecules such as O_2^- will form and upon increase in temperature it will gets modified to $2O^-$and then O_2^-. Then in the presence of acetone gas, carbon dioxide gas is released and water and electrons are gained. In oxidizing mechanism, O_2 atoms will be gained and in reduction mechanism atoms will be lost.

Mostly semiconducting gas-detecting sensors, resistance will change according to the adsorption and desorption of gas molecules on the surface of the thin film [26–29]. The ZnO nanofibers in the presence of air ambience will absorb the oxygen on the surface of the thin film. Later by collecting electrons from the conductance band, the oxygen adsorbed is changed into various chemical states. Thus, the nanofibers will show more resistance with the increment of the barrier height of e^- to move. Since the target gas acetone is a reducing gas, the oxygen which is adsorbed on the surface will undergo oxidation with acetone. In this reaction resistance of sensor will reduce when e^- enters the nanofibers as depicted in **Figure 1**.

The structure of the fibers plays a very vital role in getting sensitivity. Since the ZnO thin films will have high surface to volume ratio will have high sensing performance. Many earlier papers discussed about the ZnO thin film in sensing different gases. In this case the electrospinning technique used to synthesize nanofibers with various parameters will improve the sensor sensitivity for 5 wt% CoO-ZnO-calcinated nanofibers. Cobalt oxide is preferred in here to increase the catalytic conversion when compared with other dopants like Ni, Al, etc.

3.3 Fabrication of gas sensor

The calcinated sample in the form of thin film is fabricated to form a handheld sensor for sensing the acetone gas. Here, the thin film is placed over a glass substrate and upon the masks of desire pattern is placed as shown in **Figure 2**. The physical vapor deposition technique, a thermal evaporation process, is followed for the fabrication of the thin films. The masks were prepared by using Co_2 laser technique of comb pattern using acrylic sheet. Later the mask is placed on the thin film and these sensing materials are placed on the sample holder. Physical vapor deposition technique is followed for vaporization of the thin films where higher pressure is applied under higher temperature in a vacuum chamber. Indium electrodes were fixed by microsoldering technique, and copper leads were separated from the electrodes. The sensor is heated by placing upon a hot ceramic plate at 200°C.

Gas-sensing chamber has an inlet and an outlet chamber where the target gas is sent through an inlet nozzle. The sample is placed inside the chamber, and the two probes of electrometer were connected to the copper wires to measure the change in the resistance variation. One of the input parameters is the input voltage of about 3v; accordingly corresponding change in resistance and time variation is noted by injecting acetone of liquid form upon heating becomes gaseous state.

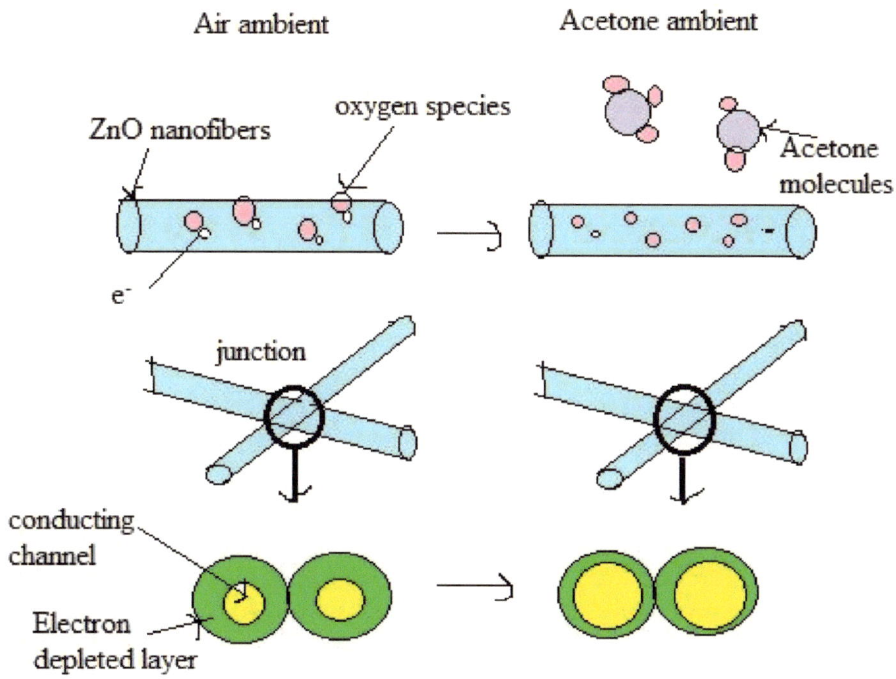

Figure 1.
Schematic sensing process of ZnO nanofibers.

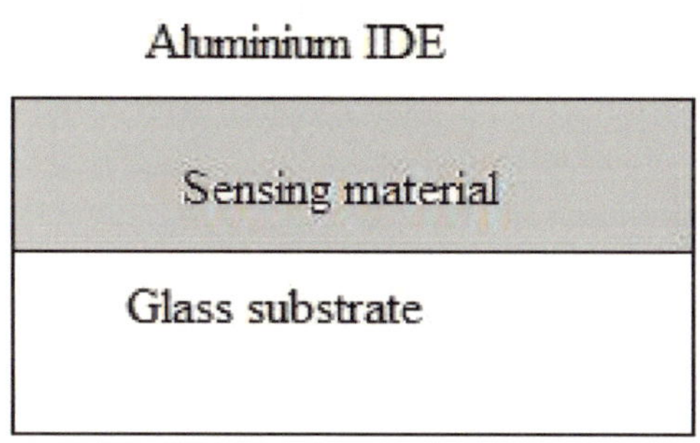

Figure 2.
Schematic of gas sensor.

The sensor resistances were calculated automatically by analysis system. The acetone liquid was injected in different concentrations of about 10, 50, 100, 150, 200, and 250 ppm. According to the volume of the chamber, this conversion is done, i.e., ml to ppm. This process is repeated at room temperature, 50°C, and 100°C by using ceramic heater to obtain the optimum temperature. The process is repeated until the gas evaporates properly from the gas chamber. The sensitivity is calculated by using the following formula:

$$S\,(\%) = \left(R_a - R_g/R_a\right) \times 100 \tag{1}$$

R_a—Resistance in air ambience.
R_g—Resistance in the presence of target gas.

4. Results and discussion

The morphological, structural, and thermal analyses are discussed in this section.

4.1 Morphological analysis of the as-prepared Co-doped ZnO nanofibers

Typical scanning electron microscopy (SEM) images of the synthesized nanofibers of pure ZnO nanofiber and cobalt-doped composite nanofibers of purity in the ratio of 1, 3, and 5 wt%, respectively, were shown in **Figure 3**. The nanofibers have a random orientation, as expected, due to the instability of the electrospinning jet. The EDX analysis (**Figure 4**) shows the presence of carbon, nitrogen, zinc, cobalt, and oxygen coming from PVP, zinc acetate, and cobalt oxide, respectively. The porosity of the electrospun mats is confirmed by detection of silicon signal

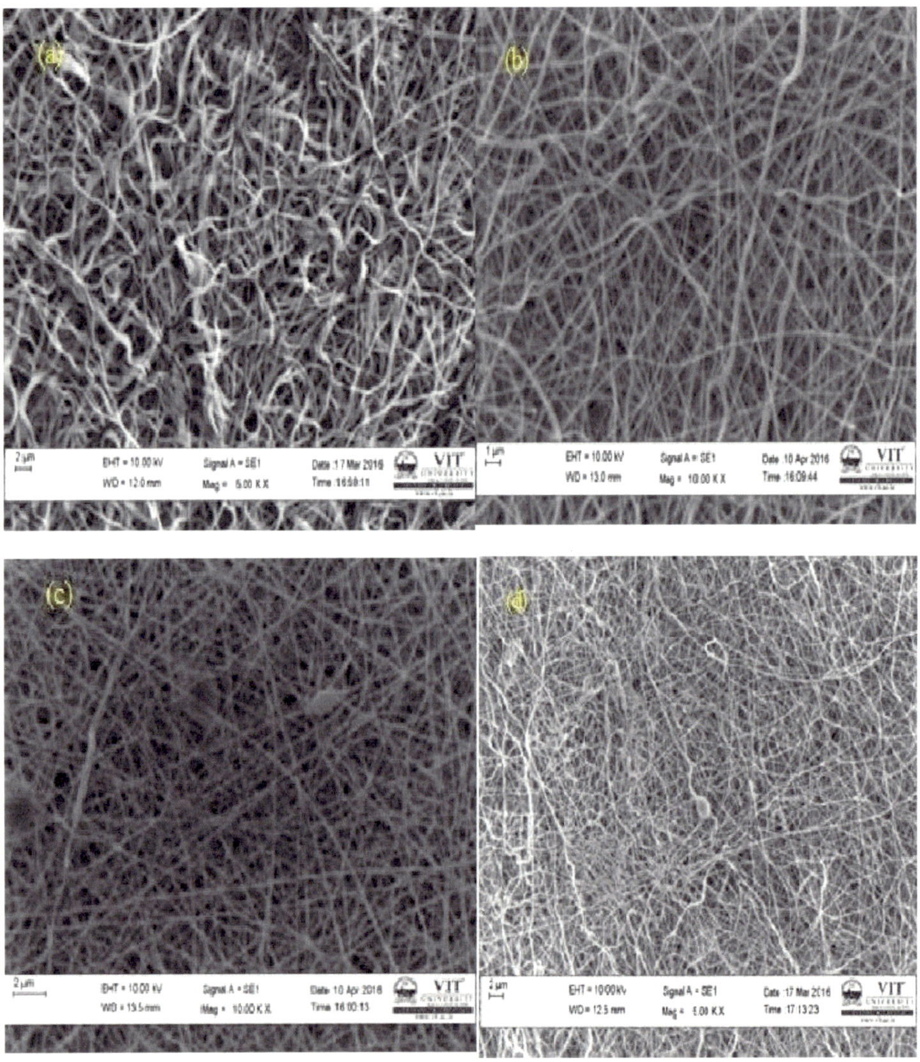

Figure 3.
(a) Pure ZnO-calcinated fiber, (b) 1 wt% CoO-ZnO-calcinated nanofibers, (c) 3 wt% CoO-ZnO-uncalcinated nanofibers, and (d) 5 wt% calcinated CoO-ZnO composite nanofiber.

coming from the substrate. The diameter of the nanofibers fabricated appears to be uniform, and it is also evident that it becomes thinner upon the increase of dopant concentration, i.e., 1 wt% CoO, 2 wt% CoO, and 5 wt% CoO, respectively. This is the result of addition of charges increased during the electrospinning process. Furthermore, a typical EDX spectrum of the Co-doped ZnO nanofibers was illustrated in **Figure 4(d)**, which shows only the peaks associated with Zn, Co, and O atoms (Cu is from copper mesh grids) were detected. Therefore, the nanofibers were indeed Co-doped ZnO materials.

4.2 Thermogravimetric analysis of the as-prepared nanofibers

Thermogravimetric (TG) analysis was performed on a NET-ZSCH STA 449 thermo-analyzer in an air atmosphere. The precursor fibers of PVP/zinc acetate/cobalt acetate composite produced after electrospinning must be annealed because of its high contents of organic polymer and ethanol solvents in these nanofibers. The thermal behavior of the precursor fibers of this composite is shown in **Figure 5**; it illustrates that most of the organics belonged to PVP and the CH_3COO group and other volatiles (H_2O, CO_2, etc.) are detached at temperature < 600°C. Beyond 600°C, there is no change in weight loss, indicating the formation of pure inorganic oxide. The XRD curve depicted in **Figure 6** shows all the peak positions of Co-doped ZnO nanofibers which suggested that Co is successfully incorporated into the crystal lattice of ZnO which is reasonable given that the ionic radii of tetrahedrally coordinated Co^{2+} and Zn^{2+} are similar.

4.3 Optical properties of the as-prepared nanofibers

Optical properties of ZnO nanofibers and Co-doped ZnO nanofibers have been examined through Raman spectrometers and photoluminescence (PL). **Figure 7** shows their typical PL spectra under an excitation wavelength of 325 nm; only a broad and strong ultraviolet (UV) luminescence with a maximum at 375 nm is observed. This ascribed to the near band edge emission of the wide band gap zinc oxide. Compared to the spectrum of ZnO nanofibers formed under the same experimental condition, the emission peak of Co-doped nanofibers is shifted about 10 nm toward longer wavelength significantly. ZnO powders are heavily doped by donors like In, and it was interpreted as mainly due to the sp-d exchange interactions between the band electrons and the localized d electrons of the Co^{2+} ions substituting Zn ions. The s-d and p-d exchange interactions lead to a negative and a positive correction to the conduction band and the valence band edges, resulting in a band gap narrowing.

To confirm that the Co ions as the dopant were successfully incorporated into the crystal lattice of ZnO, Raman scattering spectra of ZnO nanofibers and Co-doped ZnO nanofibers were measured. As observed in **Figure 8(a** and **b)**, the Raman spectrum of the ZnO nanofibers exhibits a strong peak at 437 cm^{-1}, which has been assigned to the vibration mode E_2 (H) characteristic of the ZnO with hexagonal structure. The peaks at 330 and 379 cm^{-1} are assigned to the vibration mode $2E_2$ and the A_1 (TO) mode, respectively. In comparison to the Raman spectrum of ZnO nanofibers, most modes in Co-doped ZnO nanofibers figure, expecting the A_1 (TO) mode is disappeared. The E2 (H) mode and the vibration mode 2E2 broaden asymmetrically and shift toward lower frequencies when the Co is doped. This is due to the broken symmetry induced by the incorporation of Co dopants into the ZnO structure and could be explained by the

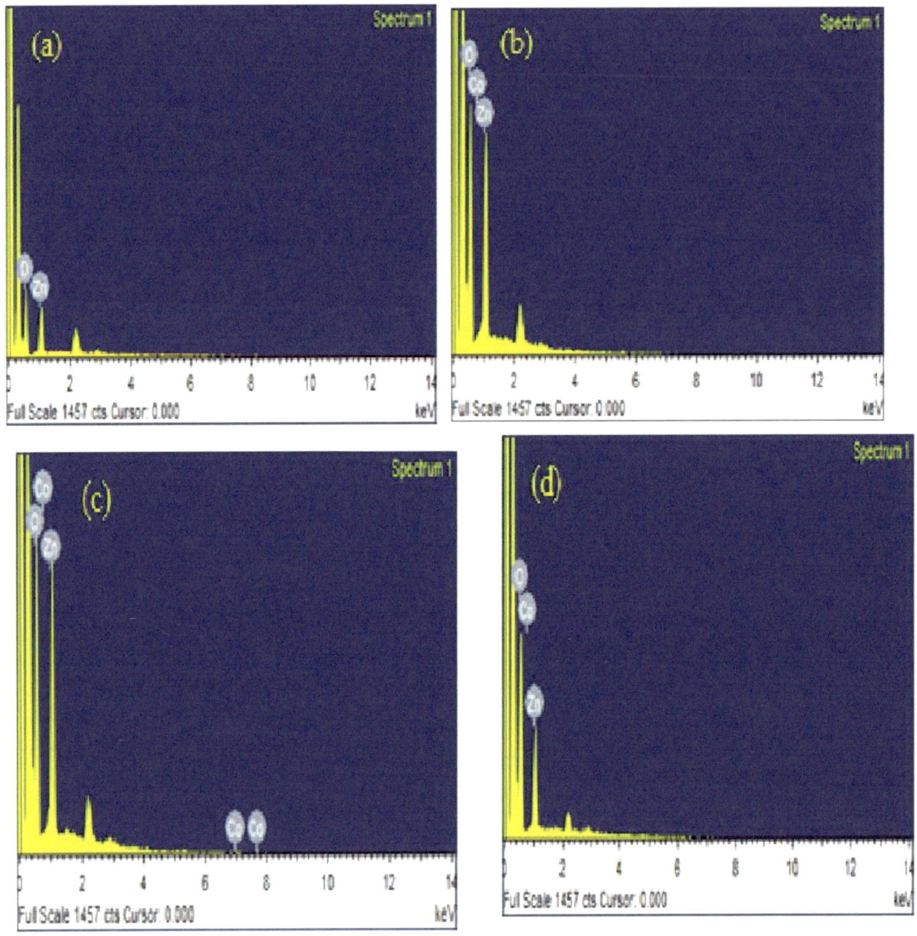

Figure 4.
EDX analysis: (a) ZnO-calcinated fiber, (b) 1 wt% CoO-ZnO-calcinated nanofibers, (c) 3 wt% CoO-ZnO-calcinated nanofibers, and (d) 3 wt% CoO-ZnO-uncalcinated nanofibers.

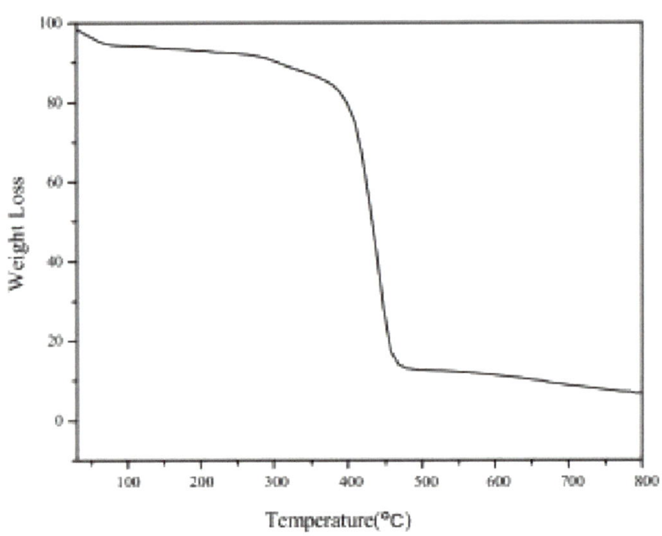

Figure 5.
TG curves of precursor fibers of PVP/zinc acetate/cobalt acetate composites.

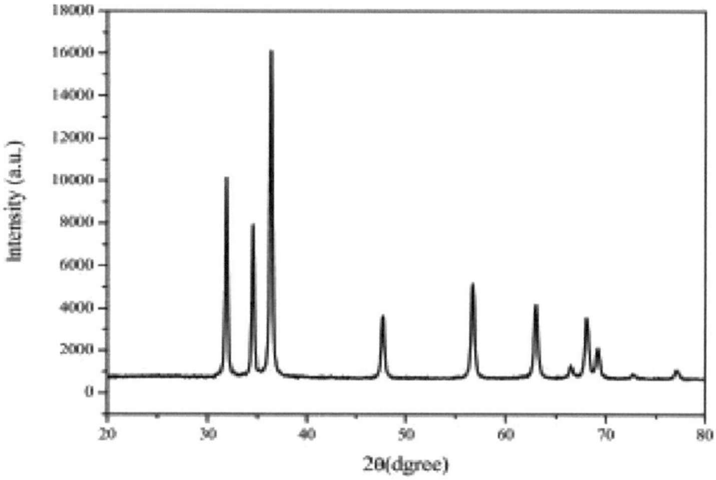

Figure 6.
XRD of Co-doped ZnO nanofibers after calcination at 600°C.

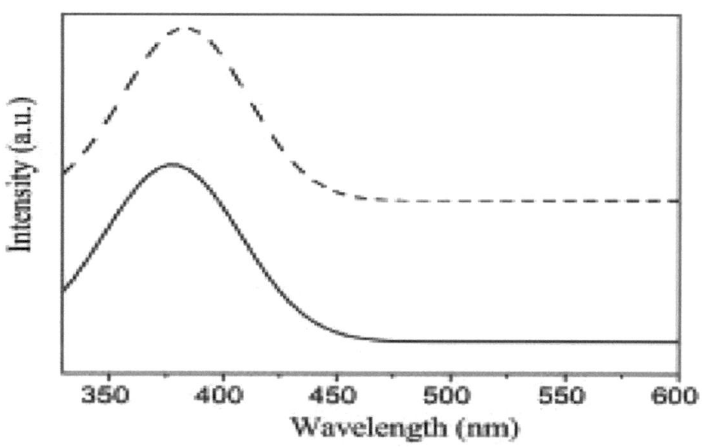

Figure 7.
PL structures of ZnO nanofibers (solid line) and Co-doped ZnO (dotted lines) after calcination at 600°C.

spatial-correlation model [30, 31]. Apart from these features, the Co-doped ZnO nanofibers exhibit an unusual peak at 540 cm^{-1}, which has been assigned to the quasi-LO phonon mode (AM1) due to the abundant shallow donor defects, such as oxygen vacancies or zinc interstitials bounded on the tetrahedral Co sites [32]. These results together confirm that Co was successfully doped into the crystal lattice of ZnO nanofibers.

4.4 Gas-sensing performance analysis

Gas-sensing performances were performed at different temperatures of ambient, 50°C, and 100°C to find out the optimum operating temperature for the detection of acetone gas. The sensitivity plots were measured for concentration of acetone gas against purity of the CoO-ZnO nanocomposite fibers fabricated. The sensitivity of the dopants was found to be increasing with increase in temperature and increases with the increase in dopant concentration when compared with the pure nanofibers.

When the operating temperature is low, the activation energy used will be reduced simultaneously, and in turn the response will be decreased. At the same time, if the operating temperature used is very high due to the increased activation energy, the adsorbed gas molecules will escape before the reaction occurs; thus, the response will decrease too. In this chapter, 5 wt% calcinated CoO-ZnO nanofibers at 100°C have high sensitivity when compared with other compositions like pure, 1, and 3 wt%. Hence this sample can be improvised to make use in many applications like diabetic's detection where the ppm range should be between 1 and 3 ppm. The base resistance, i.e., in the presence of air ambience, will increase and in the presence of acetone gas will increase. Since the acetone gas will evaporate if exposed to air temperature, the gas injected should be properly according to time to time. The ppm is calculated according to the gas chamber volume. According to the volume of the gas chamber used in this chapter, the ppm is taken as 5 ml as 50 ppm. Acetone is calculated in terms of parts per million. The main feature of this chapter is to improvise the selectivity of acetone sensing with CoO-ZnO nanofibers.

Figure 9 shows the sensor sensitivity for pure ZnO nanofibers. In this plot the sensitivity will increase linearly from 50 to 250 ppm with increase in the acetone

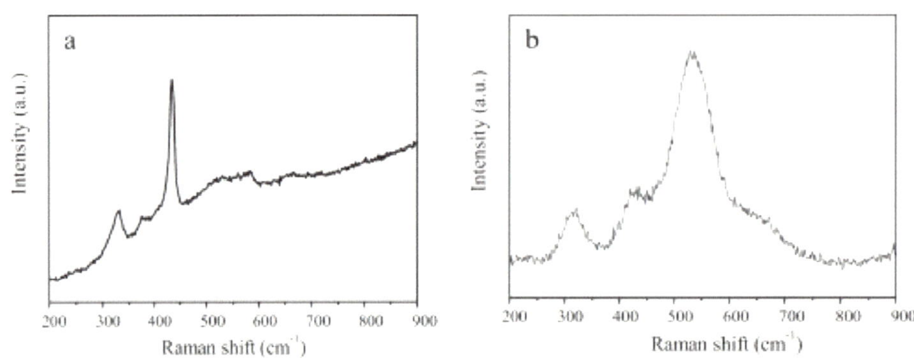

Figure 8.
Raman spectra of (a) ZnO nanofibers and (b) Co-doped ZnO nanofibers after calcination at 600°C.

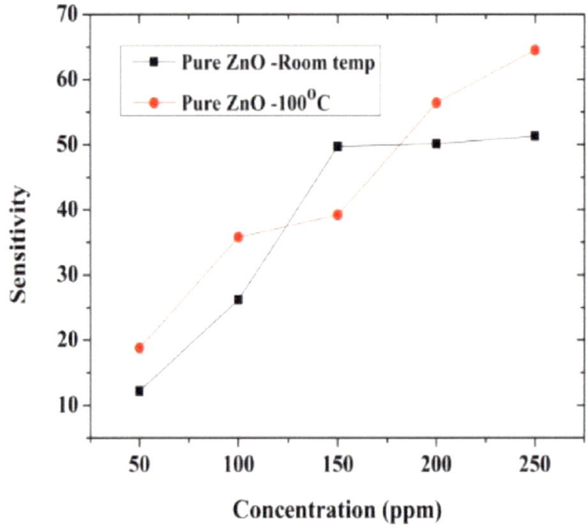

Figure 9.
Sensitivity plot for pure ZnO nanofibers.

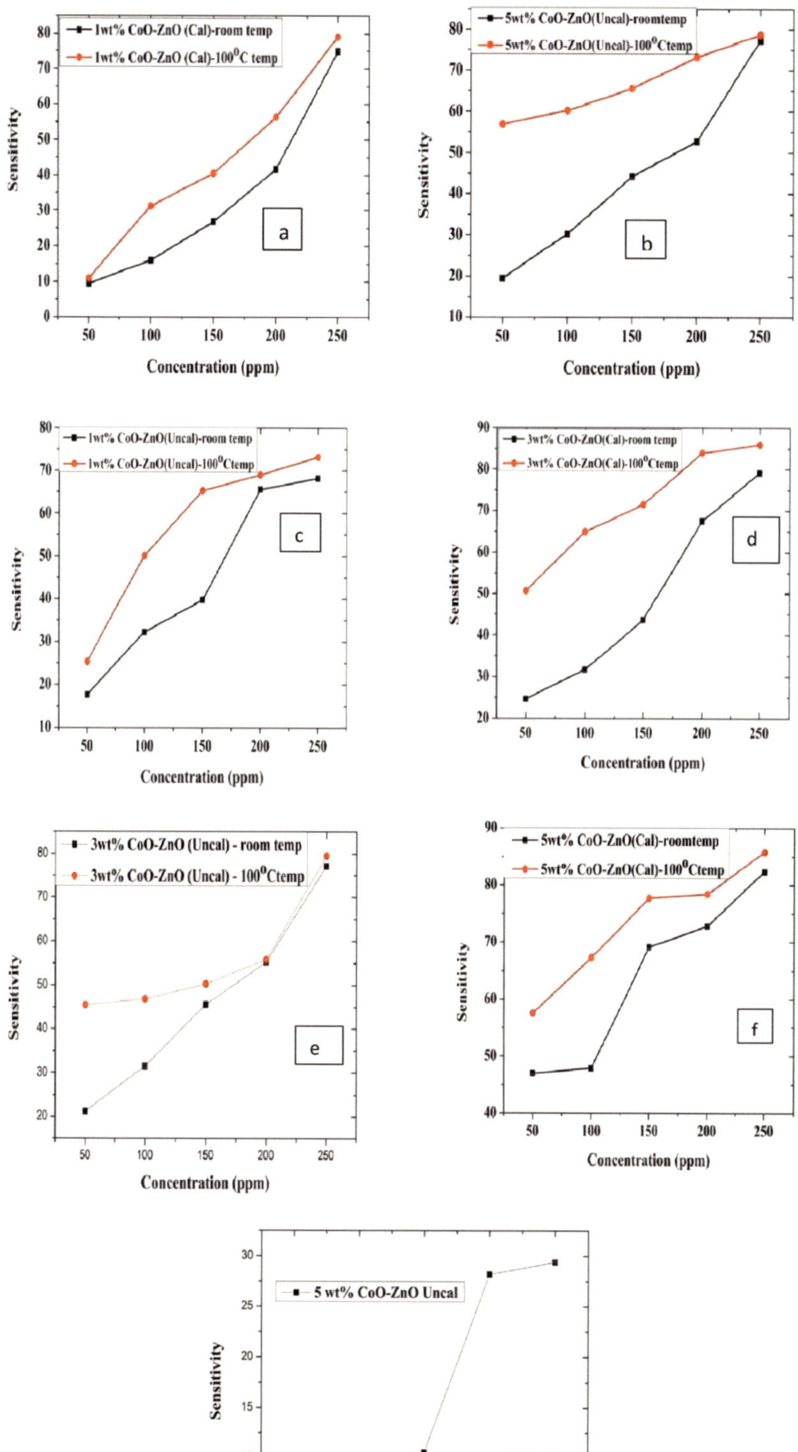

Figure 10.
Sensitivity plots: (a) 1 wt% CoO-ZnO uncal, (b) 1 wt% CoO-ZnO cal, (c) 3 wt% CoO-ZnO uncalcinated, (d) 3 wt% CoO-ZnO cal, (e) 5 wt% CoO-ZnO uncalcinated, (f) 5 wt% CoO-ZnO calcinated, and (g) 5 wt% CoO-ZnO uncalcinated.

concentration. Such linear sensitivity will be helpful in sensing gases like acetone. When compared with that of the room temperature, pure ZnO nanofibers will show high sensitivity at 100°C. The plot is between concentrations in ppm vs. sensitivity.

From **Figure 10(a–g)**, the plots for 1, 3, and 5 wt% for both calcinated and uncalcinated, the sensitivity will linearly increase with increase in concentration and also increase if the dopant concentration increases along with the operating temperature increases, i.e., at the 100°C temperature.

The above plot gives details about sensitivity plot for 5 wt% CoO-ZnO-uncalcinated nanofibers at concentration from 1 to 5 ppm where the linearity increment will be there from 3 to 5 ppm which is helpful in many applications in medical field like in diabetic's detection.

5. Conclusion

SEM analysis is used to determine the composition, structure, and surface morphologies of fibers. The amount of acetone in ppm range is detected. The sensitivity, change in resistance, and variation of calcinated and uncalcinated fibers were analyzed. The analyses which were taken from these results, that is, change in calcinated nanofibers, show more response to acetone when compared with uncalcinated nanofibers. The reason behind this kind of statement is PVP decomposed on calcination and hence it shows more resistivity, whereas in the case of uncalcinated fibers, PVP will not allow the zinc oxide to respond. From the sensitivity plots, we note that sensitivity increases when the doping concentration increases but also increases in the case of calcinated fibers. This chapter will discuss the pure, 1, 3, and 5 wt% CoO-ZnO nanofibers for acetone detection. The results give information that 5 wt% CoO-ZnO nanofibers will improve the acetone-sensing performances efficiently even if the atmosphere is complicated. These results will be helpful in many acetone gas-sensing applications widely.

Author details

Parthasarathy Panchatcharam
Research Scholar, School of Electrical Engineering, VIT University, Vellore, Tamil Nadu, India

*Address all correspondence to: arjunsarathii@gmail.com

References

[1] Lee JH. Gas sensors using hierarchical and hollow oxide nanostructures: Overview. Sensors and Actuators B: Chemical. 2009;**140**:319-336

[2] Choi SJ, Choi CY, Kim SJ, Cho HJ, Hakim M, Jeon S, et al. Highly efficient electronic sensitization of non-oxidized graphene flakes on controlled pore-loaded WO nanofibers for selective detection of H2S molecules. Scientific Reports. 2015;**5**:8067. DOI: 10.1038/srep08067

[3] Liang YC, Liao WK, Deng XS. Synthesis and substantially enhanced gas sensing sensitivity of homogeneously nanoscale Pd- and Au-particle decorated ZnO nanostructures. Journal of Alloys and Compounds. 2014;**599**:87-92

[4] Gardon M, Guilemany JM. A review on fabrication, sensing mechanisms and performance of metal oxide gas sensors. Journal of Materials Science. 2013;**24**:1410-1421. DOI: 10.1007/s10854-012-0974-4

[5] Hu J, Gao FQ, Sang SB, Li PW, Deng X, Zhang WD, et al. Optimization of Pd content in ZnO microstructures for high-performance gas detection. Journal of Materials Science. 2015;**50**:1935-1942. DOI: 10.1007/s10853-014-8758-2

[6] Wang WC, Tian YT, Wang XC, He H, Xu YR, He C, et al. Ethanol sensing properties of porous ZnO spheres via hydrothermal route. Journal of Materials Science. 2013;**48**:3232-3238. DOI: 10.1007/s10853-012-7103-x

[7] Khoang ND, Hong HS, Trung DD, Duy NV, Hoa ND, Thinh DD, et al. On-chip growth of wafer-scale planar-type ZnO nanorod sensors for effective detection of CO gas. Sensors and Actuators B: Chemical. 2013;**181**:529-536

[8] Zhang WH, Zhang WD, Zhou JF. Solvent thermal synthesis and gas-sensing properties of Fe-doped ZnO. Journal of Materials Science. 2010;**45**:209-215. DOI: 10.1007/s10853-009-3920-y

[9] Luo J, Ma SY, Li FM, Li XB, Li WQ, Cheng L, et al. The mesoscopic structure of flower-like ZnO nanorods for acetone detection. Materials Letters. 2014;**121**:137-140

[10] Zhang Y, Xu J, Xiang Q, Li H, Pan Q, Xu P. Brush-like hierarchical ZnO nanostructures: synthesis, photoluminescence and gas sensor properties. The Journal of Physical Chemistry C. 2009;**113**(9):3430-3435

[11] Bal AK, Singh A, Bedi RK. Characterization and ammonia sensing properties of pure andámodified ZnO films. Applied physics A. 2011;**103**(2):497-503

[12] Subannajui K, Wongchoosuk C, Ramgir N, Wang C, Yang Y, Hartel A, et al. Photoluminescent and gas-sensing properties of ZnO nanowires prepared by an ionic liquid assisted vapor transfer approach. Journal of Applied Physics. 2012;**112**(3):034311

[13] Song N, Fan HQ, Tian HL. PVP assisted in situ synthesis of functionalized graphene/ZnO (FGZnO) nanohybrids with enhanced gas-sensing property. Journal of Materials Science. 2015;**50**:2229-2238. DOI: 10.1007/s10853-014-8785-z

[14] Tarwal NL, Rajgure AV, Patil JY, Khandekar MS, Suryavanshi SS, Patil PS, et al. A selective ethanol gas sensor based on spray-derived Ag–ZnO thin films. Journal of Materials Science. 2013;**48**:7274-7282. DOI: 10.1007/s10853-013-7547-7

[15] Zhang ZY, Li XH, Wang CH, Wei LM, Liu YC, Shao CL. ZnO hollow nanofibers: Fabrication from facile single capillary electrospinning and applications in gas sensors. The Journal of Physical Chemistry C. 2009;**113**:19397-19403

[16] Yamazoe N. New approaches for improving semiconductor gas sensors. Sensors and Actuators B: Chemical. 1991;**5**(1-4):7-19

[17] Chen QL, Lai X, Yan ML, Tang GR, Luo JY, Chen J, et al. Preparation of tungsten oxide nanoplate thin film and its gas sensing properties. In Advanced Materials Research. Vol. 774. Trans Tech Publications; 2013. pp. 687-690

[18] Teo WE, Ramakrishna S. A review on electrospinning design and nanofibre assemblies. Nanotechnology. 2006;**17**(14):R89

[19] Alali KT, Liu T, Liu J, Liu Q, Fertassi MA, Li Z, et al. Preparation and characterization of ZnO/CoNiO2 hollow nanofibers by electrospinning method with enhanced gas sensing properties. Journal of Alloys and Compounds. 2017;**702**:20-30

[20] Sun Y, Zhao Z, Li P, Li G, Chen Y, Zhang W, et al. Er-doped ZnO nanofibers for high sensibility detection of ethanol. Applied Surface Science. 2015;**356**:73-80

[21] Zhao M, Wang X, Ning L, Jia J, Li X, Cao L. Electrospun Cu-doped ZnO nanofibers for H2S sensing. Sensors and Actuators B: Chemical. 2011;**156**(2):588-592

[22] Park JA, Moon J, Lee SJ, Lim SC, Zyung T. Fabrication and characterization of ZnO nanofibers by electrospinning. Current Applied Physics. 2009;**9**(3):S210-S212

[23] Zhao Y, Li X, Dong L, Yan B, Shan H, Li D, et al. Electrospun SnO2–ZnO nanofibers with improved electrochemical performance as anode materials for lithium-ion batteries. International Journal of Hydrogen Energy. 2015;**40**(41):14338-14344

[24] Ma L, Ma SY, Kang H, Shen XF, Wang TT, Jiang XH, et al. Preparation of Ag-doped ZnO-SnO2 hollow nanofibers with an enhanced ethanol sensing performance by electrospinning. Materials Letters. 2017;**209**:188-192

[25] Moghaddam HM, Malkeshi H. Self-assembly synthesis and ammonia gas-sensing properties of ZnO/Polythiophene nanofibers. Journal of Materials Science: Materials in Electronics. 2016;**27**(8):8807-8815

[26] Senthil T, Anandhan S. Structure–property relationship of sol–gel electrospun ZnO nanofibers developed for ammonia gas sensing. Journal of Colloid and Interface Science. 2014;**432**:285-296

[27] Egashira M, Kanehara N, Shimizu Y, Iwanaga H. Gas-sensing characteristics of Li+−doped and undoped ZnO whiskers. Sensors and Actuators B: Chemical. 1989;**18**:349-360

[28] Franke ME, Koplin TJ, Simon U. Metal and metal oxide nanoparticles in chemiresistors: Does the nanoscale matter? Small. 2006;**2**:36-50

[29] Barsan N, Koziej D, Weimar U. Metal oxide-based gas sensor research: How to? Sensors and Actuators B: Chemical. 2007;**121**:18-35

[30] Parayanthal P, Pollak FH. Raman scattering in alloy semiconductors: "Spatial correlation" model. Physical Review Letters. 1984;**52**(20):1822-1825

[31] Phan TL, Yu SC, Vincent R, Bui HM, Thanh TD, Lam VD, et al. Influence of Mn doping on structural, optical, and magnetic properties of Zn1 − x Mnx O

nanorods. Journal of Applied Physics. 2010;**108**(4):044910-044916

[32] Wang X, Xu J, Yu X, Xue K, Yu J, Zhao X. Structural evidence of secondary phase segregation from the Raman vibrational modes in in Zn1 − x Cox O (0 < x < 0.6). Applied Physics Letters. 2007;**91**(3):031908-031910

Chapter 6

Metal Organic Frameworks-Based Optical Thin Films

Cheng-an Tao, Jianfang Wang and Rui Chen

Abstract

Metal-organic frameworks (MOFs) are organic-inorganic hybrid materials with ordered pore structures assembled by metal centers and organic ligands through coordination bonds, but conventionally they are mostly in powder form. In recent years, metal-organic framework films have received increasing attention due to their potential applications in nanotechnology and nanodevices. The intrinsic ultrahigh porosity of MOFs may lead to a low refractive index of MOF materials. In addition, over 70,000 types of MOFs exist, and their properties can be tuned through the adoption of different metal motifs, organic ligands, or crystal morphologies. These characteristics make MOFs a potential new generation of optical thin film materials. In this chapter, the fabrication methods of MOF thin films, the optical properties of MOF optical thin films, and their application in optical sensors were described.

Keywords: metal-organic frameworks, thin film, fabrication method, optical property, refractive index, sensor, naked eye detection

1. Introduction

An optical film refers to a type of optical dielectric material that is attached to a surface of an optical device and is composed of a thin and uniform layered medium that propagates a light beam through an interface [1]. Commonly used optical films include: reflective films, antireflection films, polarizing films, interference filters, and beamsplitters [2]. Although there are many mature technologies and commercial products, with the increasing requirements of modern optics and optoelectronic devices, the exploration of new optical films is still the focus of current research.

Metal-organic frameworks (MOFs) are a class of porous materials that have developed rapidly in the past 20 years [3, 4]. They are crystalline materials that are composed of inorganic metal ions or metal-oxo cluster nodes and organic bridging ligands through coordination bonds. Due to the diversity of metal ions and organic bridging ligands, and the diversity of coordination linkages, there are theoretically unlimited types of MOFs, and more than 70,000 species have been reported so far, which are still increasing. MOFs have many inherently excellent properties, such as rich and tunable pore structure (pore size, pore shape, and pore opening), large specific surface area, regulatable chemical microenvironment, and good thermal stability [5–7]. They also can show special luminescent properties, magnetic properties, electrical conductivity, catalytic properties, etc. by selecting a particular metal ion or ligand. These properties make MOFs materials of great potential applications in many fields, such as gas adsorption and storage [8], separation [9], catalysis [10, 11], drug delivery and sustained release [12], medical imaging [12], luminescent materials, sensing [13–16], and so on.

In the field of optical films also, MOFs can be useful. Its almost infinite variety, tunable optical properties, and some of the inherently superior properties described above make it an excellent material for building optical films. However, MOFs are usually prepared in bulk or in powder form. Preparation of MOFs into a flat and uniform optical film is a prerequisite for their application. The optical properties of optical films are their core properties. How to effectively control their optical properties is a key issue in the study of MOF optical films. Therefore, in this chapter, firstly, five major methods for preparing MOF optical films—spin-coating, dip-coating, self-assembly, direct growth, and step-by-step liquid phase epitaxy (LPE) methods—were introduced, and a comparison of them was made under the consideration of the same MOF as the model. Then, the method of determination of refractive index of MOF films was described. The change of the linkers and post-modification of bridged ligands to control the optical properties of MOF optical films were discussed. At last, MOF thin films used as optical sensors were presented, including monolayer MOF thin films, MOF film-based 1DPC, and MOF film-combined optical fiber.

2. Fabrication methods of MOF thin films

At present, a series of methods have been developed to prepare MOF films [17, 18], but the flatness of most of the resulting films cannot meet the requirements of optical films. To date, there are mainly five typical methods to produce MOF optical films with high quality.

2.1 Spin-coating method

Spin coating is widely used in microfabrication of functional oxide layers on glass or single crystal substrates using sol-gel precursors, where it can be used to create uniform thin films with nanoscale thicknesses [19]. Hinterholzinger et al. [20] reported the fabrication of one-dimensional photonic crystals (1DPCs, also called Bragg stacks) based on zeolitic imidazolate framework (ZIF)-8 by spin coating for the first time. 1DPC was composed of multilayer of alternative ZIF-8 layers and titanium dioxide layers. In 2013, Lotsch's group [21] fabricated MOF thin films by spin-coating a MOF nanoparticle suspension onto a flat substrate. Two kinds of MOFs have been explored: one is copper trimesate ($Cu_3(BTC)_2$, also known as HKUST-1, HKUST = Hong Kong University of Science & Technology), which contains Cu (II)-paddlewheel-type nodes and trimesate struts, and the other one is the isoreticular metal-organic framework-3 (IRMOF-3, zinc amino-terephthalate).

Our group [16, 22–24] successfully applied this method to construct many other MOF thin films, such as iron (III) terephthalate MIL-88B (MIL = Materials of Institut Lavoisier), aluminum terephthalate MIL-53, MIL-101(Cr), and their analogs. We [16] reported the first example of flexible MOF optical thin film fabricated by spin coating (**Figure 1**). The flexible NH_2-MIL-88B(Fe) was chosen as the model, and nanorods of NH_2-MIL-88B(Fe) were prepared by hydrothermal method, and then MOF optical films prepared by spin-coating. The thickness of the film was nearly proportional to both suspension concentration and rotation cycles, and inversely proportional to the spin speed. And the rod-shaped MOFs tended to be aligned parallel to the substrate due to rotational shear forces. It provides the basis for subsequent application of properties.

Figure 1.
Photograph of a series of NH_2-MIL-88B films deposited on silicon wafers with various concentrations (from left to right: 0.5, 1, 2, 3, 4, 5, and 6 wt%) by spin-coating method. Reproduced from Ref. [16] with permission from The Royal Society of Chemistry.

2.2 Dip-coating method

As a popular alternative to spin coating, dip-coating methods are frequently employed to produce thin films from sol-gel precursors for research purposes, where it is generally used for applying films onto flat or cylindrical substrates [25]. In 2009, Gérard Férey et al. [26] first synthesized a colloidal dispersion of MIL-89, and then prepared a MIL-89 optical film by a dip-coating method. The film with a surface roughness of approximately 3 nm is very uniform. Subsequently, the team used microwave method to obtain nanoscale MIL-101(Cr) [27], MIL-100(Cr) [28], and MIL-100(Fe) [28], and prepared optical films using them by a similar dip-coating method. The MOF films are deposited from MOF nanoparticles naturally, and their surface roughness is about half of the size of MOF nanoparticles. The premise of this method is to obtain MOF nanoparticles with better dispersibility. MIL-100(Al) particles prepared by the same method cannot be used to fabricate optical thin films because of their large particle size (>500 nm) [28].

2.3 Self-assembly

Self-assembly is a process in which a disordered system of preexisting components forms an organized structure or pattern as a consequence of specific, local interactions among the components themselves, without external direction. The building blocks of self-assembly are not only molecules, but span a wide range of nano- and mesoscopic structures. In 2014, Prof. Huo [15] proposed a method for self-assembly of MOF nanoparticles. First, the nanoparticles of UiO-66 (UiO = University of Oslo) were prepared; then, their surface was modified with polyvinylpyrrolidone (PVP), and finally dispersed in mixture of water and ethanol (v:v = 1:1) to obtain a dispersion. A glass substrate was inserted into the water in a culture dish, and a small amount of the above dispersion was dropped on the surface of the water. Then, the MOF nanoparticles were rapidly spread to the surface of the water, and a 2% sodium dodecyl sulfonate (SDS) solution was dropped to make the MOF dispersion closely arranged into a single layer film. The film was transferred out to obtain a monolayer MOF film on the glass substrate. The MOF multilayer films can be obtained by repeating the process.

2.4 Direct in situ growth

MOFs prepared by direct in situ growth usually produce thick and rough membranes or dispersive particles. Lu et al. [14] successfully constructed ZIF-8-based optical thin films via a direct growth method that functioned as selective sensors for chemical vapors and gases. They immersed a glass substrate or silicon wafer in a newly prepared ZIF-8 growth solution (2-methylimidazole and $Zn(NO_3)_2$ in methanol) at room temperature. After 30 min of growth, a uniform ZIF-8 film of about 50 nm was obtained. More importantly, this process can be repeated, and the thickness of each subsequent growth was increased by about 100 nm, and the resulting film exhibited a very bright structural color. The color changes as the thickness increases, as shown in **Figure 3**. The method has the advantages of mild reaction conditions, fast growth rate, easy control of thickness, and easy removal of solvent in MOFs after reaction.

2.5 Step-by-step liquid phase epitaxy (LPE)

Step-by-step liquid phase epitaxy alternately deposits metal and organic ligand precursors on a functionalized surface, and the two are assembled by layer-by-layer growth to obtain a MOF film (also called surface-anchored MOF, SurMOF). The MOF prepared by this method has precisely controlled thickness and growth orientation, and the thickness can be regulated by the number of cycles of growth. In 2007, Wöll et al. [29] successfully used this method to prepare crystalline MOF films, such as two-component MOFs, HKUST-1, which alternately deposit copper acetate and BTC (BTC = trimesate) ligands on –COOH or –OH-modified gold substrate. The method is further used to prepare ternary layered MOFs composed of two organic ligands, such as [$Cu(bdc)_2(dabco)_2$] (bdc = 1,4-benzenedicarboxylic acid; dabco = 1,4-diazabicyclo[2.2.2]octane) [30], [Zn_2(+/−)$(cam)_2$dabco](cam = (1R,3S)-(+)-camphoric acid) [31], Fe(pz)[$Pt(CN)_4$](pz = pyrazine) [32, 33], DA-MOF($M(L)_2(P)_2$)((M = Cu or Zn; L = naphthalene dicarboxylate or 2,3,5,6-tetrafluoroterephthalic acid); P = dabco) [34], and so on.

The orientation of SurMOF can be regulated by different functional groups on the surface. For example, modifying the carboxyl or hydroxyl groups on the surface of the substrate allows the growth direction of HKUST-1 to be [111] and [100], respectively, under suitable conditions [29]. On the pyridyl- or carboxyl-modified surface, the growth direction of [$Cu(bdc)_2(dabco)_2$] is [100] and [001], respectively [30].

Unfortunately, it usually takes a long time to obtain a thicker film of MOFs due to the layer-by-layer growth mechanism. To overcome this shortcoming, researchers further combined this method with the spray method [35, 36], the spin-coating method [37], and the dip-coating method [38], and thereby developed a few improved methods for rapidly preparing the MOF films. Wöll et al. [36] alternately sprayed a solution of $Cu_2(CH_3COO)_4{\cdot}H_2O$ and BTC on a substrate modified with a coordinating group by a high-pressure carrier gas to obtain a HKUST-1 film. The thickness of the cyclic growth can reach 10 nm, and the growth rate of the method is increased by two orders of magnitude compared with the original method. In 2016, Eddaoudi group [37] combined the LPE method with the spin-coating method to achieve the growth of MOF films by spin-coating the metal salt solution and the organic ligand solution on the substrate, respectively. $Cu_2(bdc)_2{\bullet}xH_2O$, $Zn_2(bdc)_2{\bullet}xH_2O$, HKUST-1, and ZIF-8 films were successively prepared on the surface of substrates such as gold and alumina, which is more effective than traditional LPE methods. MOF films with thicknesses ranging from nm to μm are available. However, compared with the conventional LPE, the roughness of the obtained MOF film is remarkably increased. Particularly, the MOF films having too much growth cycle or growing on a porous substrate are difficult to use as optical films.

In 2014, Benes team [38] at the University of Twente developed a step-by-step dip-coating method that yielded thicker, dense MOFs in a single cycle. They dipped the silicon wafer vertically into the $ZnCl_2$ solution for 30 min, and then pulled it out at a certain rate (0.1–4 mm/s). After washing, it was immersed in the 2-methylimidazole solution for 30 min, then pulled out again, and then washed again. ZIF-8 film can be obtained by drying naturally in a culture dish, wherein the rate of each pulling is controlled to be the same, wherein the thickness of the film can be controlled by the pulling speed. The thickness of the ZIF-8 film is about 100 nm at

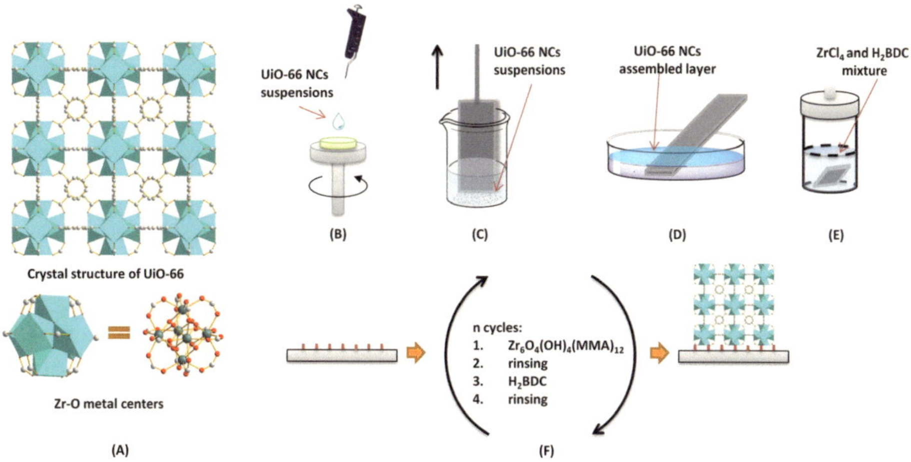

Figure 2.
(A) Crystal structure of UiO-66, consisting of Zr-O metal centers connected by terephthalate linkers. Zr, cyan; C, gray; O, red; H, omitted. (B–F) Illustration of five fabrication methods of MOF optical thin film: (B) spin coating, (C) dip coating, (D) self-assembly, (E) direct in situ growth, (F) LPE method [39].

a pulling rate of 1 mm/s. If the rate is too large or too small, the film thickness will increase. This method has not been extensively studied and has not been reported for the suitability of other types of MOFs.

2.6 Comparison

The above-mentioned five methods have shown success in preparation of MOF thin films, but there is still lack of systematic research on the performance and potential application of the MOF films from the aspects of optical films. Moreover, for the same MOF system, there is a lack of detailed comparison on these preparation methods. Recently, our group [39] chose UiO-66 as a model to prepare UiO-66 thin films by the above-mentioned five methods, respectively (**Figure 2**). The resultant thin films were denoted as OTF-SP, OTF-DP, OTF-SA, OTF-DG, and OTF-LPE, respectively. The qualities of the films were quantitatively analyzed using surface roughness. The arithmetic average roughness, Ra, is the arithmetic average value of filtered roughness profile and the root mean squared roughness. The OTF-SP film has the best flatness, while the OTF-DG has the worst roughness (Ra = 40.3 nm). The roughness of MOF thin films made of octahedral UiO-66 nanocrystals is generally higher than that of spherical MOF nanoparticles-based thin films [23, 24, 27]. For the dip-coated films, the more the depositions, the larger the roughness due to accumulation. For the self-assembled films, the trend is just the opposite. Their roughness decreases with the increase of depositions, thanks to the interleaved filling of following nanocrystals.

3. Optical properties of MOF optical thin films

At present, the research on optical properties of MOF optical films is still relatively preliminary, mainly focusing on the determination of their optical constants, especially the refractive index, and the regulation of optical constants.

3.1 Determination of optical constants

Refractive index is one of the important properties and parameters of optical films (such as antireflective coatings, high reflectivity coatings, polarizing films, long/short wave filters, etc.). The main method for determining the optical constant of an optical film of MOFs is an ellipsometry method. Férey et al. [26] reported the optical film of MIL-89 prepared by the dip-coating method and determined its refractive index at 700 nm. Since MIL-89 is a flexible MOF, its refractive index is larger after absorption of water vapor. The change was reduced to 1.45. They used an ellipsometer to monitor the change in refractive index with the water vapor pressure. They also measured the refractive indices of MIL-101(Cr) [27], MIL-100(Cr) [28], and the MIL-100(Fe) [28] optical films at 700 nm, which are 1.18, 1.17, and 1.43, respectively. The optical film prepared by the spin-coating method inevitably has interparticle pores, so the determined values of the refractive index are the effective refractive indices of the MOF optical films, not the intrinsic refractive indices of the MOF materials themselves. Ranft et al. [40] also measured the refractive index of the HKUST-1 optical films prepared by spin coating using ellipsometry, and found it was only 1.20, while Redel et al. [41] studied the refractive index of HKUST-1 optical film prepared by LPE method, and found it was 1.39 at 750 nm, which was much larger than that of the spin-coated HKUST-1 film, and close to the intrinsic refractive indices of HKUST-1, benefiting from the compactness of the optical film prepared by LPE method.

The calculation of the interference fringes on the upper and lower surfaces of the optical film is another method to determine refractive index of an optical film. Lu et al. [14] determined the refractive index of the ZIF-8 film prepared by the direct growth method using this method. The value is 1.39, which is quite different from the value of 1.20 obtained by Ranft et al. [40], mainly because the ZIF-8 film prepared by Ranft et al. is composed of ZIF-8 nanoparticles, while for the dense ZIF-8 film reported by Lu et al. prepared by direct in situ growth, the value (1.39) of refractive index is closer to the that of the ZIF-8 bulk material [14].

Redel et al. [41] prepared two copper-carboxylic acid MOF Cu-BDC films on Si substrate by LPE method. The refractive index of the film was directly measured by spectroscopic ellipsometry, and the refractive index of a series of MOF films with the same Cu-BDC topology but different lengths of organic ligands was predicted. This study also showed the advantages of MOFs in optical applications: compared to the optical properties of traditional inorganic materials, MOF can be used as an optical film active material to adjust the optical properties by changing the composition or structure of the MOF.

For the UiO-66 films prepared by different methods, in general, their refractive index has this order: OTF-SP < OTF-DP < OTF-SA < OTF-LPE. The refractive index of OTF-LPE has the highest value, which can be considered as the intrinsic refractive index of UiO-66 due to the compactness and integrity of the film. The nanocrystal-based films can achieve a refractive index lower than 1.23, which is important for application as antireflection films. Such a low refractive index benefits from the cracks and voids between the UiO-66 nanocrystals. Assuming the intrinsic refractive index of UiO-66 was 1.512, the index of air is 1, the void percent in the film was estimated according to the effective medium theory [7, 23, 42]. The voids of OTF-SP, OTF-DP, and OTF-SA films are about 25, 30, and 23%, respectively.

3.2 Tuning of optical constants

The optical constants of MOF optical films can be regulated in various ways, such as changing the metal ion species, ligand type, crystal type, adsorbing guest molecules, and changing metal ions or ligands by post-modification. But many strategies have not been realized. The way of adsorbing guest molecules is more used in sensing. Here, the manners of changing linkers and post-modification of ligands are introduced to tune the optical constants of optical films.

3.2.1 Tuning through changing of the linkers

Our group [23] reported the regulation of the optical properties of MOF optical films by changing linkers with different functional groups. Five different ligands were selected and synthesized with chromium ions to obtain five MOFs of the same MIL-101 configuration, which were respectively recorded as MIL-101(Cr), NH_2-MIL-101(Cr), NO_2-MIL-101(Cr), OH-MIL-101(Cr), and $(NO_2)_2$-MIL-101(Cr), and the corresponding optical film was produced by spin coating. The refractive index and extinction coefficient of the MOF optical film changed with the change of linkers. The average refractive indices of MIL-101(Cr), NH_2-MIL-101(Cr), OH-MIL-101(Cr), and $(NO_2)_2$-MIL-101(Cr) optical films were 1.306, 1.268, 1.223, and 1.250, respectively. NO_2-MIL-101(Cr) has the lowest refractive index of only 1.208, which is lower than 1.22, making it have great potential in the preparation and application of antireflection coatings [43, 44].

The effect of ligand on the refractive index of MOF materials was further studied by eliminating the effect of porosity of the optical films. The order of the intrinsic refractive indices of MOFs is: $(NO_2)_2$-MIL-101(Cr) > NO_2-MIL-101(Cr) > NH_2-MIL-101(Cr) > OH-MIL-101(Cr) > MIL-101(Cr). The intrinsic refractive index of the MOF material in the same topology increases with the increase of the atomic density of the ligands.

The change of the extinction coefficient *k* is related to the electron-absorbing and electron-donating states of the ligands. It was found that the substitution of the electron-donating group was good at increasing the value of *k*, and the electron-withdrawing group would decrease *k* value.

3.2.2 Tuning through post-modification

Post-modification of ligands is also an effective means to change the optical constants of MOF optical films benefiting from abundant organic chemical reactions. Our group [24] reported the first example of successful regulation of optical films by post-modification of NH_2-MIL-53(Al) (see **Figure 3**). Propionaldehyde, valeraldehyde, and heptaldehyde with different carbon chain lengths were chosen as the modifier. After modification with propionaldehyde and valeraldehyde, the n_{eff} of the MOF optical films became larger as the carbon chain length increased, and the refractive index increased from 1.292 to 1.371 (propionaldehyde modification) and 1.424 (valeraldehyde modification). After heptaldehyde modification, the refractive index changed only slightly, from 1.292 to 1.299 (**Figure 3F**). This is because propionaldehyde and valeraldehyde have small molecular size and can effectively modify MOF, while heptaldehyde has a larger molecular size (10.7 Å) and finds it difficult to enter into the pores of MOF (8.5 Å), so the grafting rate is not high.

In addition, the extinction coefficient (k) is the imaginary part of the complex refractive index and is another important parameter of the optical film. After post-modification, the extinction coefficient of the MOF optical film also changed significantly. This is due to the change in the electronic structure inside the MOFs after post-modification [41]. The stop band width of MOFs after modification decreased from 2.74 to 2.71 eV (propionaldehyde modification) and 2.68 eV (valeraldehyde modification), which is consistent with the change in k. After grafting, the optical properties of the material itself are tuned both due to changing the internal electronic structure of NH_2-MIL-53 (Al) and occupying its internal pore structure.

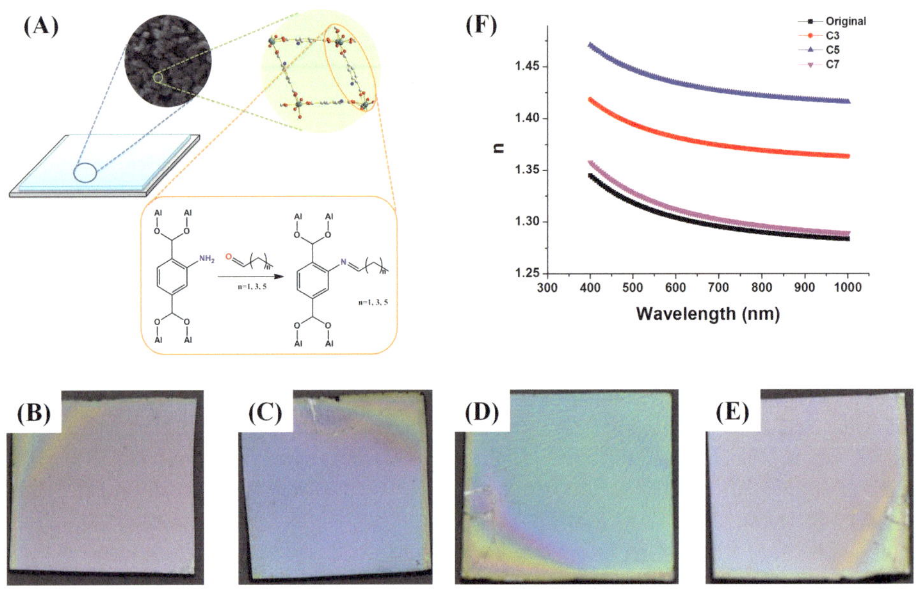

Figure 3.
(A) Scheme of tuning the optical properties through post-modification. Photographs of MOF optical thin film without modification (B), modified with propionaldehyde (C), pentanal (D), and heptanal (E). (F) The effective refractive index (n) of MOF optical thin films [24].

4. The application of MOF films in optical sensors

The application of MOF optical films in optical sensing is also essentially using the changes in optical properties, typically changes in the reflectance spectra (or colors) caused by the stimulation of target molecules.

4.1 Monolayer MOF thin films

Lu and Hupp [14] constructed a ZIF-8 monolayer film-based Fabry-Pérot device, which can serve as a selective sensor for chemical vapors and gases. The red shift of the reflection spectrum was observed within 1 min due to adsorption of propane. The sensor can distinguish n-hexane and cyclohexane due to their different sizes. It can also detect the ethanol from water/ethanol system with a limit as low as 0.3 vol%, corresponding to an ethanol vapor concentration of ca. 100 ppm.

Compared with the rigid pore structure of ZIF-8, the flexible pore structure of MOFs can change more significantly after adsorbing water or organic vapors. Our group [16] chose flexible NH_2-MIL-88B as a model MOF to fabricate a Fabry-Pérot device with a vivid color by spin-coating method. The NH_2-MIL-88B photonic film displayed high chemical selectivity, for example, acetone induced 380-nm redshifts, while water only led to a redshift of about 50 nm, and the color would change accordingly after absorbing the water or organic vapors (**Figure 4**), which can be observed by the naked eye. Depending on the nature of the organic solvent and their interaction with NH_2-MIL-88B, the selective breathing behavior of NH_2-MIL-88B promotes the excellent selectivity of the optical films.

4.2 MOF thin film-based one-dimensional photonic crystals

A one-dimensional photonic crystal (1DPC), also called Bragg stack, is a periodic nanostructure with a refractive index distribution along one direction [45].

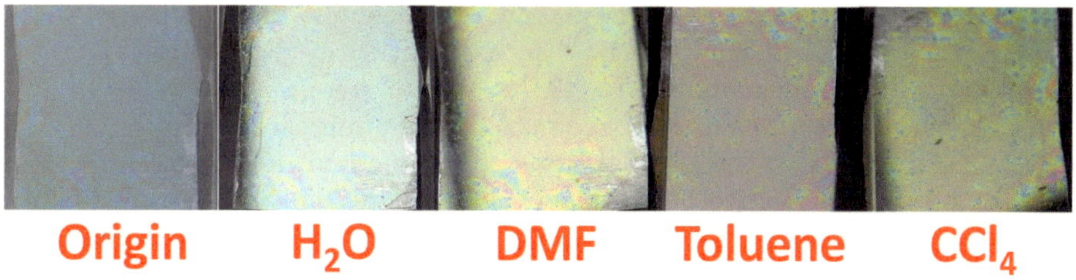

Figure 4.
Photographs of the NH_2-MIL-88B optical film upon exposure to various organic vapors. Reproduced from Ref. [16] with permission from The Royal Society of Chemistry.

Hinterholzinger et al. [20] presented the fabrication of 1DPC based on ZIF-8 and mesoporous titanium dioxide (TiO_2) for the first time. ZIF-8 is intended to impart molecular selectivity, and mesoporous TiO_2 is used to ensure high refractive index contrast and to guarantee molecular diffusion within the 1DPC. The 1DPC is sensitive and selective toward a series of chemically similar solvent vapors due to different sorption behavior of the photonic material to the solvent vapors.

In the consideration of replacing the rigid MOF with a flexible MOF, we [22] also prepared a 1DPC by alternately spin-coating NH_2-MIL-88B and TiO_2, wherein the TiO_2 layer functioned as a high-refractive index contrast. The optical properties (color, refractive index, etc.) of the 1DPCs can be adjusted by varying the number of depositions, and the thickness of the film. Benefiting from the flexible pore structure of NH_2-MIL-88B, the 1DPCs exhibited a highly selective response to different organic vapors, including dimethyl formamide, isopropanol, methanol, acetone, and ethanol

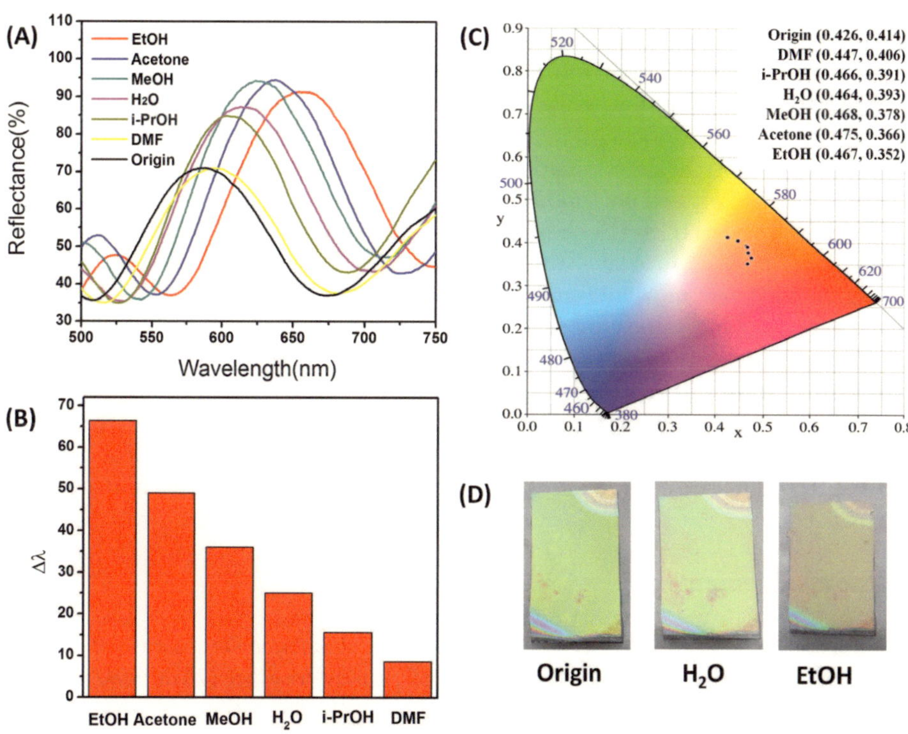

Figure 5.
(A) Reflection spectra, (B) shift of the photonic stop band, (C) chromaticity coordinates, and (D) typical images of the NH_2-MIL-88B/TiO_2 1DPCs upon exposure to various organic vapors. Reproduced from Ref. [22] with permission from The Royal Society of Chemistry.

(**Figure 5A–C**). Typically, the color of the 1DPCs in air and saturated vapors of water and EtOH were different, which can be observed by naked eye (**Figure 5D**). The 1DPC also can detect ethanol from 0 to 43.8 vol% in a water-ethanol mixture.

Liu et al. [46] described the fabrication of 1DPC by alternate deposition of HKUST-1 and indium tin oxide (ITO) layers by LPE method and sputtering method, among which HKUST-1 was used as low-refractive index layer, and ITO as high-refractive index layer. The 1DPC had a red shift response of only 38 nm for the adsorption of toluene because of the change in refractive index brought about by the adsorbing guest molecules.

4.3 MOF thin film on fiber surfaces

MOF thin films can be deposited not only on the flat surface but also on the nonplanar optical components, which further expands their use as optical sensors. Kim et al. [47] recently reported ZIF-8 films grown on linear fiber surfaces to create optical gas sensors, which are more sensitive to the target gases than Fabry-Pérot devices or 1DPCs based on normal incident light. The resultant sensors showed high sensitivity and selectivity to CO_2 gas relative to other small gases such as H_2, N_2, O_2, and CO. They also exhibited rapid (<tens of seconds) response time and excellent reversibility, which may be ascribed to the physisorption of gases into a nanoporous MOF. A refractive index-based sensing mechanism for the MOF-integrated optical fiber platform was proposed.

5. Conclusions

There are five major methods for preparing MOF optical films with high-quality spin coating, dip coating, self-assembly, direct growth, and LPE. Under the consideration of the same MOF (UiO-66) as the model, the MOF film prepared by spin coating has the best flatness, while the one prepared by direct growth is the worst one. The research on optical properties of MOF optical films is still in its early stage, mainly focusing on the determination and regulation of optical constants. The change of the linkers and post-modification of bridged ligands to control the optical properties of optical films have been realized. The application of MOF thin films in optical sensors can induce label-free optical sensors with variable colors exposed to various organic vapors. MOF films not only on the flat surface but also on the non-planar surface (for instance fiber surface) can be served as optical sensors. They can have great selectivity due to different interactions between the target molecules with MOFs. In the future, the quality of optical films will be further improved, and a wide range of chemical modification methods will be applied to control the properties of optical films, which will further meet the needs of specific optical devices. MOF optical films are expected to become the next generation of optical materials.

Acknowledgements

This research was funded by the National Natural Science Foundation of China (21573285) and Natural Science Foundation of Hunan Province (2018JJ3597).

Conflict of interest

The authors declare no conflict of interest.

Author details

Cheng-an Tao*, Jianfang Wang and Rui Chen
College of Liberal Arts and Science, National University of Defense Technology, Changsha, China

*Address all correspondence to: taochengan@nudt.edu.cn

References

[1] Wen X, Xiong Q. A large scale perfect absorber and optical switch based on phase change material ($Ge_2Sb_2Te_5$) thin film. Science China Materials. 2016;**59**:165-172. DOI: 10.1007/s40843-016-0129-7

[2] Epstein LI. The design of optical filters. Journal of the Optical Society of America. 1952;**42**:806-810. DOI: 10.1364/josa.42.000806

[3] Sindoro M, Yanai N, Jee A-Y, Granick S. Colloidal-sized metal–organic frameworks: Synthesis and applications. Accounts of Chemical Research. 2013;**47**:459-469. DOI: 10.1021/ar400151n

[4] Bai Y, Dou Y, Xie L-H, Rutledge W, Li J-R, Zhou H-C. Zr-based metal–organic frameworks: Design, synthesis, structure, and applications. Chemical Society Reviews. 2016;**45**:2327-2367. DOI: 10.1039/C5CS00837A

[5] Serre C, Mellot-Draznieks C, Surblé S, Audebrand N, Filinchuk Y, Férey G. Role of solvent-host interactions that lead to very large swelling of hybrid frameworks. Science. 2007;**315**: 1828-1831. DOI: 10.1126/science.1137975

[6] Cavka JH, Jakobsen S, Olsbye U, Guillou N, Lamberti C, Bordiga S, et al. A new zirconium inorganic building brick forming metal organic frameworks with exceptional stability. Journal of the American Chemical Society. 2008;**130**:13850-13851. DOI: 10.1021/ja8057953

[7] Park KS, Ni Z, Côté AP, Choi JY, Huang R, Uribe-Romo FJ, et al. Exceptional chemical and thermal stability of zeolitic imidazolate frameworks. Proceedings of the National Academy of Sciences. 2006;**103**:10186-10191. DOI: 10.1073/pnas.0602439103

[8] Murray LJ, Dincă M, Long JR. Hydrogen storage in metal–organic frameworks. Chemical Society Reviews. 2009;**38**:1294-1314. DOI: 10.1039/B706207A

[9] Kang Z, Fan L, Sun D. Recent advances and challenges of metal–organic framework membranes for gas separation. Journal of Materials Chemistry. 2017;**5**:10073-10091. DOI:10.1039/C7TA01142C

[10] Chen J, Shen K, Li Y. Greening the processes of metal–organic framework synthesis and their use in sustainable catalysis. ChemSusChem. 2017;**10**: 3165-3187. DOI:10.1002/cssc.201700748

[11] Shen C, Mao Z, Xu H, Zhang L, Zhong Y, Wang B, et al. Catalytic mof-loaded cellulose sponge for rapid degradation of chemical warfare agents simulant. Carbohydrate Polymers. 2019;**213**:184-191. DOI:10.1016/j.carbpol.2019.02.044

[12] Horcajada P, Chalati T, Serre C, Gillet B, Sebrie C, Baati T, et al. Porous metal–organic-framework nanoscale carriers as a potential platform for drug delivery and imaging. Nature Materials. 2010;**9**:172-178. DOI: 10.1038/nmat2608

[13] Kreno LE, Leong K, Farha OK, Allendorf M, Van Duyne RP, Hupp JT. Metal–organic framework materials as chemical sensors. Chemical Reviews. 2011;**112**:1105-1125. DOI: 10.1021/cr200324t

[14] Lu G, Hupp JT. Metal−organic frameworks as sensors: A ZIF-8 based Fabry−Pérot device as a selective sensor for chemical vapors and gases. Journal of the American Chemical Society. 2010;**132**:7832-7833. DOI: 10.1021/ja101415b

[15] Cui C, Liu Y, Xu H, Li S, Zhang W, Cui P, et al. Self-assembled

metal-organic frameworks crystals for chemical vapor sensing. Small. 2014;**10**:3672-3676. DOI: 10.1002/smll.201302983

[16] Hu Z, Tao C-a, Liu H, Zou X, Zhu H, Wang J. Fabrication of an NH_2-MIL-88B photonic film for naked-eye sensing of organic vapors. Journal of Materials Chemistry A. 2014;**2**:14222-14227. DOI: 10.1039/c4ta01916d

[17] Bétard A, Fischer RA. Metal-organic framework thin films: From fundamentals to applications. Chemical Reviews. 2012;**112**:1055-1083. DOI: 10.1021/cr200167v

[18] Zacher D, Shekhah O, Wöll C, Fischer RA. Thin films of metal-organic frameworks. Chemical Society Reviews. 2009;**38**:1418-1429. DOI: 10.1039/b805038b

[19] Hanaor D, Triani G, Sorrell C. Morphology and photocatalytic activity of highly oriented mixed phase titanium dioxide thin films. Surface and Coating Technology. 2011;**205**:3658-3664. DOI: 10.1016/S0254-0584(02)00138-4

[20] Hinterholzinger FM, Ranft A, Feckl JM, Rühle B, Bein T, Lotsch BV. One-dimensional metal–organic framework photonic crystals used as platforms for vapor sorption. Journal of Materials Chemistry. 2012;**22**:10356-10362. DOI: 10.1039/c2jm15685g

[21] Ranft A, Betzler SB, Haase F, Lotsch BV. Additive-mediated size control of MOF nanoparticles. CrystEngComm. 2013;**15**:9296-9300

[22] Hu Z, C-a T, Wang F, Zou X, Wang J. Flexible metal–organic framework-based one-dimensional photonic crystals. Journal of Materials Chemistry C. 2015;**3**:211-216. DOI: 10.1039/C4TC01501K

[23] Yin W, Tao C-a, Wang F, Huang J, Qu T, Wang J. Tuning optical properties of MOF-based thin films by changing the ligands of MOFs. Science China Materials. 2018;**61**:391-400. DOI: 10.1007/s40843-017-9143-5

[24] Yin W, Tao C-a, Zou X, Wang F, Qu T, Wang J. The tuning of optical properties of nanoscale MOFs-based thin film through post-modification. Nanomaterials. 2017;7:242. DOI: 10.3390/nano7090242

[25] Scriven L. Physics and applications of dip coating and spin coating. MRS Online Proceedings Library Archive. 1988;**121**:717-729

[26] Horcajada P, Serre C, Grosso D, Boissiere C, Perruchas S, Sanchez C, et al. Colloidal route for preparing optical thin films of nanoporous metal–organic frameworks. Advanced Materials. 2009;**21**:1931-1935. DOI: 10.1002/adma.200801851

[27] Demessence A, Horcajada P, Serre C, Boissière C, Grosso D, Sanchez C, et al. Elaboration and properties of hierarchically structured optical thin films of MIL-101(Cr). Chemical Communications. 2009;**0**:7149-7151. DOI: 10.1039/B915011K

[28] Marquez AG, Demessence A, Plateroprats AE, Heurtaux D, Horcajada P, Serre C, et al. Green microwave synthesis of MIL-100(Al, Cr, Fe) nanoparticles for thin-film elaboration. European Journal of Inorganic Chemistry. 2012;**2012**: 5165-5174. DOI:10.1002/ejic.201200710

[29] Shekhah O, Wang H, Strunskus T, Cyganik P, Zacher D, Fischer R, et al. Layer-by-layer growth of oriented metal organic polymers on a functionalized organic surface. Langmuir. 2007; **23**:7440-7442. DOI: 10.1021/la701148z

[30] Zacher D, Yusenko K, Bétard A, Henke S, Molon M, Ladnorg T, et al. Liquid-phase epitaxy of multicomponent layer-based porous

coordination polymer thin films of [M(L)(P)0.5] type: Importance of deposition sequence on the oriented growth. Chemistry—A European Journal. 2011;**17**:1448-1455. DOI: 10.1002/chem.201002381

[31] Liu B, Shekhah O, Arslan HK, Liu J, Wöll C, Fischer RA. Enantiopure metal–organic framework thin films: Oriented SURMOF growth and enantioselective adsorption. Angewandte Chemie International Edition. 2012;**51**:807-810. DOI: 10.1002/anie.201104240

[32] Otsubo K, Haraguchi T, Sakata O, Fujiwara A, Kitagawa H. Step-by-step fabrication of a highly oriented crystalline three-dimensional pillared-layer-type metal–organic framework thin film confirmed by synchrotron X-ray diffraction. Journal of the American Chemical Society. 2012;**134**:9605-9608. DOI: 10.1021/ja304361v

[33] Haraguchi T, Otsubo K, Sakata O, Fujiwara A, Kitagawa H. Remarkable lattice shrinkage in highly oriented crystalline three-dimensional metal–organic framework thin films. Inorganic Chemistry. 2015;**54**:11593-11595. DOI: 10.1021/acs.inorgchem.5b02207

[34] So MC, Jin S, Son H-J, Wiederrecht GP, Farha OK, Hupp JT. Layer-by-layer fabrication of oriented porous thin films based on porphyrin-containing metal–organic frameworks. Journal of the American Chemical Society. 2013;**135**:15698-15701. DOI: 10.1021/ja4078705

[35] Shekhah O, Wang H, Zacher D, Fischer RA, Wöll C. Growth mechanism of metal–organic frameworks: Insights into the nucleation by employing a step-by-step route. Angewandte Chemie International Edition. 2009;**48**:5038-5041. DOI: 10.1002/anie.200900378

[36] Arslan HK, Shekhah O, Wohlgemuth J, Franzreb M, Fischer RA, Wöll C. High-throughput fabrication of uniform and homogenous MOF coatings. Advanced Functional Materials. 2011;**21**:4228-4231. DOI: 10.1002/adfm.201101592

[37] Chernikova V, Shekhah O, Eddaoudi M. Advanced fabrication method for the preparation of MOF thin films: Liquidphase epitaxy approach meets spin coating method. ACS Applied Materials & Interfaces. 2016;**8**:20459-20464. DOI: 10.1021/acsami.6b04701

[38] Cookney J, Ogieglo W, Hrabanek P, Vankelecom I, Fila V, Benes NE. Dynamic response of ultrathin highly dense ZIF-8 nanofilms. Chemical Communications. 2014;**50**:11698-11700. DOI: 10.1039/C4CC04754K

[39] Huang Y, Tao CA, Chen R, Sheng L, Wang J. Comparison of fabrication methods of metal-organic framework optical thin films. Nanomaterials. 2018;**8**:676. DOI: 10.3390/nano8090676

[40] Ranft A, Niekiel F, Pavlichenko I, Stock N, Lotsch BV. Tandem MOF-based photonic crystals for enhanced analyte-specific optical detection. Chemistry of Materials. 2015;**27**:1961-1970. DOI: 10.1021/cm503640c

[41] Redel E, Wang Z, Walheim S, Liu J, Gliemann H, Wöll C. On the dielectric and optical properties of surface-anchored metal-organic frameworks: A study on epitaxially grown thin films. Applied Physics Letters. 2013;**103**:091903. DOI: 10.1063/1.4819836

[42] Sihvola A. Mixing rules with complex dielectric coefficients. Subsurface Sensing Technologies and Applications. 2000;**1**:393-415. DOI: 10.1023/A:1026511515005

[43] Schulze M, Lehr D, Helgert M, Kley E-B, Tünnermann A. Transmission enhanced optical lenses with self-organized antireflective subwavelength structures for the UV range. Optics

Letters. 2011;**36**:3924-3926. DOI: 10.1364/OL.36.003924

[44] Wongcharee K, Brungs M, Chaplin R, Hong Y, Pillar R, Sizgek E. Sol-gel processing by aging and pore creator addition for porous silica antireflective coatings. Journal of Sol-Gel Science and Technology. 2002;**25**:215-221. DOI: 10.1023/A:1020243527580

[45] Shen H, Wang Z, Wu Y, Yang B. One-dimensional photonic crystals: Fabrication, responsiveness and emerging applications in 3d construction. RSC Advances. 2016;**6**:4505-4520. DOI: 10.1039/C5RA21373H

[46] Liu J, Redel E, Walheim S, Wang Z, Oberst V, Liu J, et al. Monolithic high performance surface anchored metal–organic framework Bragg reflector for optical sensing. Chemistry of Materials. 2015;**27**:1991-1996. DOI: 10.1021/cm503908g

[47] Kim K-J, Lu P, Culp JT, Ohodnicki PR. Metal–organic framework thin film coated optical fiber sensors: A novel waveguide-based chemical sensing platform. ACS Sensors. 2018;**3**:386-394. DOI: 10.1021/acssensors.7b00808

Chapter 7

Nanoscale Optical Patterning of Amorphous Silicon Carbide for High-Density Data Archiving

Tania Tsvetkova

Abstract

The work presented here is related to some developments in providing a new generation ultrastable (>100 years), ultrahigh density (>1 Tbit/sq.in.) data storage materials for archival applications. The chosen material to write nanoscale data by finely focused ion beams is hydrogenated amorphous silicon carbide (a-SiC:H) films. Wide bandgap a-SiC:H has been chosen for its appropriate optical, chemical and mechanical properties. Ga^{+} was prefered as the implant species for the focused ion beam (FIB) implantation due to its widespread uses in FIB equipment and its modifying effects on the amorphous silicon carbide target. A range of a-SiC:H film samples have been FIB patterned under different implantation conditions for this study. The emphasis in these investigations was the influence of different substrate temperatures on the patterning process. The effects of further annealing of room temperature implanted samples were also studied. The FIB patterned samples under different conditions were analysed using near-field techniques, like atomic force microscopy (AFM), to define optimum implantation parameters for archival data storage applications. Using the established optimal conditions for the FIB patterning process of a-SiC:H films, it is expected to achieve the aimed ultrahigh density and stability with this novel data storage method for archival applications.

Keywords: optical data storage, nanoscale data archiving, focused ion beams

1. Introduction

The archival sector is becoming increasingly important, due in part to the introduction of new legal requirements governing the storage of governmental and commercial data but also as a consequence of the ever-increasing amount of digital data generated by all aspects of our everyday life. Indeed, it has been estimated [1] that the total archival capacity required worldwide is expected to exceed 1000 exabytes by 2020, generating market values of in excess of $50 billion. For archival applications reliability, data integrity and media longevity feature much more prominently than in other storage sectors. The main recognized archival storage media currently in use are magnetic tape and optical disks, although it is estimated that around 50% of the archive data for commercial organizations is in fact held on magnetic hard disk drives [2]. While optical disks for professional archiving are usually 'guaranteed' a lifetime of at least 50 years, magnetic tapes and disks have in general a much shorter predicted lifespan. The lifetime of magnetic hard disk media in particular is reducing as the storage density increases due to the superparamagnetic effect, with lifetimes of current high-capacity drives limited to around 10 years. Magnetic tape, while a very good choice for high-capacity short-term storage such as backup and disaster recovery, is yet rather delicate contact media, which can become physically damaged with use, and so can degrade with time, while also the magnetic data can be easily lost when exposed to magnetic fields. Hence, to retain tape data available over extended periods of time, tapes must be frequently re-tensioned and data be periodically refreshed (read and rewritten)—a potentially complex and costly task with serious consequences if not managed properly. It is clear therefore that archival storage is an exceedingly important application and it is equally clear that conventional archival storage media are quite limited in terms of lifetime and, particularly for optical disks and magnetic tapes, in terms of storage density (bits per square inch). It is in this context that the present work has aimed to develop novel, ultra-stable and ultrahigh density storage media for archival applications. Indeed, the approach of ion implantation in SiC can potentially lead to extremely long (many hundreds of years) data lifetimes, which may find application in the long-term preservation of important scientific and cultural assets.

2. Nanoscale optical patterning of amorphous silicon carbide films by Ga^+-focused ion implantation

To realize our goal of 'permanent' high-density storage, we have chosen amorphous hydrogenated silicon carbide (a-SiC:H) as a storage medium. a-SiC:H is an exciting material with a wide range of useful optical and electrical properties (e.g., high transparency in the visible region due to its wide optical band gap from 1.8 to 3.0 eV), as well as mechanical durability and chemical inertness [3]. These properties, together with ease of fabrication and a remarkable thermal stability, make SiC a very promising candidate for various optoelectronic, photonic and other device applications (e.g., solar cells, light-emitting diodes, imaging sensors, high-power microwave devices, high-energy radiation detectors), particularly in adverse environments [3, 4]. The relative immunity of SiC to environmentally induced degradation, in particular its high thermal stability (stable to temperatures in excess of 1500°C), makes it attractive for data storage applications, and promising results have been achieved by using ion implantation to write micro- and nanoscale marks in SiC films [5–11].

Ion-beam implantation is a standard technique for controlling electrical, band gap and optical properties of semiconductors (see, e.g., [12]), and some data exist for the SiC materials. Recently the ion-beam implantation route has attracted renewed interest due to the development of computer-controlled focused-ion-beam systems that enable the fabrication of sophisticated ion-implanted structures [11, 13, 14]. The focused-ion-beam diameter can be <10 nm, allowing the modification of the dielectric and optical properties of materials on this same nanoscale (10 nm bit sizes is roughly equivalent to a data density of 10 Tbit/sq.in.). Such nanoscale property changes in optical transmission and reflection offer ultrahigh density yet permanent optical data archives. To readout such high-density optical data, a super-resolution (sub-diffraction) optical technique is required. In the laboratory environment and for research purposes, this can be provided by scanning near-field optical microscopy (SNOM) [15]. SNOM techniques have been successfully applied for studies of ion-beam modifications in silicon carbide in infrared [16] and visible [8] spectral ranges. Of course imaging by SNOM is slow (and expensive), but for practical applications, other super-resolution readout techniques, such as the use of solid-immersion lens (SIL) optics [17], the so-called super-RENS technique [18] or arrays of scanning near-field apertures, might ultimately be used to achieve the necessary readout data rates (and required system cost).

2.1 Ga^+ ion beam-induced optical contrast in amorphous silicon carbide

Several ion-implanted species have already been investigated for the modification of the optical properties of a-SiC:H thin films, including Ar^+, ions of group IV elements [5, 7–9] and, most recently, Ga^+. The use of gallium is attractive since it is available in standard FIB machines and in addition has been shown to be capable of generating large optical contrasts [6]. Investigations were carried out further in order to study the structural and chemical bonding modifications of thin a-SiC:H films when using high-dose Ga^+ ion implantation, resulting in considerable changes of the optical band gap and the electronic properties of the material [6, 7].

The Ga^+ ion implantation has yielded in a considerable optical effect as registered by the results of transmission, reflection and PDS measurements, used further to derive the absorption coefficient spectra, shown in **Figure 1**. This effect becomes more markedly pronounced as the dose increases. As seen from the figure, the optical absorption edge moves to lower photon energies while also being accompanied by an increase of the absorption coefficient in the measured energy range, as a result of the Ga^+ ion implantation.

The optical absorption edge shift increases with the ion dose, signifying a corresponding change in the density of states distribution in the optical gap. It could be assumed, therefore, that the observed changes are related to the breaking of bonds due to ion bombardment and modification of the chemical bonding arrangements of the host atoms due to the presence of the implanted Ga^+ ions. This assumption is also supported by the registered changes in the absorption spectra in the IR region shown in **Figures 2** and **3**. The well-defined peak at about 2080 cm^{-1}, related to the stretching vibration mode of Si-H bonds in a-SiC:H films for the non-implanted case, shows well-expressed tendency to decrease. Notably, it is also shifting towards lower energies with the increase of the ion dose (**Figure 2**).

This shift could be attributed to an underlying mechanism of interaction between the ions of the added element with the Si atoms in the host material, resulting in a change of the Si-H stretching mode. This behaviour could be explained remembering the lower electronegativity of the implanted Ga atoms than the C host atoms in the target material, so that Ga will be substituting for the C atom in the C-Si-H bond, as in the previously reported case of Ge and Sn ion implantation in a-SiC:H films [19–21]. The observed accompanying decrease of the peak with the ion dose is further confirmed for the modifying effect of the Ga ion bombardment on the bonding configuration of the host atoms and is an evidence for the breaking of the Si-H bonds which is accompanied by considerable loss of hydrogen.

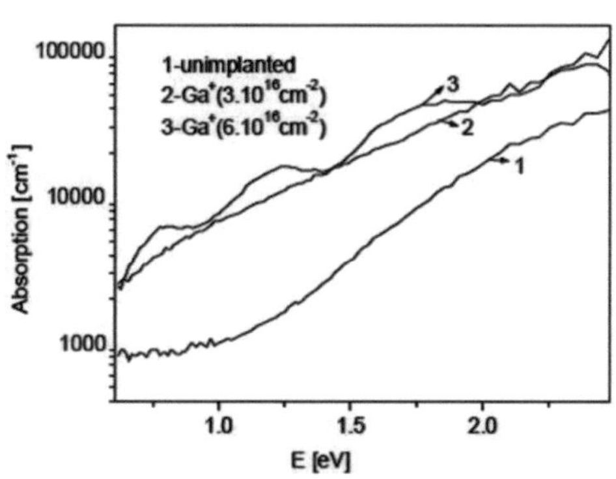

Figure 1.
Optical absorption coefficient α of a-$Si_{0.85}C_{0.15}$:H films: unimplanted (1) and Ga^+ implanted at doses (in cm^{-2}) 3×10^{16} (2) and 6×10^{16} (3) [7].

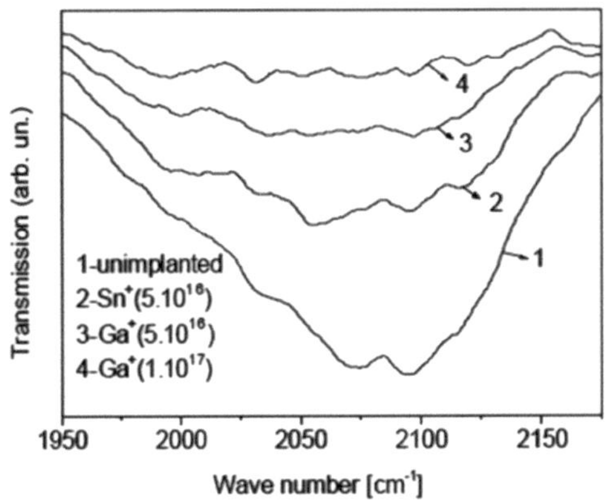

Figure 2.
IR transmission spectra in the region 1950–2200 cm^{-1} of an unimplanted a-$Si_{0.85}C_{0.15}$:H film (1) and a-$Si_{0.85}C_{0.15}$:H films implanted with Sn^+ at dose D = 5 × 10^{16} cm^{-2} (2) and Ga^+ with doses D = 5 × 10^{16} cm^{-2} (3) and D = 1 × 10^{17} cm^{-2} (4) [7].

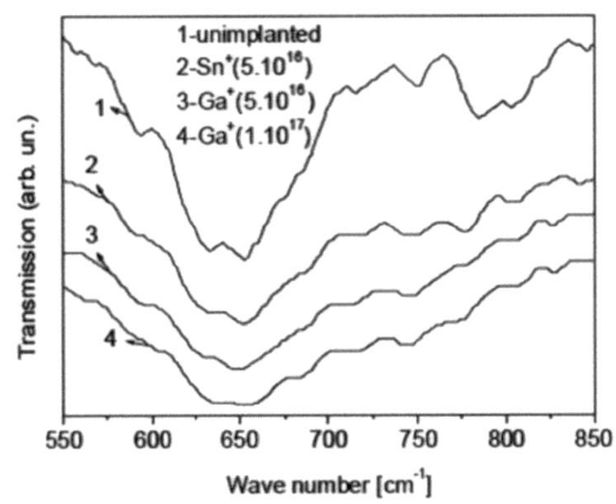

Figure 3.
IR transmission spectra in the region 550–850 cm^{-1} of an unimplanted a-$Si_{0.85}C_{0.15}$:H film (1) and a-$Si_{0.85}C_{0.15}$:H films implanted with Sn^+ at dose D = 5 × 10^{16} cm^{-2} (2) and Ga^+ with doses D = 5 × 10^{16} cm^{-2} (3) and D = 1 × 10^{17} cm^{-2} (4) [7].

As also demonstrated in **Figure 2**, the modification effect of Ga^+ ion implantation on the bonding configurations of the host atoms in the a-SiC:H films is very similar to the one of Sn^+ ion implantation in a-SiC:H films [22]. Apparently, less explicit but still observable are the changes in the region 550–800 cm^{-1} shown in **Figure 3**. In the same figure, the observed change of the Si-C bond stretching mode as shown at ~780 cm^{-1} [23, 24] signifies the modification of the Si-C bonding arrangements in the a-SiC:H host material, the mechanism including considerable Si-C bond breaking as a result of the Ga^+ ion bombardment.

The Raman measurement results presented in **Figure 4** could also be considered as confirmation of the above reasoning. Again in this case, there is a registered well-expressed effect of increased disordering of the silicon network, similar to the earlier results of Sn^+ ion implantation in a-SiC:H films [22]. Here this is demonstrated in the figure by a considerable decrease of the Si-Si bond related peak in the spectra near 500 cm^{-1}.

The investigation results presented here for the case of Ga^+ broad-beam implanted a-SiC:H films have shown various structural and chemical modification changes, resulting in an effective optical bandgap decrease. The underlying mechanism of these modifications comprises breaking of Si-H and C-H bonds

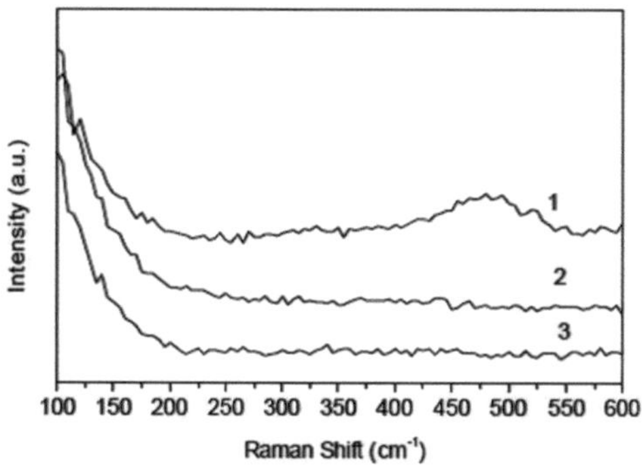

Figure 4.
Raman spectra of an unimplanted a-$Si_{0.85}C_{0.15}$:H film: (1) and a-$Si_{0.85}C_{0.15}$:H films implanted with Sn^+ (2) and Ga^+ (3) at a dose D = 5 × 10^{16} cm^{-2} [7].

and accompanying considerable loss of H in the target material as a result of the ion bombardment. The IR and Raman investigations also reveal increased Si-C bond breaking and formation of Si-Ga bonds, bearing in mind the lower electronegativity of the implanted Ga atoms, compared with the C host atoms, and assuming Ga will substitute for the C atom in the C-Si-H bond [5, 7]. Yet, due to the very low melting point (T_m = 29.8°C) of Ga, some part of the implanted Ga ions are inevitably embedded as Ga clusters and are not directly bonded chemically to the host atoms [25].

2.2 Ga^+-focused ion beam patterning of amorphous silicon carbide

The particular attention that has been focused on Ga^+ as the implant species is due to the fact of its widespread use in FIB systems. Computer-generated Ga^+ FIB pattern in an a-SiC:H film is demonstrated in **Figure 5** [6]. The figure represents an optical micrograph of an a-SiC:H thin film with a variety of Ga^+ FIB-created patterns. The combined pattern consists of a number of chess-board-like (CBL) and series-of-line (SL) patterns, implanted with different doses. The optical density of the irradiated areas is increased, due to ion bombardment, as a result of which they are seen as optically darker regions in the picture.

The computer-generated FIB-irradiated pattern, presented in the figure, comprises several patterns of line series and chess-board-type fields with varying sizes and doses of implanted Ga^+. In this picture, the shorter line series patterns LS1 (size 2.5/2.5/20 μm) and the longer line series patterns LS2 (size 5/5/40 μm) were FIB written with Ga^+ ion doses in the range D = 5 × 10^{15} ÷ 2 × 10^{17} cm^{-2}. The big chess-board-type field (BCF) (size 20 μm) and the smaller chess-board-type fields SCF1 and SCF2 (both size 10 μm) were made with Ga^+ ion doses D_1 = 2.5 × 10^{16} cm^{-2} (BCF and SCF1) and D_2 = 4 × 10^{16} cm^{-2} (SCF2). The picture shows the potential of this method for applications in high-density optical data storage, as well as for direct writing of submicron lithographic masks for further uses in micro- and optoelectronics.

3. Near-field analysis of FIB-patterned amorphous silicon carbide

Near-field techniques were used for the study of the FIB-patterned amorphous silicon carbide films. The topography of the FIB-patterned samples was analysed with an atomic force microscope (dimension 3000 digital instruments). Scanning near-field optical microscope was employed to study and compare the changes in

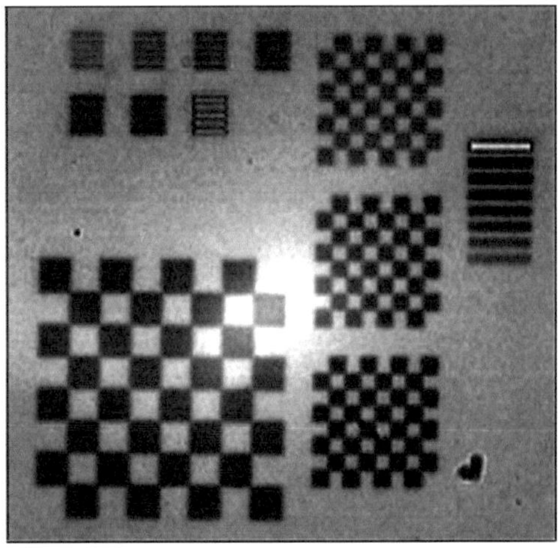

Figure 5.
Micron-scale combined pattern on an a-SiC:H film written with a FIB system by using a 15 pA Ga^+-focused ion beam. The size of the big chess field is 20 × 20 μm² [6].

the local optical absorption of the irradiated (patterned by design) areas relative to the nonirradiated ones.

3.1 Atomic force microscopy (AFM) studies of FIB-patterned a-SiC:H

Before proceeding with the SNOM experiments, topographic images of the FIB-patterned areas were recorded and studied with AFM [8]. The AFM results are presented in **Figure 6** for the CBL and SL structures, and apparently the irradiated areas are topographically lower than nonirradiated ones. Presumably this is due to the ablation of the thin film caused by the ion irradiation. A cross-sectional view of the AFM image of a SL is shown in **Figure 6b**. The ion dose is highest for the left-hand side (3.2×10^{16} cm^{-2}), and it is decreasing towards the right-hand side (0.8×10^{15} cm^{-2}) of the SL pattern. These results suggest that the higher ion doses cause larger topography variations, which should be expected.

Further important result observed in **Figure 6b** is that each line of the SL pattern is as large as ~11 μm, even though the beam diameter of the Ga^+ ion beam was set to be ~50 nm. This unexpectedly large line width of the SL pattern is probably the result of imperfect charge neutralization which leads to severe charge buildup, causing distortion and defocusing of the Ga^+ ion beam during the irradiation.

3.2 Scanning near-field optical microscopy analysis of FIB-patterned amorphous silicon carbide

The implanted set of patterns was characterized by SNOM measurements using a custom-built microscope. The details of the SNOM measurements are described elsewhere [8, 9]. In short, an unpolarized beam from a He-Ne laser (633 nm) is scanned over the substrate side of the sample. The laser beam is kept defocused keeping the laser intensity over the scanned area as uniform as possible. The light passing through the patterned area is then collected using a sharp, uncoated optical fibre probe from one point to another for mapping the optical images obtained by the near-field method. The gap between the probe end and the sample surface is kept

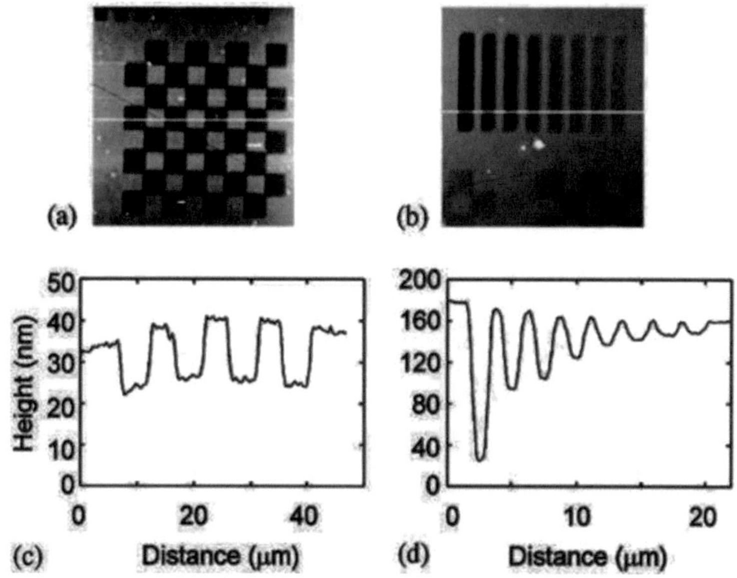

Figure 6.
AFM images of CBS (a) and SSL (b) patterns written in an a-SiC:H film with Ga^+ FIB. The cross-sections indicated in (a) and (b) are shown in (c) and (d), respectively [8].

constant by non-optical shear-force technique and the feedback from the shear-force regulation scheme, providing the required topographic information. Thus, topography mapping and local optical mapping are both registered simultaneously.

The SNOM experiments can provide an optical image of the corresponding topographic image simultaneously, as already mentioned. A series of CBL patterns with different doses have been investigated with SNOM in order to demonstrate its potential. The obtained topographic and corresponding optical images with SNOM are represented in **Figure 7**. When compared with unpatterned areas, e.g., in the left-hand side of the image or in some parts of the CBS pattern, the areas irradiated with ion beam are topographically lower, while they appear also optically darker in the SNOM image. The observed trend of the topographic features of the irradiated

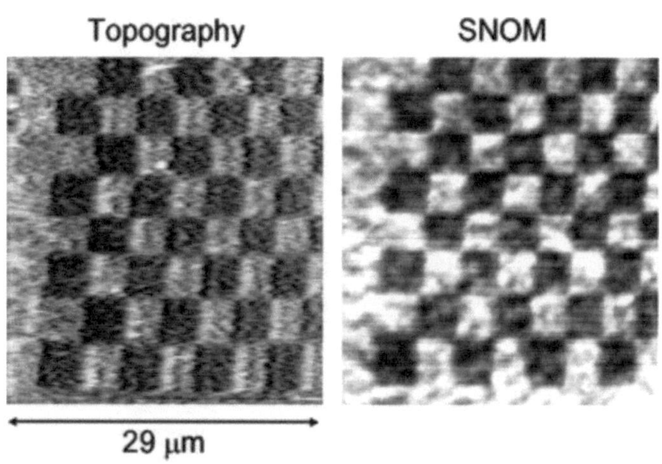

Figure 7.
Topographic image (left) and corresponding SNOM image (right) of a CBS pattern created on an a-SiC:H thin film by irradiation with Ga^+-focused ion beam [8].

areas appears to be the same as that obtained with the AFM. The trend of the optical contrasts observed with SNOM is qualitatively the same as the one obtained in the conventional optical microscopy case (e.g., see **Figure 5**).

4. Ion-implantation temperature and post-implantation annealing effects on the ion beam-induced optical patterning of a-SiC:H

Gallium has also previously been used to make both embedded thin films and layers rich in Ga nanoparticles for broadband high reflectivity devices (such as sunglasses). For the interpretation of these earlier studies in optical glasses, the reflectivity control by the Ga implantation was linked to the fact that Ga has a very low melting point (T_m = 29.8°C) and a rather unusual feature of volume contraction upon melting. These factors were assumed to contribute to Ga incorporation as dispersed clusters and/or small nanoparticles [26, 27]. There was also preferential diffusion and loss of Ga from the implanted region. Any such changes not only influence the implant zone but in turn can modify the interzone reference regions both from changes in the H content, lateral in-diffusion of Ga and induced stress due to the volume changes associated with gallium nanoparticles. Such changes are very sensitive to the implant temperature. It was noted that Ga precipitation into nanoparticles can vary dramatically (in terms of particle size) with concentration and small changes in surface implant temperature [28]. Overall, these factors will contribute to the maximum optical contrast (readout signal) between implanted and unimplanted regions and will also play a significant role in determining the lifetime of the recorded information.

The precise role of implantation temperature effects of target temperature during Ga^+ ion irradiation on the forms of Ga incorporation in the SiC:H film, and hence on the optical contrast obtainable, has therefore formed a key part of this research. This could allow optimization of the process not only from the point of view of the implantation itself (i.e., maximize processing speed and minimize implant costs) but also from the point of view of data storage density, readout contrast and data longevity. Experience also suggests that appropriate post-implantation annealing treatments could offer further benefits in reducing the required ion dose [29] and enhancing contrast, thus increasing the cost-effectiveness of the FIB bit-writing method. Hence, we have also investigated a range of post-implantation treatments in order to determine the most appropriate route for dose reduction while maintaining suitable readout contrast.

4.1 Investigation of the thermal effects on the Ga^+ ion beam-induced optical contrast and structural modification of amorphous silicon carbide

The effects of implantation temperature and post-implantation thermal annealing on the Ga^+ ion beam-induced optical contrast formation in hydrogenated silicon-carbon alloy films have been further studied [30, 31]. For these studies, Ga^+ broad-beam ion implantation in a-SiC:H samples was chosen to be carried out at different film substrate temperatures (T_1 = RT (room temperature), T_2 = LN_2 (liquid nitrogen) and T_3 = +50°C). Then some of the RT implanted samples were thermally annealed post-implantation at higher temperatures. The benefits expected for the optical data storage method, relying on a readout of reduced optical transmission as detected by SNOM [15, 26], were as follows: (1) lower implantation temperature should result in an increased amount of defects which will be leading to an absorption increase, i.e., further decrease in transmission (hence greater readout contrast); (2) higher implantation temperatures, or high-temperature post-implantation annealing, were expected to lead to a possible increase in the optical reflectivity and hence decrease in transmission, due to Ga clusters coalescing into bigger light-reflecting Ga colloid particles.

A range of in a-SiC:H samples have been implanted with Ga^+ broad-beam ion implantation at different substrate temperatures (from liquid nitrogen (LN_2) temperatures to around room temperature (RT) and also at a higher temperature T = +50°C). Some of the RT implanted samples were further post-implantation thermally annealed at several different temperatures in vacuum. The whole range of implanted and annealed samples were optically studied, using optical transmission spectroscopy in the UV–VIS range, in order to define the optimum implantation conditions for archival data storage applications. Using the established optimal dose, $D = 2D_3 = 5 \times 10^{16}\ cm^{-2}$ [30], the transmission (T), the reflection (R) and the absorption coefficient α of Ga^+ implanted a-$Si_{1-x}C_x$:H ($x_1 = 0.18$) films at different temperatures (T_1 = RT, $T_2 = LN_2$ (liquid nitrogen) and T_3 = +50°C) were studied (**Figure 8**).

For further post-implantation annealing studies, some of the RT implanted a-$Si_{1-x}C_x$:H ($x_1 = 0.18$) samples were annealed at higher temperatures (T_1 = +50°C, T_2 = +120°C and T_3 = +250°C) and optically characterized by measurements of the transmission T, the reflection R and the absorption α (**Figure 9**). The obtained results do not represent any appreciable transmission change, even for the higher dose $D = 2D_3 = 5 \times 10^{16}\ cm^{-2}$ and for the highest annealing temperature T_3 = +250°C. Even though in this case there is some slight increase in reflectivity detected, compared to the RT implanted sample, the ion beam-induced thermal annealing of the irradiation defects, which results in absorption decrease, compensates this effect, resulting in similar transmission change as the one for the RT implantation case.

The presented results have indicated the optimum conditions for the new optical data recording method which uses the reading near-field technique SNOM, chosen to be working in optical transmission mode. The results demonstrate that the greatest effect on the optical transmission T is achieved for the RT implanted samples. In the lower temperatures case, the expected ion beam-induced increased defect introduction, resulting in absorption α increase, does occur. However, the simultaneous ion beam-induced thickness decrease due to sputtering compensates this effect, resulting in lower transmission change than the RT implantation case. Likewise, Ga^+ implantation at increased temperatures does result in increased reflectivity R of the implanted films, presumably due to Ga^+ nanoscale particles coalescing into bigger, light-reflecting clusters. Yet, the simultaneously occurring thermal annealing of the ion beam-induced radiation defects in the films decreases the absorption α, resulting in a lower transmission change than the RT implantation case.

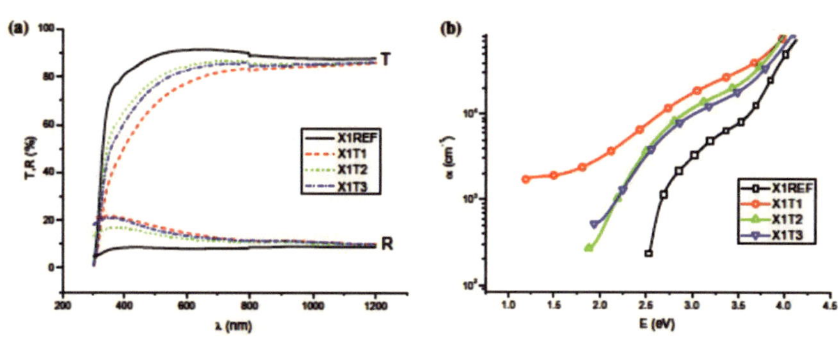

Figure 8.
Transmission (T) and reflection (R) (a), and absorption coefficient α (b), of Ga^+-implanted a-$Si_{1-x}C_x$:H ($x_1 = 0.18$) films with a dose $D = 2D_3 = 5 \times 10^{16}\ cm^{-2}$, at different temperatures: T_1 = RT (room temperature), $T_2 = LN_2$ (liquid nitrogen) and T_3 = +50°C [30].

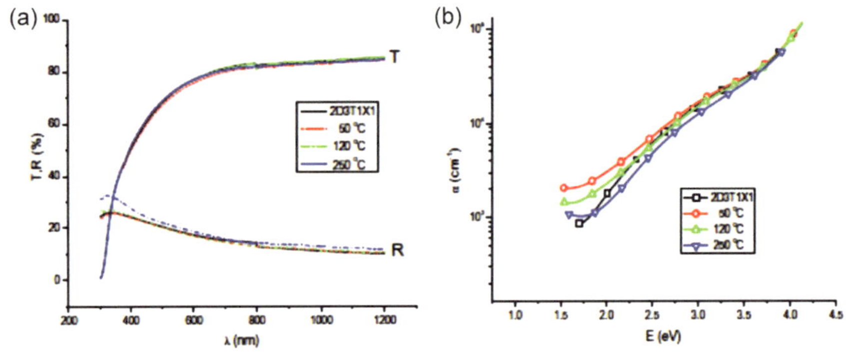

Figure 9.
Transmission (T) and reflection (R) (a) and absorption coefficient α (b), of Ga^+ implanted a-$Si_{1-x}C_x$:H (x_1 = 0.18) films with dose D = $2D_3$ = 5 × 10^{16} cm^{-2} at RT and then annealed at different temperatures: T_1 = +50°C, T_2 = +120°C and T_3 = +250°C [30].

Summarizing, studying the effects of implantation temperature and post-implantation thermal annealing on the Ga^+ ion beam-induced optical contrast formation in hydrogenated silicon-carbon alloy films has led to some practical conclusions [30]. The reasoning about the benefit of lower implantation temperature for the new optical data storage method, relying for optical data readout on reduced optical transmission as detected by SNOM working in transmission mode, was that a lower implantation temperature would lead to an increased amount of introduced defects resulting in an absorption increase, i.e., further decrease in transmission (hence greater data readout contrast). The expected effect was indeed registered but was apparently overridden by an effect of increased ion beam-induced sputtering, which greatly reduced the film thickness and thus increased the transmission in the irradiated films. These results are consistent with those in the case of the Ga^+ FIB patterning of a-SiC:H films, where an increase of both the depth and the width of the individual line patterns within the combined written patterns at the lower temperature case has also been observed as a result of an increased ion beam-induced sputtering [32].

On the other hand, the expected advantage of higher implantation temperatures, or high-temperature post-implantation annealing processing, was that in both cases, the higher temperatures would lead to an increase in the optical reflectivity and hence decrease in transmission, due to Ga clusters effectively coalescing into bigger, optically reflecting Ga colloid particles, resulting in a higher readout contrast. The predicted effect of optical reflectivity increase was indeed observed but was overridden in this case by an apparent decrease in absorption (hence increase in transmission), both in the high-temperature implanted or annealed samples, due to a decrease in the ion beam-introduced defect concentration, as a result of thermally enhanced self-annealing. Thus, it can be concluded that the best conditions for optical data storage for archival storage applications when using Ga^+ ion implantation in a-SiC:H films were found to be at room temperatures with an optimal dose. The conclusion that the optimum implantation temperature for storage applications has to be that of room temperature appears advantageous since it means that the required ion-implantation process can be simplified and reduced in cost.

4.2 Ion-implantation temperature effects on the nanoscale optical patterning of amorphous silicon carbide by Ga^+-focused ion beams

Focused Ga^+ FIB ion implantation was used to write nanoscale data into hydrogenated amorphous silicon carbide films. As was previously reported,

Ga precipitation into nanoparticles can vary greatly (in terms of the particle sizes) with Ga concentration and small variations in the sample surface temperature [25]. Therefore, the precise role of the effects of the implantation temperature, i.e., the target surface temperature during Ga^+ ion implantation and appropriate post-implantation annealing treatments of the FIB-patterned samples at different temperatures, were studied with respect to optical contrast achieved [30] and obtained data storage densities/feature size (present chapter section) [32], in order to further optimize the cost-effectiveness of the FIB bit-writing method for data archiving.

For this purpose, a range of wide bandgap a-SiC:H thin film samples has been prepared by Ga^+-focused ion beam patterning [32]. The samples have been FIB-patterned under different implantation conditions, with emphasis on the different substrate temperatures (from 0°C temperature to around room temperature). Some of the RT implanted samples were further annealed at +250°C in vacuum. The FIB-patterned samples were then analysed using near-field techniques, like atomic force microscopy, to define the optimum implantation conditions and their implications for archival data storage applications.

Focused Ga^+ implantation in these samples was performed with the FIB system simultaneously employing a charge neutralizer (electron-beam shower) to implant a series of FIB patterns with different ion doses in the range 1×10^{15}–1.25×10^{17} ions cm^{-2}. This choice of the Ga^+ ion doses range has been prompted by our earlier Ga^+ broad-beam implantation results, where the optimized range of ion doses for Ga^+ and other elements has been established to be the same so to yield an optimal optical contrast. The chosen type of the combined implanted FIB pattern for this experiment consisted of combination of four individual patterns: a full square, an open square and two sets of parallel lines. They were situated around a central cross-feature (designed for an eye-guide in the AFM microscope to help finding the area of the nanoscale pattern at the start of each measurement) (**Figure 10**).

In these measurements atomic force microscope (AFM), type Dimension 3000 Digital Instruments [8, 9], was used to analyse the topography of the designed FIB-patterned samples. A lab-built scanning near-field optical microscope [8, 9, 15] was also employed to study the sample morphology and the existing film non-homogeneities in the as-deposited a-SiC:H films.

Initially, the as-deposited and some broad-beam Ga^+ implanted samples were analysed with high-resolution optical imaging using SNOM. The AFM imaging was carried out with a range of different wavelengths (λ = 500÷800 nm), including the wavelength of the He-Ne laser used for the further SNOM analysis of the FIB-patterned samples (λ = 633 nm).

The obtained results showed that both the as-deposited and the ion-implanted samples exhibit sub-μm scale nonuniformities, with regions of lower and higher optical absorptions, presumably due to variations in the samples stoichiometry at the sub-μm scale. The SNOM image of these nonuniformities is shown in **Figure 11** for the case of the a-$Si_{1-x}C_x$:H (x_2 = 0.35) sample implanted with Ga^+ with a dose $D = 5 \times 10^{16}$ cm^{-2}.

The preliminary designed actual patterns consisted of a combination of the above described individual patterns, each of which was ion implanted at four different Ga^+ ion doses: 1×10^{15}, 5×10^{15}, 2.5×10^{16} and 1.25×10^{17} cm^{-2}. For creating these patterns, a specially designed programme was used with the Ga^+ FIB equipment. The length of the individual lines in each pattern, as well as the side of the full and open-squares pattern, was chosen to be 5 μm, while all line widths were fixed to be 200 nm.

While some low temperature (0°C) implanted samples were prepared, a greater number of RT implanted samples were designed with view of the planned following post-implantation annealing treatments for the investigation of their effect on

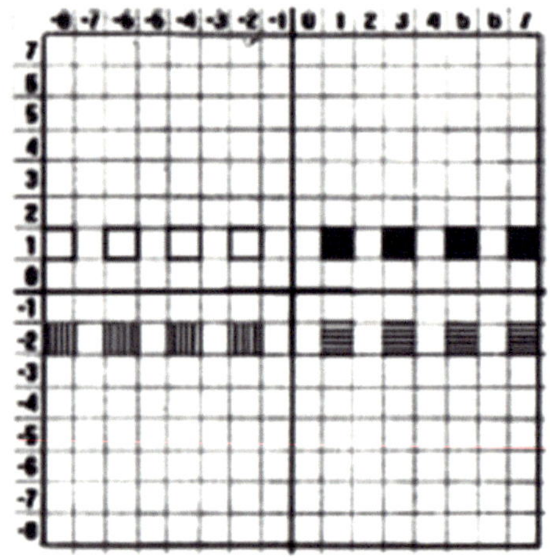

Figure 10.
Combined FIB pattern consisting of individual square patterns [32].

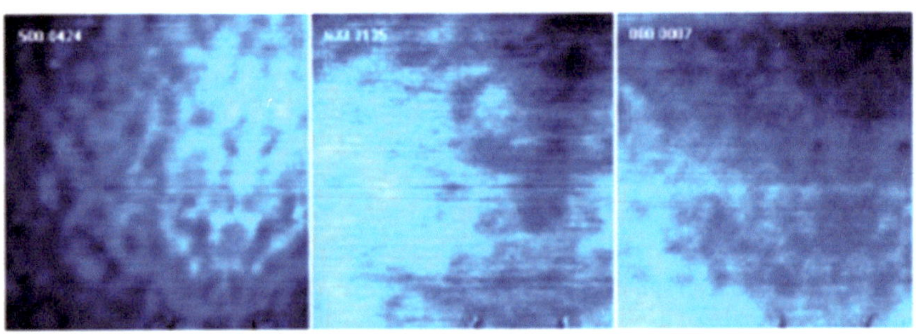

Figure 11.
Images of high-resolution SNOM optical mapping (25 × 25 μm²) of an a-$Si_{1-x}C_x$:H (x_2 = 0.35) sample broad-beam ion implanted with Ga^+ with a dose D = 5 × 10^{16} cm^{-2} [32].

the recorded optical pattern characteristics, to be further studied using near-field techniques (AFM).

The AFM topographic analysis of the FIB-patterned samples has been focused mainly on the open-square patterns (example shown in **Figure 12**). The AFM results for RT and 0°C implanted samples revealed an increased ion beam-induced sputtering yield for the 0°C implanted samples as compared to the RT implanted ones, similarly to the results in the Ga^+ broad-beam implantation studies. Here, in the Ga^+ FIB case, this resulted in an increase of both the depth and the width of the individual lines within the FIB patterns (see **Table 1**).

For studying the post-implantation annealing effects, some of the RT Ga^+ FIB-implanted a-$Si_{1-x}C_x$:H samples have been thermally annealed in vacuum at T = 250°C. The investigations with AFM topographic analysis of the RT FIB-implanted and the thermally annealed samples have been focused again on the open-square patterns (**Figure 13**, showing the examples for the highest doses in the a-$Si_{0.65}C_{0.35}$:H samples).

The presented AFM results for the RT implanted and the thermally annealed samples revealed slight changes in the depth of the individual lines within the FIB written patterns with the annealing. There is a noticeable decrease of the depth of the individual lines after thermal annealing, denoting slight thickness increase in the irradiated areas of the films, which should contribute to their increased optical absorption and hence result in an increased optical contrast as compared to the

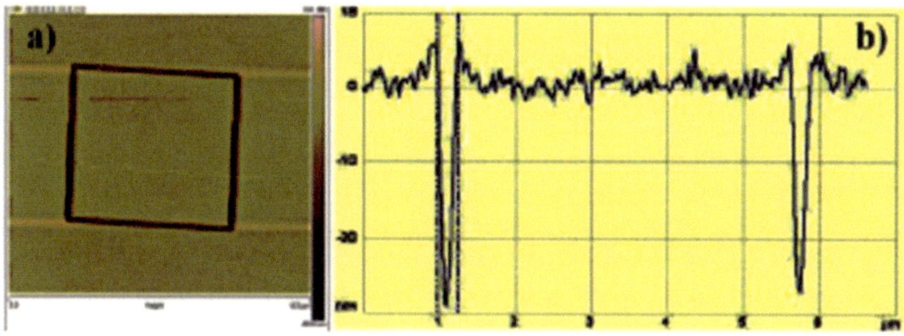

Figure 12.
AFM topographic image (a) and corresponding cross-section (b) of a FIB written open-square pattern in an a-$Si_{1-x}C_x$:H (x_2 = 0.35) sample with a Ga^+ ion dose D = 1.25 10^{17} cm^{-2} [32].

a-$Si_{1-x}C_x$:H		$D_1 = 1 \times 10^{15}$ cm^{-2}		$D_2 = 5 \times 10^{15}$ cm^{-2}		$D_3 = 2.5 \times 10^{16}$ cm^{-2}		$D_4 = 1.25 \times 10^{17}$ cm^{-2}	
		h_1 [nm]	w_1 [μm]	h_2 [nm]	w_2 [μm]	h_3 [nm]	w_3 [μm]	h_4 [nm]	w_4 [μm]
$x_1 = 0.18$	RT	3	0.25	4	0.30	4	0.70	4.5	0.85
	0°C	3	0.59	9	0.69	11	0.79	12	0.99
$x_2 = 0.35$	RT	2.8	0.15	3	0.16	8	0.17	10	0.19
	0°C	9	0.22	15	0.25	18	0.27	22	0.50

Table 1.
Depth (h) and width (w) of the lines in the Ga^+ FIB written open-square patterns in a-$Si_{1-x}C_x$:H [32].

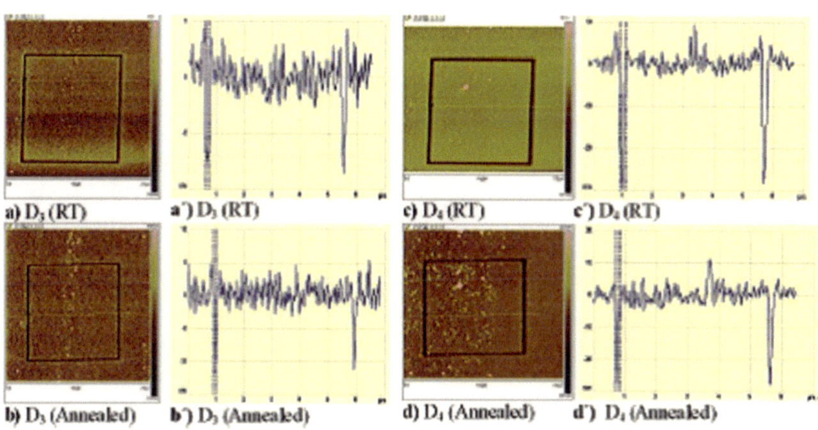

Figure 13.
A set of AFM topographic images of Ga^+ FIB-patterned open-square patterns in a-$Si_{1-x}C_x$:H (x = 0.35) samples: Implanted at a dose D_3 = 2.5 × 10^{16} cm^{-2}, before (a) and after annealing (b), and implanted at a dose D_4 = 1.25 × 10^{17} cm^{-2}, before (c) and after annealing (d); also shown are their corresponding cross-sections (a', b', c' and d') [32].

nonirradiated film areas. This effect is expected to be overridden though by the reduced absorption in the irradiated areas, as a result of thermal annealing of the ion irradiation-induced defects, similarly to the Ga^+ broad-beam implanted case [30].

As demonstrated, the AFM analysis of the Ga^+ FIB-implanted a-$Si_{1-x}C_x$:H samples at RT and at 0°C has revealed an increase of both the depth and the width of the individual lines within the FIB written patterns at lower temperature. This is presumably due to an increased ion beam-induced sputtering yield, in good agreement with the previous results for the case of Ga^+ broad-beam implantation in a-$Si_{1-x}C_x$:H, presented in a previous chapter section. Hence, the present results again suggest that the preferable conditions for optical data storage for archival storage uses would be choosing Ga^+ FIB implantation in a-SiC:H films with an optimal dose at room temperatures. The AFM data also confirm that no advantage would be gained by applying post-implantation annealing treatments.

5. Conclusions

- The present study has shown the existence of optimal conditions for Ga ion implantation of a-SiC:H films in order to achieve large ion beam-induced optical contrast with less topographic change. Low ion irradiation doses are not only sufficient but also preferable for achieving strong modification of the optical absorption, which is required for applications in optical data storage and for creating permanent optical archives.

- With view of optimizing the FIB optical patterning of a-SiC:H, studying the effects of the implantation temperature and post-implantation thermal annealing on the Ga^+ broad ion beam-induced optical contrast formation in the implanted films has been undertaken. The expected benefit for the optical data storage method, which relies on achieving reduced optical transmission detected by scanning near-field optical microscopy, was that a lower implantation temperature would result in an increased amount of ion beam-induced defects, leading to an absorption increase, i.e., further decrease in transmission (greater readout contrast). This expected effect was indeed observed but was overridden by an additional effect of increased ion beam-induced sputtering yield, which greatly reduced the film thickness and hence increased the optical transmission.

- The expected effect of the use of higher implantation temperatures, or high-temperature post-implantation annealing, was that there would be an increase in the optical reflectivity and hence decrease in transmission, since there was expected Ga clusters coalescing into bigger light-reflecting Ga colloid particles. This effect of reflectivity increase was indeed observed but was overridden by a decrease in absorption (transmission increase) in the high-temperature implanted or annealed samples, due to a decrease in the defect concentration as a result of self-annealing.

- In the case of the Ga^+-focused ion beam writing of optical patterns in a-SiC:H films, an increase of both the depth and the width of the individual lines within the FIB written patterns at the lower implantation temperatures, as a result of an increased ion beam-induced sputtering, has been observed. These results are also in good agreement with the previous data for the broad-beam implantation case.

- The AFM results for the RT implanted and the thermally annealed samples revealed slight changes in the depth of the individual lines within the FIB written patterns. The noticeable decrease of the depth of the individual lines after the thermal annealing, and hence obtained slight thickness increase in the irradiated areas of the films, which should contribute to an increased optical absorption thereof, and hence should result in an increased optical contrast as compared to the nonirradiated film areas, is overridden by the reduced absorption in the irradiated areas, as a result of the thermal annealing of the irradiation-induced defects, in agreement with the results for the broad-beam implantation case.

- The AFM analysis of the Ga^+ FIB-patterned a-$Si_{1-x}C_x$:H samples at RT and at 0°C showed an increase of both the depth and the width of the individual lines within the written patterns at the lower temperature case, due to an increased ion beam-induced sputtering. This data is in good agreement with previous results for the case of Ga^+ broad-beam implantation and again suggests that the preferable conditions for the optical data storage method would be using gallium ion implantation in amorphous silicon carbide films at room temperatures with an optimal gallium ion dose. Similarly, the AFM data also confirm that no advantage is expected to result from further post-implantation annealing treatments.

- In conclusion, we can assume that the best conditions for optical data storage for archival storage applications, when using Ga^+ ion implantation in a-SiC:H films, would be employing an optimal gallium ion dose at room temperatures. The fact that the optimum implantation temperature for optical data storage applications appears to be that of room temperature is fortuitous since the implications are that the required ion-implantation processing could be simplified and reduced in cost.

Acknowledgements

Throughout this work, the support of the Bulgarian Academy of Sciences and Maria Curie-Sklodowska University in Lublin, Poland, has been invaluable. Also gratefully acknowledged is the help of the staff of the Ion Beam Centre at the Helmholtz-Zentrum Dresden-Rossendorf e.V., Germany, for performing the focused ion beam implantation.

Author details

Tania Tsvetkova
Institute of Solid State Physics, Bulgarian Academy of Sciences, Sofia, Bulgaria

*Address all correspondence to: tsvet@issp.bas.bg

References

[1] Enterprise Strategy Group (ESG) Research Report. Digital Archiving: End-User Survey & Market Forecast; 2006-2010

[2] Enterprise Strategy Group (ESG) Archive Multi-Client Study; September 2006

[3] Powell JA, Matus L. In: Harris GL, Yang CYW, editors. Amorphous and Crystalline Silicon Carbide. Berlin: Springer; 1989

[4] Kanicki J. Amorphous and Microcrystalline Semiconductor Devices. Boston, London: Artech House; 1991

[5] Tsvetkova T, Tzenov N, Tzolov M, Dimova-Malinovska D, Adriaenssens GJ, Pattyn H. Optical properties and chemical structure of ion implanted a-SiC:H. Vacuum. 2001;**63**:749

[6] Bischoff L, Teichert J, Kitova S, Tsvetkova T. Optical pattern formation in a-SiC:H films by Ga^+ ion implantation. Vacuum. 2003;**69**:73

[7] Tsvetkova T, Angelov O, Sendova-Vassileva M, Dimova-Malinovska D, Bischoff L, Adriaenssens GJ, et al. Structural and optical properties modification of a-SiC:H by Ga^+ ion implantation. Vacuum. 2003;**70**:467

[8] Tsvetkova T, Takahashi S, Zayats A, Dawson P, Turner R, Bischoff L, et al. Near-field optical mapping of ion-implanted patterns fabricated in amorphous silicon carbide. Vacuum. 2005;**79**:94-99

[9] Tsvetkova T, Takahashi S, Zayats A, Dawson P, Turner R, Bischoff L, et al. Fabrication of nano-scale optical patterns in amorphous silicon carbide with focused ion beam writing. Vacuum. 2005;**79**:100-105

[10] Kalbitzer S. Semiconductors for optical memories. Current Opinion in Solid State & Materials Science. 2002;**6**:271

[11] Müller G. The contribution of ion-beam techniques to the physics and technology of amorphous semiconductors. Nuclear Instruments and Methods in Physics Research Section B. 1993;**957**:80-81

[12] Hirvonen JK. Ion Implantation and Ion Beam Processing of Materials. Amsterdam, North Holland; 1984

[13] Böhringer K, Jousten K, Kalbitzer S. Development of a high-brightness gas field-ionization source. Nuclear Instruments and Methods in Physics Research Section B. 1988;**30**:289

[14] Ruttensberger B, Krotz G, Muller G, Derst G, Kalbitzer S. Crystalline-amorphous contrast formation in thermally crystallized SiC. Journal of Non-Crystalline Solids. 1991;**137-138**:635

[15] Richards D, Zayats AV, editors. Nano-optics and near-field microscopy. Philosophical Transactions of the Royal Society. 2004;**362**:699-812

[16] Ocelic N, Hillenbrand R. Subwavelength-scale tailoring of surface phonon polaritons by focused ion-beam implantation. Nature Materials. 2004;**3**:606

[17] Birukawa M, Ito E, Narumi K, Kojima R, Yamada N, Mizukuki T, Nakaoki A, et al. Rewritable dual-layer recording with near-field coupled SIL system. Proc. of EPCOS. 2008. Prague; 2008. Available from: www.epcos.org

[18] Liu Q, Fukaya T, Cao S, Guo C, Zhang Z, et al. Study on readout durability of super-RENS disk. Optics Express. 2008;**16**:213

[19] Tzenov N, Dimova-Malinovska D, Marinova T, Krastev V, Tsvetkova T. X-ray

photoelectron study of Sn^+ implanted $a\text{-}Si_{1-x}C_x$:H thin films. Materials Science and Engineering B. 1995;**29**:165

[20] Tzenov N, Dimova-Malinovska D, Marinova T, Krastev V, Tsvetkova T. Modification of magnetron sputtered $a\text{-}Si_{1-x}C_x$:H films by implantation of Ge^+. Nuclear Instruments and Methods in Physics Research Section B. 1996;**112**:342

[21] Tzenov N, Dimova-Malinovska D, Tsvetkova T. Modification of magnetron sputtered $a\text{-}Si_{1-x}C_x$:H films by implantation of Sn^+. Materials Research Society Symposium Proceedings. 1996;**396**:243

[22] Tsvetkova T. Ion beam modification of amorphous silicon-carbon alloys. In: Singh J, Copley S, Mazumder J, editors. Beam Processing of Advanced Materials. Metals Park: ASM International; 1996. p. 321

[23] Mohr W, Tsai C, Street R. Properties and local structure of plasma-deposited amorphous silicon-carbon alloys. Materials Research Society Symposium Proceedings. 1986;**70**:319

[24] Herremans H, Grevendonk W, van Swaaij RACMM, van Sark WGJHM, Berntsen AJM, Arnold Bik WM, et al. Structural, compositional and optical properties of hydrogenated amorphous silicon-carbon alloys. Philosophical Magazine B. 1992;**66**:787

[25] Hole DE, Townsend PD, Barton JD, Nistor LC, Landuyt J. Gallium colloid formation during ion implantation of glass. Journal of Non-Crystalline Solids. 1995;**180**:266-274

[26] Takahashi S, Dickson W, Pollard R, Zayats AV. Near-field magneto-optical analysis in reflection mode SNOM. Ultramicroscopy. 2004;**100**:443

[27] Karrai K, Grober RD. Piezoelectric tip-sample distance control from near field optical microscopes. Applied Physics Letters. 1995;**66**:1842

[28] Hoffmann P, Dutoit B, Salathe RP. Comparison of mechanically drawn and protection layer chemically etched optical fiber tips. Ultramicroscopy. 1995;**61**:165

[29] Pangaribuan T, Yamada K, Jiang SD, Ohsawa H, Ohtsu M. Reproducible fabrication technique of nanometric tip diameter fiber probe for photon scanning tunneling microscope. Japanese Journal of Applied Physics. 1992;**31**:L1302

[30] Tsvetkova T, Wright CD, Kitova S, Bischoff L, Zuk J. Effects of implantation temperature and thermal annealing on the Ga^+ ion beam induced optical contrast formation in a-SiC:H. Nuclear Instruments and Methods in Physics Research Section B. 2013;**307**:71

[31] Tsvetkova T, Wright CD, Craciun MF, Bischoff L, Angelov O, Dimova-Malinovska D. Thermal effects on the Ga^+ ion beam induced structural modification of a-SiC:H. Journal of Physics Conference Series. 2012;**398**:012048

[32] Tsvetkova T, Wright CD, Hosseini P, Bischoff L, Zuk J. Implantation temperature effects on the nanoscale optical pattern fabrication in a-SiC:H. Acta Physica Polonica A. 2013;**123**:952

Chapter 8

Multilayered and Chemiresistive Thin and Thick Film Gas Sensors for Air Quality Monitoring

Tynee Bhowmick, Vibhav Ambardekar, Abhishek Ghosh, Moumita Dewan, Partha Pratim Bandyopadhyay, Sudip Nag and Subhasish Basu Majumder

Abstract

Selective detection of gases such as nitrogen dioxide (NO_2), carbon monoxide (CO), carbon dioxide (CO_2), and various volatile organic components (VOCs) is necessary for air quality monitoring. Detection of hydrogen (H_2) is equally important as it is a flammable gas and poses serious threat of explosion when exposed to oxygen gas. We have studied the sensing characteristics of these gases using thin film deposited by chemical solution deposition as well as relatively thicker films deposited by atmospheric plasma spray (APS) process. The chapter starts with the sensing mechanism of chemiresistive sensors followed by the definition of gas sensing parameters. Subsequently, we have demonstrated selective NO_2 sensing characteristics of zinc oxide-graphene (ZnO-G) multilayered thin film followed by CO and H_2 sensing characteristics of ZnO thin film and SnO_2 thick film. Cross-sensitivity among CO and H_2 gases has been addressed through the analysis of conductance transients with the determination of activation energy, E_a, and heat of adsorption, Q. The concepts of reversible and irreversible sensing have also been discussed in relation to CO and H_2 gases. CO_2 sensing characteristics of $LaFe_{0.8}Co_{0.2}O_3$ (LFCO)-ZnO thin film have been elucidated. Interference from CO has been addressed with principal component analyses and the ascertaining of E_a and Q values. Additionally, the variation of response with temperature for each gas was simulated to determine distinct parameters for the individual gases. Further, VOC sensing characteristics of copper oxide (CuO) thin film and WO_3-SnO_2 thick film were investigated. Principal component analysis was performed to discriminate the gases in CuO thin film. The interaction of WO_3-SnO_2 thick film with various VOCs was found to obey the Freundlich adsorption isotherm based on which E_a and Q values were determined.

Keywords: air quality monitoring, gas sensing, thin film sensor, thick film sensor, Langmuir adsorption isotherm, Freundlich adsorption isotherm, reversible sensing, irreversible sensing

1. Introduction

A recent surge in diseases related to poor air quality has made it mandatory to effectively monitor the level of harmful pollutants in the atmosphere. Poor ventilation in indoor spaces as well as vehicular exhausts and emissions from industries poses serious threats to the quality of breathable air. Indoor air quality monitoring has become essential in order to counter the rising incidence of diseases such as sick building syndrome, which occurs in residential and office spaces due to poor ventilation. Indoor air pollutants are mostly volatile organic compounds (VOCs) which include alcohols such as ethanol and methanol and adverse pollutants such as benzene and formaldehyde. Other than this, carbon dioxide (CO_2) and nitrogen dioxide (NO_2) also pollute indoor living space [1].

Outdoor air quality is contaminated mainly due to the presence of carbon monoxide (CO), carbon dioxide, nitrogen oxides (NO_X), sulfur dioxide (SO_2), volatile organic compounds, and particulate matter. Urban sources of pollutants include vehicles and industries. In rural India, many people still use solid fuel such as wood, crop wastes, charcoal, and cow dung in open fires popularly known as "chulhas." Such cooking practices are highly damaging to human health as they emit huge quantities of CO and NO_2 and cause an array of diseases including but not limited to pneumonia, stroke, heart disease, and lung cancer [2].

Air quality monitoring is thus mandatory to check the levels of various pollutants in the atmosphere. Agencies like Occupational Safety and Health Administration (OSHA) define the permissible exposure limits (PELs) of air pollutants to protect the health of workers. **Table 1** shows the permissible limit of various air contaminants.

Apart from these pollutants, detection of hydrogen (H_2) gas is also important. It is used as fuel in space craft and rockets. It is reported that almost all CO sensing materials are cross-sensitive toward H_2 as well. It remains a major challenge to address the cross-sensitivity of CO sensor.

To detect these pollutants, one can use chemiresistive type thin or thick film oxide sensors, made from economic precursor materials. The use of oxide sensing materials in thin/thick film form has advantage over commonly used bulk Taguchi-type sensors. Thin/thick film sensors consume less power and also allow miniaturization of gas sensing systems.

The surface of oxide sensors acts as a catalyst to oxidize reducing pollutants (CO, H_2, VOCs, etc.) to relatively benign gases (viz., CO forms CO_2). Limited attempt has so far been made to study the gas sensing performance of oxide ceramics together with catalytic activity to oxidize the gas sensed [4]. For gas sensing performance, thin film sensor is preferred; however, for catalytic conversion, relatively thicker coating of oxide materials over a large area surface (often curved) is required.

Pollutant	Averaging time (h)	Level (ppm)
Carbon monoxide	8	35
Nitrogen dioxide	1	100
Carbon dioxide	8	5000
Sulfur dioxide	8	2
Ethanol	8	1000
Acetone	8	1000
Benzene	8	1
Formaldehyde	8	0.75

Table 1.
OSHA permissible limit for various air pollutants [3].

Understanding of gas–solid interaction is of utmost importance to optimize the sensing performance of chemiresistive oxide sensors. Typically, Langmuir adsorption isotherm has been utilized to understand the resistance transients recorded upon exposing the oxide sensor to test gas environment. Through these kinetic analyses, interrelation among sensing layer thickness, its morphology, response %, sensitivity, and response/recovery times can be addressed adequately [5–7]. One can also estimate the adsorption and desorption energies of the test gas and reaction products, respectively. The estimated energies could be an effective tool to differentiate various types of test gases [7]. Base line stability of the sensing materials is also important to fabricate commercial sensing elements. Interrelation among the surface morphology, response time, and test gas concentration needs to be understood to fine-tune the base line stability [6, 8].

In view to the above, the primary goal of the present chapter is to describe a comprehensive approach to model the resistance/conductance transients based on Langmuir and Freundlich adsorption isotherms to address cross-sensitivity. Modeling the resistance/conductance transients allows us to estimate sensing parameters (such as response %, response/recovery time, etc.) and adsorption/desorption energies of test gas and reaction products. The estimated parameters help us to better understand the effect of the receptor (adsorption of gas molecule on sensing surface), transduction (electronic conduction pathway), and utility functions (effect of thickness and morphology of sensing materials) controlling the sensitivity, stability, and selectivity criteria of thin film type chemiresistive sensors.

The chapter is organized as follows: In the following sections, we have elucidated the mechanism of gas sensing followed by defining various sensing parameters. Subsequently we have briefed the dominant material characteristics influencing the sensing performance of thin film chemiresistive type gas sensors. Next, we have outlined the synthesis procedure of thin as well as thick film sensing elements. The synthesized materials were characterized in terms of their phase purity and microstructure evaluation. Finally, we have outlined the details of static as well as dynamic flow gas sensing measurements to investigate the sensing performance of the synthesized materials.

We have thus prepared thin and thick film sensors to detect NO_2, CO, H_2, CO_2, and VOCs (ethanol, acetone, and isopropanol). Gas sensing mechanism has been described for each gas. Cross-sensitivity has been addressed through conductance transient analysis. The gas–solid interaction has been observed to follow either Langmuir or Freundlich adsorption characteristics. Based on this approach, the activation energy, Ea, and heat of adsorption, Q, were calculated to discriminate the gases. Stability of the gas sensors was also studied with respect to thin and thick films.

2. Salient features of semiconducting oxide gas sensor

Semiconducting metal oxide (SMO) sensors are attractive for lower cost, smaller size, simpler operation principle, durability, and ease of fabrication together with their low concentration of gas detection limit [9]. These sensors change their conductivity when exposed to test gases of different concentrations [9]. However, the sensor operating temperature (~300°C) needs to be lowered, and cross-sensitivity toward multiple gases needs to be minimized for its wide commercial adaptation. The operating principle and salient features of SMO-based chemiresistive sensors are briefly described as follows:

2.1 Principal of operation

For reducing CO gas sensing, **Figure 1** schematically illustrates the operative sensing mechanism. At elevated temperature, when the sensing film is exposed to air, the sensor surface provides active sites for the chemiadsorption of oxygen ions. Thus, oxygen adsorbs on to the sensor surface by accepting electrons from the conduction band of the sensing material. In the operating temperature ranging 100–500°C, O_2^-, O^-, and O^{2-} are chemiadsorbed [10] forming an electron depletion layer (EDL). The EDL impedes grain to grain electron migration increasing the surface resistance for an "n"-type sensing material. When the sensor is exposed to a reducing gas such as CO or H_2, the adsorbed oxygen ions react with the test gas to release the electrons back to the conduction band of the sensing material. As a result the EDL width is reduced to decrease the surface resistance of the sensor.

In the case of a p-type sensor, since the majority carriers are holes, the oxidation reaction of the reducing test gas and the oxygen ions on the sensor surface would decrease the concentration of holes, increasing the sensor resistance. Assuming n-type and p-type sensors have similar morphologies, the response of p-type and n-type sensor is related by the following relation [11].

$$S_p = \sqrt{S_n} \tag{1}$$

where S_p is the gas response of p − type sensor and S_n is the gas response of n − type sensor.

2.2 Characteristic sensor parameters

Figure 2 illustrates the characteristic sensor parameters one can estimate from the recorded resistance transient of an n-type sensor used for reducing gas sensing.

Response (S) (%): Response is defined by the expression $(R_a - R_g)/R_a \times 100$ or R_a/R_g where R_a and R_g are the measured resistances of the sensor exposed to air and test gas, respectively.

Response time (τ_{res}) (s): Time required for a sensor to reach 90% of the maximum response on exposure to test gas [12].

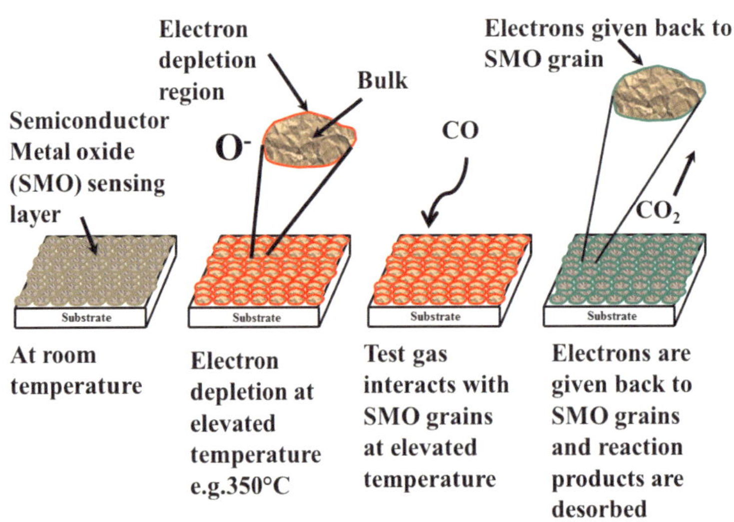

Figure 1.
Gas sensing mechanism for a semiconducting metal oxide sensor.

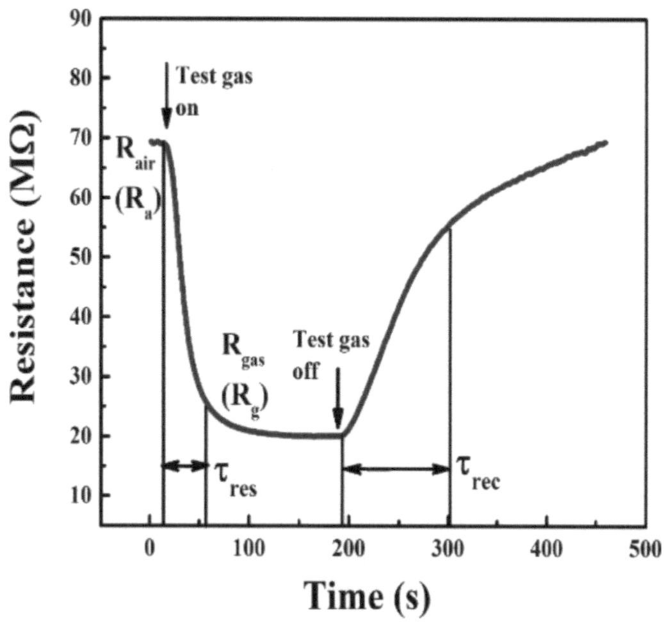

Figure 2.
Resistance transient of an n-type gas sensor showing stable resistance in air (R_a), stable resistance in gas (R_g), response time (τ_{res}), and recovery time (τ_{rec}).

Recovery time (τ_{rec}) (s): Time required for a sensor to reach 90% of the original baseline resistance upon removal of the test gas.

Optimum temperature (T_{opt}) (°C): The temperature at which the sensor exhibits maximum response.

Selectivity: The ability of a sensor to sense a selective gas preferentially among other test gases.

Stability: Minimum variation of sensor base resistance over a period of time.

Sensitivity: Change of measured signal (resistance, voltage or current) per analyte concentration unit [13].

2.3 Factors affecting gas sensing performance of thin film sensor

The sensing performance of SMO-based sensor is primarily influenced by the receptor, transducer, and utility functions of thin film sensing elements. The receptor function is influenced by the crystallite size of the sensing material, as at elevated temperature oxygen ions are chemiadsorbed on these crystallites forming a charge-depleted layer. The receptor function controls the interaction of test gases with the sensing elements. The transducer function controls the flow of charges between the electrodes. Finally the utility function controls the diffusion of gases. The parameters that regulate the receptor function of the sensor include composition of the material, surface to volume ratio, and size of the pores and grains. The transducer function is influenced by the grain to grain contact and the phase purity of the material. The composition of the material influences the defect chemistry which in turn changes the nature and concentration of the defects in the sensing material. The surface to volume ratio, pore size and grain size, and grain to grain contact of the sensing material are dependent on the growth condition.

Usually, the sintered block as well as thick film sensors consists of loosely sintered secondary particles made of innumerous tiny primary particles. Generally, the dimension of pores between the primary particles is in microns (termed as macropores), whereas those in between the primary particles are in the order of nanometers (<50 nm) (termed as mesopores). When the sensor is aged at elevated temperature, oxygen gas is diffused through both the macro- (corresponds to molecular diffusion) and mesoporous (corresponds to Knudsen diffusion) regions. Sensor particles are charge depleted due to chemiadsorption of oxygen. During gas sensing, the test gas diffuses through macro- and mesoporous regions and reacts with chemiadsorbed oxygen. The diffusion of gases through the mesoporous regions is known to play an important role during reducing gas sensing using semiconducting oxide-based sensing elements. For compact thin films, the gas-solid interaction mostly takes place at the geometric surface and to a limited extent along grain boundaries. In contrast, the active surface of the porous thin film is much larger than the compact thin films. In both these thin film structures, the key parameters that control the receptor function of the sensing element include its composition, nature of metal-oxygen bonding, specific surface area, porosity, and crystallite size. The transducer function is influenced by grain to grain contact and phase purity of the sensing element, and the utility function will be controlled by the gas diffusion either through grain boundaries or pores. For thin film sensing elements with pore size distribution in the range of 2–50 nm, Knudsen-type gas diffusion seems to be operative. Since the Knudsen diffusion coefficient is related to operating temperature, pore radius, molecular weight of the diffusing gas, and diffusion distance are related to film thickness; it is quite obvious that the sensing performance of a thin film sensor is related to the operating temperature, film thickness, and molecular weight of the test gas. Very limited attempt has so far been made to theoretically investigate the gas transport phenomena within the sensing film and correlate it with its sensing performance.

In the subsequent sections, we have outlined the sensing performance of ZnO-graphene multilayered thin films for NO_x sensing. Subsequently, CO and H_2 sensing performance of ZnO thin film and SnO_2 thick film were investigated. This was followed by CO_2 sensing performance of 0.5% LFCO-ZnO thin film and VOC sensing characteristics of copper oxide (CuO) thin film and WO_3-SnO_2 thick film.

At constant sensor operating temperature, we assumed monolayer coverage of the reducing gas. Assuming Langmuir adsorption kinetics, we have theoretically predicted the response and recovery transients during gas sensing using thin and thick film sensing elements. From an application point of view, marginal base resistance drift of these thin film sensing elements is desirable during repeated response and recovery cycles. Often a significant drift in base line resistance is observed due to the partial recovery of thin and thick film sensing elements. For such partial recovery, the sensing is termed irreversible, whereas for reversible sensing, the base resistance is fully recovered. For a wide range of test gas concentrations and operating temperatures, the response transients of thin film sensing elements are modeled using Langmuir-Hinshelwood reaction mechanism. It is predicted that for irreversible-type sensing, the response time is reduced with the increase in test gas concentration, whereas for reversible sensing, response time is independent of the test gas concentration [12]. The WO_3-SnO_2 thick film sensor was modeled with Freundlich adsorption isotherm in which the concentration-dependent conductance was ascertained to be nonlinear.

3. Experimental

An economic wet chemical synthesis route was adopted to synthesize thin film gas sensing elements. Wet chemical synthesis is also energy efficient, and relatively cheaper precursor materials are used in wet chemical synthesis. We have synthesized the ZnO-graphene (G) multilayer to study its NO_2 sensing characteristics. ZnO thin film sol was prepared by dissolving stoichiometric amount of zinc acetate dihydrate ($Zn(CH_3COO)_2.2H_2O$) in 2-methoxyethanol (2-MOE) solvent with monoethanolamine (MEA) as stabilizer. The final concentration of the precursor solution was ~0.4 M/L. The precursor solution was heated at ~80°C for complete dissolution and stabilization. Graphene sheet from crystalline graphite was exfoliated by micromechanical cleavage using facile ultrasonication synthesis [14]. For this purpose, 5 mg of graphite powder was mixed in 5 ml of 2-MOE solution. N-methyl-2-pyrrolidine (NMP) solution was added for stabilization of the graphite sol. This suspension was then ultrasonicated for 3 h at room temperature and aged for 48 h. Multilayered ZnO-G thin films were fabricated by alternate deposition of ZnO and exfoliated graphite sol. Prior to sol deposition, the glass substrates were cleaned in trichloroethylene for 5 min followed by ultrasonic cleaning in the presence of acetone and deionized water. This was followed by drying of the substrates in flowing air. Details of deposition process have been reported elsewhere [14]. The multilayered films were annealed at 500°C for 1 h in air. The sensing performance of ZnO-G multilayered films were characterized using NO_2 (0.5–5.0 ppm), CO (50 and 500 ppm), H_2 (50 and 500 ppm), and i-C_4H_{10} (50 and 500 ppm) as test gases. The operating temperature was maintained in the range 100–200°C during gas sensing measurements. Additionally, mixed gas sensing behavior for the ZnO-G multilayered films has also been investigated using NO_2/CO (5 ppm/50 ppm), NO_2/H_2 (5 ppm/50 ppm), and NO_2/i-C_4H_{10} (5 ppm/50 ppm) gas mixtures.

We have also studied the CO and H_2 sensing behavior of ZnO thin film and SnO_2 thick film. ZnO thin film was synthesized following the method described above. SnO_2 thick film was synthesized using a plasma spray deposition process. Commercial SnO_2 powder was used as feedstock material for the plasma spray process. Nitrogen (sans hydrogen) was used as the plasma forming gas and compressed air was used for cooling. Details of the process and equipment can be found elsewhere [7]. Alumina substrates (Ants Ceramic, India) were used for deposition as these are unaffected by the high temperatures generated during the atmospheric plasma spray process.

We have also studied the CO_2 sensing behavior of 0.5% LFCO ($LaFe_{0.8}Co_{0.2}O_3$)-ZnO thin film composite. Lanthanum nitrate hexahydrate ($La(NO)_3.6H_2O$), iron (III) nitrate nonahydrate ($Fe(NO)_3.9H_2O$), and cobalt (II) nitrate hexahydrate (Co $(NO)_2.6H_2O$) were dissolved in 2-MOE to prepare LFCO precursor sol with a final solution of 0.2 M via continuous stirring at room temperature. LFCO and ZnO sol were mixed in appropriate proportions to make composites of the composition xLFCO-(1−x) ZnO. The resulting sol was then spin coated on quartz substrate. Each layer was baked at 300°C with an intermittent heating interval of 10 min between layers to burn out the organic moieties. The coating and baking processes were repeated to yield films about 300nm thick. The resultant films were then annealed at 600°C for 2 h.

VOC sensing characteristics of CuO thin film and WO_3-SnO_2 thick film were measured in a static gas sensing system. CuO sol (0.25 M) was synthesized using copper (II) acetate monohydrate ($Cu(COOCH_3).H_2O$) and 2-MOE as precursor materials. Glass substrates used for deposition were cleaned using trichloroethylene, acetone, and deionized water as described earlier. CuO sol was spin coated on glass substrates. Each layer was baked at 300°C for 5 min. The resultant film which was about 200 nm thick was annealed at 600°C for 1 hour to achieve complete crystallization. WO_3-SnO_2 thick film composite was deposited on alumina substrates using a plasma spray deposition process. The feedstock material consists of WO_3 and SnO_2 powders in the weight ratio ~25:75. This ratio was found to be optimum in a series of experiments with bulk pellets with various weight ratios (~25:75, 50:50, and 75:25) of WO_3 and SnO_2 [15].

The resistance change of the thin film sensing elements was measured using gold interdigitated electrode which was sputter coated onto the thin films using a dc sputter coater (Cressington 108, Ted Pella, Inc., USA). High-temperature silver paste was used as electrodes for the thick coatings. An indigenously developed dynamic gas sensing setup was used to measure the gas sensing characteristics of the thin films and thick plasma sprayed coatings. The gas sensing setup consists of a gas chamber connected to mass flow controllers (PR 4000, MKS Instruments, Germany) which control the flow rate of gases. The flow rate is calculated by the relation [12, 16].

$$C_{mixed\ gas} = \left[C_{test\ gas} \times \frac{dV_{test\ gas}}{dt}\right] / \left[\frac{dV_{test\ gas}}{dt} + \frac{dV_{carrier\ gas}}{dt}\right] \tag{2}$$

where $C_{test\ gas}$ is the concentration of test gas, $\frac{dV_{test\ gas}}{dt}$ is the flow rate of test gas, and $\frac{dV_{carrier\ gas}}{dt}$ is the flow rate of carrier gas.

The temperature of the sensor inside the dynamic chamber is controlled with a temperature controller. An electrometer (6517 A, Keithley, USA) is connected to the sensor via gold-coated silver probes. Necessary voltage for electron migration between the grains of the sensor is controlled through the electrometer. The entire gas sensing setup is interfaced with a PC equipped with an RS-232 interface, GPIB card (National Instruments, USA), and LabVIEW 8.5 (National Instruments, USA) software [17]. A schematic of the mentioned gas sensing setup may be found in [12].

A static gas sensing set was employed to study the VOC (ethanol, acetone, and isopropanol) sensing characteristics of CuO thin film and WO_3-SnO_2 thick film. Such a setup (also indigenous) consists of a rectangular chamber housing the sensor and the heater which is connected to the temperature controller [18]. The sensor is connected to a microcontroller which in turn is connected to the PC. Details of the system schematic may be found elsewhere [15]. Data (in this case, resistance) is collected at the rate of 1 bit/sec with the help of terminal emulation software. The sensor is heated to the temperature of operation and aged at that temperature to obtain a stable resistance (R_a) which is followed by injection of gas. The resistance at which the sensor stabilizes in the presence of gas is termed R_g. The response of the sensor to the test gas in measured in terms of response percent which is defined as [19, 20].

$$Response\ (S)\% = \frac{|R_a - R_g|}{R_a} \times 100 \tag{3}$$

4. Investigations on the gas sensing performance for air quality monitoring

Air quality monitoring involves the monitoring and quantification of various gases present in the atmosphere. Such gases include NO_X [21], CO [22], CO_2 [23], and VOCs [24]. In addition, flammable gases such as H_2 also need detection in order to prevent major disasters [25]. Apart from the detection of low concentration test gases, addressing the cross-sensitivity of the sensors also remains a major research issue. In addition to this, cost of air quality monitoring system is high, and efforts are needed to make it economic for wide uses. Keeping these considerations in mind, we have developed sensors to monitor these gases selectively. For cross-sensitive sensing elements, necessary protocols are developed to address their selectivity. The sensing performances of synthesized sensing elements to detect test gases pertinent to air quality monitoring are described as follows.

4.1 NO_2 sensing

The main sources of NO_2 emissions are the incomplete burning of fuel and vehicular exhausts. According to a report in *The Hindu* [26], one of India's premier newspapers, winter smog is caused in many parts of northern India due to burning of crop residue. An estimated 240 million tons of NO_X (NO and NO_2) is generated because of this practice. **Figure 3** shows percentage of NO_X emissions in India from various sources. Other than this, vehicular exhausts also contribute to NO_X formation. NO_X, along with other pollutants such as SO_2, VOCs, and particulate matter, forms smog as well as acid rain. These are markers of extremely polluted air and can have adverse health effect on humans.

Traditionally, semiconducting metal oxides such as SnO_2 [27], WO_3 [28], ZnO [29], etc. have been used for NO_2 sensing. Recently, composites such as Fe_2O_3-ZnO [30] has also gained prominence for selective NO_2 sensing. In addition, metal oxide (MO)-graphene (G) composites such as 0.5 wt%G-WO_3 [31] and SnO_2-RGO [32] have also been reported as these sensors show high response at low operating temperature. These sensors are also highly selective.

Based on literature study (**Table 2**), we realized that MO-graphene (G) sensor will be an effective candidate for NO_2 sensing study. We have chosen the MO to be ZnO as it is an extremely stable sensor and can be easily modulated with graphene to operate at a low temperature. We have thus prepared a ZnO-G multilayer system

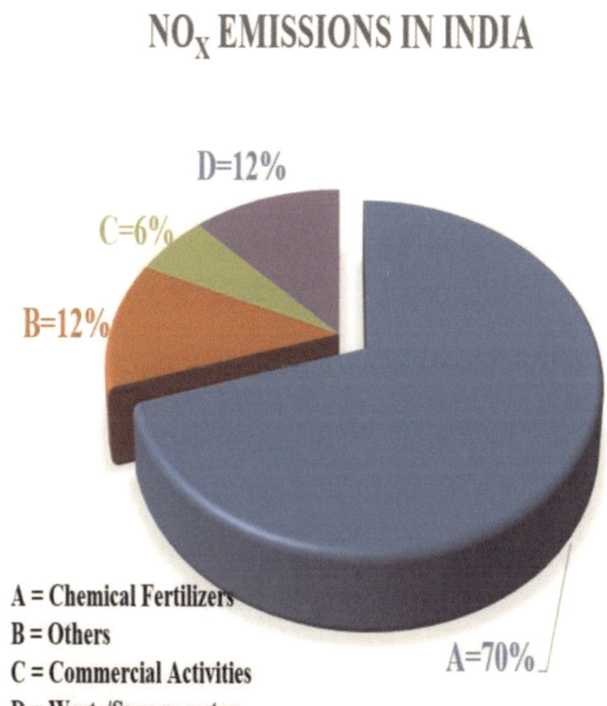

Figure 3.
NO_X emissions in India from various sources. Most of NO_X emissions in India come from the use of chemical fertilizers in agriculture.

for NO_2 detection. We have measured NO_2 sensing with ZnO which is an n-type semiconductor. It shows optimum response of 26% to 5 ppm NO_2 at 150°C when the test temperature range is 100–200°C. Below 150°C, the response to NO_2 is almost negligible. However, with the addition of graphene to ZnO layers, the sensing response significantly increases to 894% [14]. These results are presented in **Figure 4**.

The response of ZnO-G multilayer thin film was also measured for 500 ppm H_2, CO, and i-butane gases in the said temperature range. It was found that the response to 5 ppm NO_2 was far superior to any of the other gases. We have defined the selectivity (κ) as

$$\kappa = \frac{S_{NO2}}{S_{gas}} \tag{4}$$

where S_{NO2} is the response % for NO_2 and S_{gas} is the response % for other gas (H_2, CO, and i-butane).

The selectivity plot for NO_2 with respect to H_2, CO, and i-butane has been shown in **Figure 5**. For all the gases, the selectivity factor κ has been found to be greater than 1. Thus we could detect NO_2 selectively in the presence of other gases.

Sensor material	Form	Gas	T_{opt} (°C)	S%	Selectivity	τ_{res}, τ_{rec} (s)	Conc. (ppm)	Ref.
SnO_2-RGO	Nano powder	NO_2	30	4.63 (R_g/R_a)	Yes	177, 510	5	[32]
SnO_2	Hollow sphere	NO_2	160	—	Yes	90, 25	5	[27]
0.5 wt%G-WO_3	Nanopowder	NO_2	RT, 250	1.96, 130 (R_g/R_a)	No	25–200	1, 5	[31]
ZnO	Nano powder	NO_2	350	1.8 (R_g/R_a)	Yes	180, —	1	[29]
WO3	Thin film	NO_2	300	70	No	—	10	[28]
Fe_2O_3-ZnO	Thin film	NO_2	150	—	No	—	2.5	[30]
WO_3	Thin film	NO_2	300	87	Yes	300, 360	2	[33]
ZnO-G	Thin film	NO_2	150	894	Yes	150, 315	5	[14]

Table 2.
Comparison of the NO_2 sensing characteristics of various metal oxides.

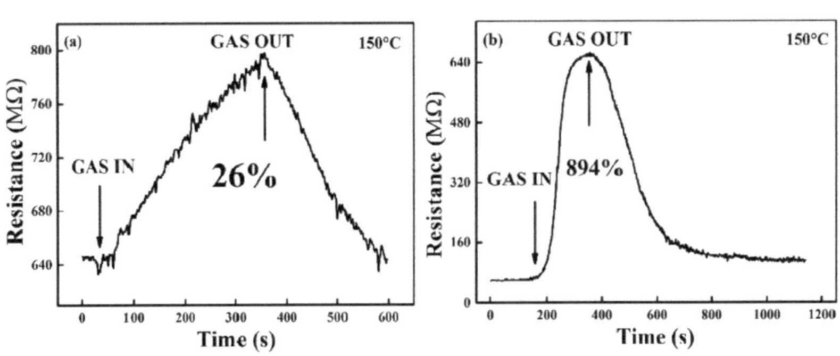

Figure 4.
Resistance transients of (a) ZnO and (b) ZnO-G multilayer at 150°C. an enormous increase in response is observed on addition of graphene (Reprinted from [14] with the permission of AIP publishing).

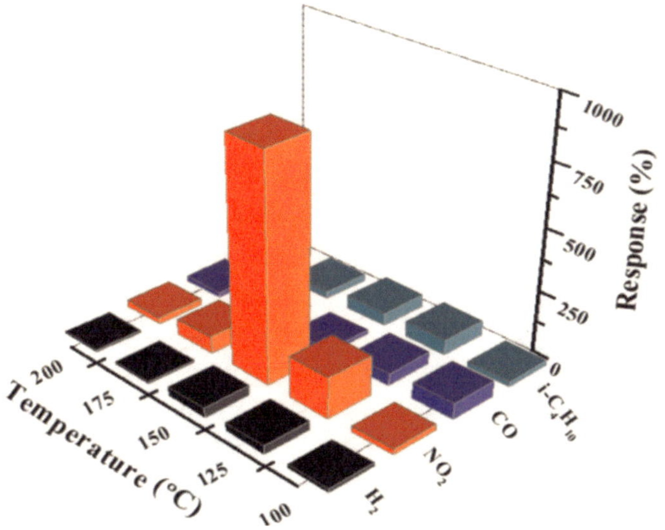

Figure 5.
Response % for 500 ppm i-butane, 500 ppm CO, 5 ppm NO_2, and 500 ppm H_2 gases. Figure clearly shows the selective detection of NO_2 at 150°C (Reprinted from [14] with the permission of AIP publishing).

4.1.1 NO_2 sensing mechanism

The significant enhancement in NO_2 sensing may be attributed to the formation of p-n junction between ZnO and graphene. In ZnO-G composite, graphene adheres to ZnO surface to increase the interaction with NO_2. Due to the formation of p-n junction, electron flows from graphene to ZnO resulting in an electron-depleted surface. On exposure to air, majority of $\boldsymbol{O_2^-}$ (dominant species at 150°C) species adsorb on to the ZnO-G surface, leading to the formation of electron-depleted layers. NO_2 reacts with this surface to form NO which is explained via the following equations:

$$NO_2(gas) + e^- \leftrightarrow NO_2^-(ads) \tag{5}$$

$$NO_2^-(ads) + O_2^- \leftrightarrow 2NO\ (gas)\uparrow + 2O_2^- \tag{6}$$

The interaction of reducing gases (H_2, CO and i-butane) with O_2^- is poor at 150° C, thus leading to selective detection of NO_2. A detailed figure of the sensing mechanism may be found in [14].

4.1.2 Mixed gas sensing

We felt it important to sense NO_2 selectively in a mixed gas environment. Thus, the sensing performance of NO_2 was investigated in the presence of 5 ppm NO_2 and 50 ppm CO, 5 ppm NO_2 and 50 ppm i-butane, and 5 ppm NO_2 and 50 ppm H_2. The response % histograms are presented in **Figure 6**. Principal component analyses (PCA) were used to address the cross-sensitivity. PCA involves the fast Fourier transforms (FFT) of the resistance transients of each gas (accomplished using MATLAB®, Mathworks, USA) which are fed into this unsupervised pattern recognition algorithm using STATISTICA-9, StatSoft, USA.

The plots between PC #1 and PC#2 are plotted separately in **Figure 7**. As shown in the inset of **Figure 7a**, distinct cluster forms among individual test gases (H_2, CO, and iC_4H_{10}). Also for NO_2-CO, NO_2-i-butane, and NO_2-H_2 mixtures, distinct

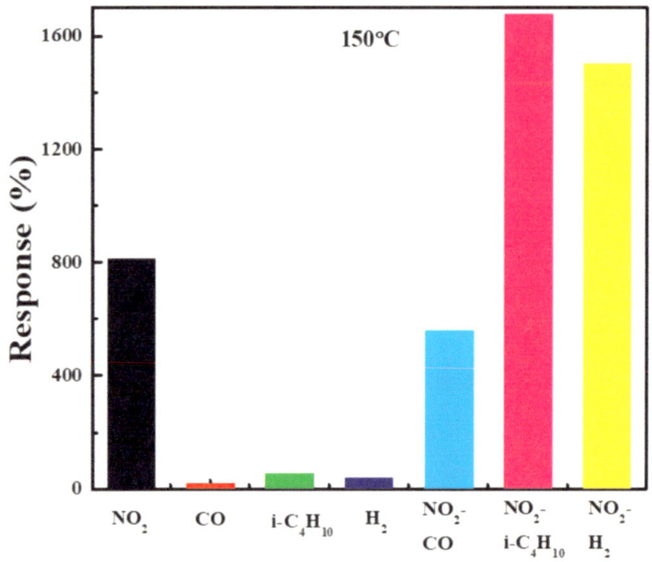

Figure 6.
Gas sensing responses of individual NO_2, CO, i-C_4H_{10}, and H_2 gases and mixed gas sensing for NO_2/CO, NO_2/i-butane, and NO_2/H_2 (Reprinted from [14] with the permission of AIP publishing).

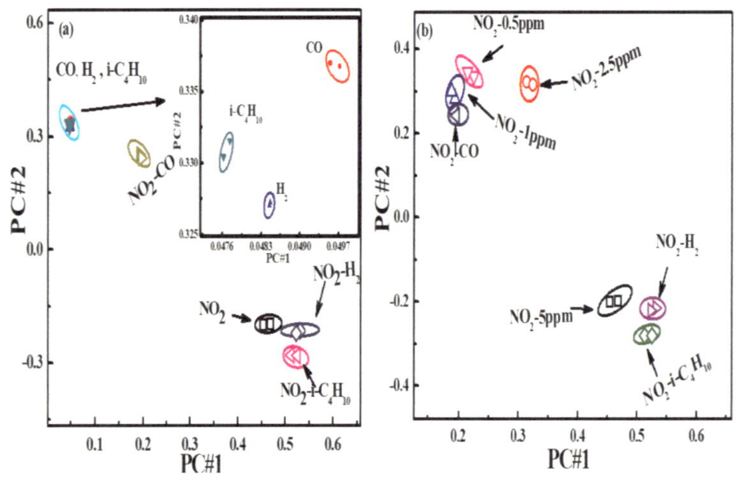

Figure 7.
Principal component analysis of (a) NO_2, CO, H_2, i-C_4H_{10}, NO_2-CO, NO_2-H_2, and NO_2-i-C_4H_{10} gases and (b) mixed NO_2/CO, NO_2-H_2, and NO_2-i-C_4H_{10} gases and NO_2 gas varying NO_2 contents in the range 0.5–5 ppm measured at 150°C (Reprinted from [14] with the permission of AIP publishing).

clustering is readily identified in **Figure 7a**. PCA is also performed in mixed gases varying NO_2 contents in the range of 0.5–5 ppm (**Figure 7b**). Note that distinct clustering is also achieved in mixed gases with variable NO_2 contents, which eventually leads to successful differentiation between them.

4.2 CO and H_2 sensing characteristics of thin and thick films

Carbon monoxide is a major air pollutant. Open burning of wastes as well as fuel and biomass burning mainly in rural India is the main source of carbon monoxide. Additionally, fuel adulteration, especially in auto-rickshaws, also adds to the CO in air. On breathing CO-rich polluted air, one might end up inhaling CO as opposed to oxygen. This CO then easily combines with the hemoglobin in blood forming carboxyhemoglobin, which reduces the oxygen-carrying capacity of blood. Additionally, tobacco smoke and heating appliances add to CO pollution of indoor air.

Hydrogen (H_2) is a colorless, odorless, and tasteless gas. H_2 is also highly flammable with a low ignition energy (0.017 mJ) and low ignition temperature (~560° C). Hydrogen detection is critical to maintaining the safety of nuclear reactors where the reaction of water with high temperature core or cladding gives rise to evolution of hydrogen in radioactive waste tanks [34]. A small amount of H_2 is also produced as a result of methane explosions in coal mines [35]. H_2 is used as a liquid propellant for rockets as it is a zero emission fuel. It is also used in the reduction of metallic ores and for methanol production. H_2 has a lower explosive limit (LEL) of 4 vol % which makes its detection fairly crucial.

Semiconducting metal oxides such as SnO_2, ZnO, WO_3, etc. have demonstrated huge research potential when it comes to sensing of reducing gases such as hydrogen, carbon monoxide, hydrogen sulfide, ammonia, and VOCs. However, such materials exhibit poor selectivity, and efforts are on to achieve selectivity either by doping metal oxides or synthesis of composites. A brief review of literature (**Table 3**) suggests that selectivity has not been addressed for CO and H_2 sensing of metal oxides. We have synthesized ZnO thin film via spin coating process and SnO_2 thick film via plasma spray deposition process and studied their gas sensing characteristics with respect to H_2 and CO gases. We have analyzed the issue of selectivity through conductance transient analyses. The analyses have been performed for both thin and thick films for comparison. Additionally, we have discussed how stability varies in thin and thick film gas sensors when the test gases under consideration are the same.

ZnO thin film and SnO_2 thick film were investigated for H_2 and CO sensing characteristics in the temperature range 350–200°C. The concentration of each gas was fixed at 500 ppm. A maximum response of 68% for 500 ppm H_2 and 49% for 500 ppm CO was recorded at 300°C by ZnO thin film. SnO_2 thick film recorded maximum response to 500 ppm H_2 (81%) and 500 ppm CO (73%) at 250°C. Careful inspection of **Figure 8** circumstantiates cross-sensitivity of H_2 and CO gases at all temperatures for both ZnO thin film and SnO_2 thick film sensors. This necessitates

Sensor material	Geometry	Gas	T_{opt} (°C)	S%	Selectivity (w.r.t H_2/CO)	τ_{res}, τ_{rec} (s)	Conc. (ppm)	Ref.
Sm_2O_3	Thick film	CO	250	1.4 (R_g/R_a)	No	35, 110	5	[36]
CuO	Nano-tube	CO	175	3.27 (R_g/R_a)	No	29, 37	1000	[37]
In_2O_3	Thin film	CO	350	45, 85 (R_g/R_a)	No	10– 50	1000	[38]
Pd-ZnO	Thin film	CO	150	—	No	17, 23	500	[39]
ZnO	Thin film	H_2	300	21	No	58, 70	200	[40]
In-ZnO	Thin film	H_2	300	65	No	35, 64	500	[41]
SnO_2	Thin film	H_2	350	83	No	23, 139	265	[42]
ZnO	Thin film	H_2	400	23	Yes	110	1200	[43]
ZnO	Thin film	H_2	300	68	Yes	53, 67	500	[9]
SnO_2	Thick film	H_2	250	80	Yes	30, 118	500	[7]

Table 3.
Comparison of the CO and H_2 sensing characteristics of various semiconducting metal oxides.

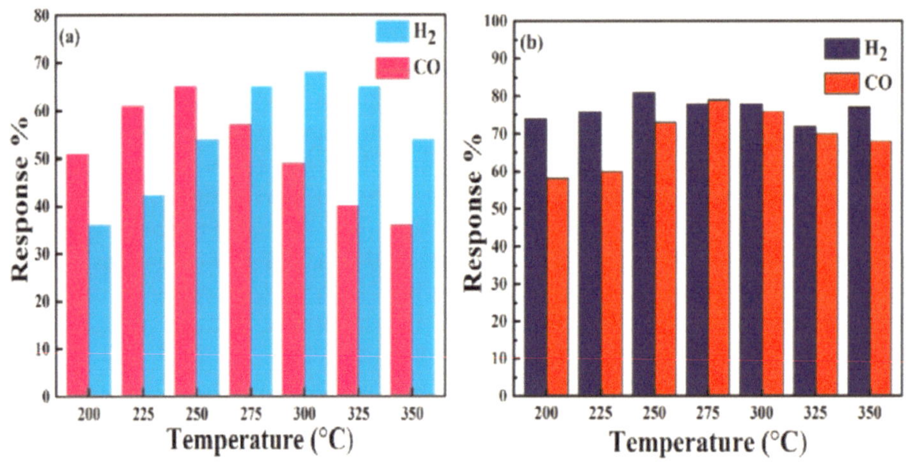

Figure 8.
Response % of (a) ZnO thin film and (b) SnO_2 thick film in response to 500 ppm H_2 and 500 ppm CO.

further analysis of the sensing data obtained. Analyses of conductance transients are a well-proven coherent technique to address the problem of cross-sensitivity. Although cross-sensitivity is shown by the sensors at all temperatures, we have performed mathematical analyses only for the optimum temperature of each kind of film. For H_2 and CO sensing, the recorded conductance transients of both ZnO (thin film) and SnO_2 (thick film) sensors were modeled using Langmuir adsorption kinetics [5].

4.2.1 Analyses of conductance transients

In a preceding section, we have already outlined the sensing mechanism of n-type semiconducting sensor toward reducing gas sensing. Thus, ambient oxygen physiadsorbs on the sensor surface at ambient temperature (Eq. (7)). At elevated temperatures, the physiadsorbed oxygens are chemiadsorbed (Eq. (8)) by extracting electrons from semiconductor oxide grains. This creates an electron depletion layer (EDL) at the sensor surface [5, 10, 44, 45].

$$\frac{1}{2}O_2 + sensor_{surface} \leftrightarrow O_{ad-surface}(physi - adsorption) \tag{7}$$

$$O_{ad-surface} + e \leftrightarrow O^{-}_{ads}\ (chemi - adsorption) \tag{8}$$

On exposing the sensor to a reducing gas, the chemiadsorbed oxygen reacts with the test gas (R) to yield oxidized reaction products. The reaction schemes and desorption of reaction product (RO_{gas}) is described in Eqs. (9)–(11).

$$R + [sensor] \leftrightarrow R_{ad}\ (physi - adsorption) \tag{9}$$

$$R_{ad} + O_{ad^-} \leftrightarrow RO_{ad} + e \tag{10}$$

$$RO_{ad} \leftrightarrow RO_{gas} \uparrow + sensor_{surface} \tag{11}$$

Among the cited relations, adsorption of oxygen ions and desorption of the reaction product (Eqs. (10) and (11)) are the rate-determining steps for response and recovery, respectively.

As discussed earlier, when the sensor is exposed to air, oxygen chemisorbs onto the sensor surface which results in the formation of an electron depletion layer (EDL). The EDL formed impedes the transfer of charge carriers from one grain to another. The relation between the height of this potential barrier and the conductance G of the sensor is given by the following Eq. (5):

$$G = G_0 \exp\left(-\frac{eV_s}{kT}\right) \tag{12}$$

where eV_s is the Schottky barrier, k is the Boltzmann constant, and T is the absolute temperature of operation of the sensor.

As the sensor is exposed to a reducing gas like CO or H_2, the chemisorbed oxygen ions react with the gas to reduce the potential barrier between the grains, thus allowing the flow of electrons [46]. This process increases the conductance of the sensor.

The adsorption process of gas molecules on a solid surface was studied by Langmuir who developed an empirical model based on the surface coverage θ. The Langmuir model assumes that (i) the surface on which the gas adsorbs is homogeneous, (ii) a specific number of sites are available for gas adsorption, and once all these sites are occupied, further adsorption is not possible. This is called monolayer adsorption, and (iii) all sites on which gas is adsorbed are equivalent in terms of heat of adsorption. Assuming the above, Langmuir formulated the equation below [5, 15].

$$\theta(t) = \frac{N(t)}{N^*} = A\{1 - \exp\left(-\frac{t}{t_A}\right)\} \tag{13}$$

where $N(t)$ is the number of gas molecules adsorbed at time t and N^* is the total adsorption sites available, and

$$A \equiv \frac{P}{P+b} \tag{14}$$

where P is the gas pressure and b is related to heat of adsorption Q and is defined by the following relation:

$$\mathrm{b} = \mathrm{b}_0 \exp\left(-\frac{\mathrm{Q}}{\mathrm{RT}}\right) \tag{15}$$

$$\mathrm{b}_0 = \frac{\upsilon}{\mathrm{K}_0.\mathrm{S}_{\mathrm{area}}.\mathrm{N}_\mathrm{A}} \times \sqrt{2\pi \mathrm{MRT}} \tag{16}$$

where b_0 is a constant, Q is the heat of adsorption ($Q = E_d - E_a > 0$), E_a and E_d are the adsorption and desorption activation energies of gas molecules, R is the gas constant, and T is the operating temperature, T_{opt}. The pre-exponential constant b_o depends on υ (frequency of oscillation of adsorbed molecules), K_o (fraction of those molecules with energy > E_A, adsorbed on surface), S_{area} (the surface area of a single adsorbed molecule), N_A (Avogadro number), M (molar mass of the gas), and R (the gas constant).

The amount of gas adsorption is dependent on the nature of the gas sensing surface, the operating temperature, and pressure of the gas. An adsorption isotherm is defined as the study of the amount of gas adsorption as a function of gas pressure. The simplest adsorption isotherm was given by Langmuir in which he assumes no interaction between adsorbed atoms or molecules. At equilibrium, the rate of adsorption is equal to the rate of desorption. Thus, when t $\rightarrow \infty$,

$$\theta = \frac{N(t)}{N^*} = A = \frac{P}{P + b} \tag{17}$$

Assuming a linear functional dependence between $\theta(t)$ and conductance transient ($G(t)$), the response behavior of the sensor that follows Langmuir model can be written as [5]

$$G(t)^{response} = G_0 + G_1\left[1 - exp\frac{-t}{\tau^{res}}\right] \tag{18}$$

where G_0 is the base conductance of the sensor at time t = 0, G_1 is the fitting parameter, and τ^{res} is the characteristic response time. This equation will only be valid if all three of Langmuir's criteria are satisfied.

G_1 in Eq. (18) is identical to A in Eq. (17); hence, at low gas concentration if we assume $P \approx$ gas concentration (C), we can then write

$$\frac{G_1}{1 - G_1} = \frac{C}{b} \tag{19}$$

If Langmuir adsorption isotherm is followed, then variation of $G_1/(1\text{-}G_1)$ vs. C should be a linear fit and pass through the origin. The parameter b can be estimated from the slope of the plot.

Using Eq. (18), conductance transients during response at various concentrations for SnO_2 sensing element for H_2 and CO sensing were fitted to estimate values of G_0, G_1, and τ^{res}. **Figure 9a** shows the fitting of conductance transient of SnO_2 thick film at 250°C in response to 500 ppm H_2. **Figure 9b** shows the variation of $G_1/(1\text{-}G_1)$ with test gas concentration (C) at 250°C. Since the linear fits for H_2 and CO pass through the origin, Langmuir adsorption isotherm is satisfied [5, 47, 48]. The parameter b can be obtained from the inverse of the slope of the plot. Accordingly, the estimated 'b' values for H_2 (1.5349×10^{10} Pa) and CO (1.7834×10^{11} Pa) have been determined from these plots [7]. From Eq. (15), expression for heat of adsorption (Q) can be written as

$$Q = RT[ln(b_0) - ln(b)] \tag{20}$$

The molecular area of nitrogen is approximately 16.2 Å^2 as given in [49]. We have considered this as the area per site occupied by the test gas molecule (S_{area}). Nitrogen has a molecular weight of 28 g mole^{-1} which gives $S_{area} \sim 16.2$ Å^2. A similar calculation for carbon monoxide with molecular weight *28.01* g mol^{-1} yields S_{area} to be $\sim$*16.2* Å^2. Assuming condensation coefficient K_o = *1* [47], $\upsilon = 10^{14}$ Hz [49], and R = 8.314 JK^{-1}; b_o (at T $\sim$ 523 K) can be estimated for H_2 and CO gases using (Eq. (16)).

From the estimated values of b and b_0, heat of adsorption (Q in J mol^{-1}) can be obtained for H_2 and CO sensing using Eq. (20) [7].

$$Q_{H_2} = 23.418 \times 10^3 \text{ J mol}^{-1} = 23.418 \text{ KJmol}^{-1} \tag{21}$$

$$Q_{CO} = 7.0138 \times 10^3 \text{ J mol}^{-1} = 7.0138 \text{ KJ mol}^{-1} \tag{22}$$

Figure 9c shows the plot of response time (τ^{res}) versus gas concentration (C) on a logarithmic scale. As shown in **Figure 9c**, variation of τ^{res} with C follows the following relation:

$$\tau^{res} = \tau_0 \times C^{-\beta} \tag{23}$$

The ln τ^{res} versus ln C can be fitted linearly to estimate the rate constant and β for H_2 and CO sensing where τ_0 is the fitting constant.

The temperature variation of τ^{res} can be expressed by the following relation (47):

$$\tau^{res} = \upsilon^{-1} \times \exp\left(\frac{E_a + Q}{RT}\right) \tag{24}$$

From Eqs. (23) and (24), the value of E_a can be written as

$$E_a = RT\,ln\left(\tau_0 \times C^{-\beta} \times \upsilon\right) - Q \tag{25}$$

The variation of E_a with concentration for H_2 and CO gases is shown in **Figure 9d** [7]. The activation energy decreases from 134.3 kJ/mol to 131.8 kJ/mol for 500 ppm H_2 to 50 ppm H_2, whereas the activation energy decreases from 151.05 kJ/mol to 150.07 kJ/mol and from 500 ppm CO to 50 ppm CO for *SnO_2 thick film* sensor.

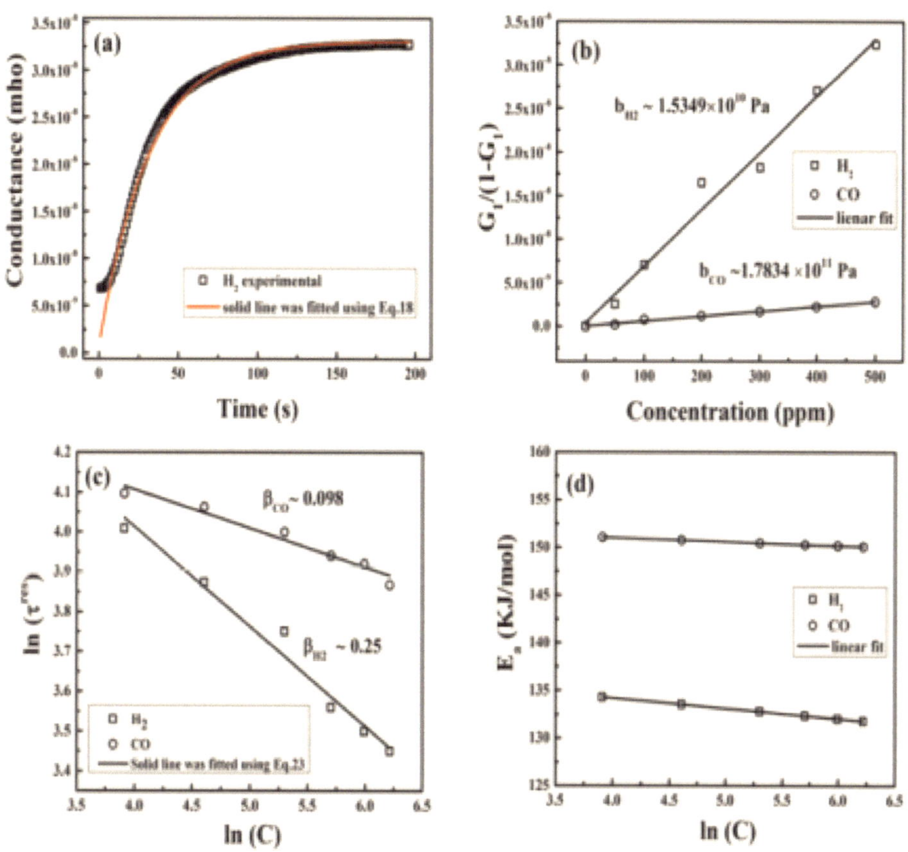

Figure 9.
*(a) Conductance transient plot for 500 ppm H_2 sensing. Experimental points are shown by symbols, and the solid line was fitted using Eq. (18), (b) linear variation of G1/(1-G1) with concentration, (c) log–log plot of τ^{res} with C. Experimental points are shown by symbols, and the solid line was fitted using Eq. (23), (d) variation of E_a with ln C for H_2 and CO sensing. For all these measurements, operating temperature was kept fixed at 250°C (**Figure 9a–d**, reprinted from [7] with permission from Elsevier).*

Similar analyses for the *ZnO thin film sensor* yield activation energy values of 138.62 and 151.87 kJ/mol for 500 ppm H_2 and CO, respectively. Activation energy increases with decreasing concentration for both the gases. The Q values for H_2 and CO are 31.48 and 18.48 kJ/mol, respectively. Thus, the activation energy, E_a, and heat of adsorption, Q, may be used to differentiate H_2 and CO gases for both thin and thick films.

4.2.2 Stability of thin and thick film gas sensors

Figure 10 shows the schematic of Langmuir-Hinshelwood mechanism [8] in which the oxygen as well as reducing test gas is initially adsorbed on the sensor surface shown by "A." An oxidized reaction product is produced as a result of collision among adsorbed oxygen and reducing test species. For efficient formation of an oxidized reaction product "RO_{gas}," reducing test gas "R_{gas}" (having concentration "Cg") and oxygen "O" should cross the energy barrier height shown by "TS" (transition state).

During this investigation, the sensing element was exposed to the test gas (during response) and air (during recovery), respectively. After removal of the test gas, it is expected that the sensor should switch back to the initial base resistance in air. Considering this fact, gas sensing can be classified into reversible and irreversible type. In reversible sensing, the base resistance is fully recovered, whereas for irreversible sensing, the base resistance recovers partially. The mechanism of reversible and irreversible sensing was reported elsewhere [8]. The irreversible sensing can be better understood from the gas–solid interaction. During gas sensor response process, CO and H_2 gases are oxidized to CO_2 and H_2O. These oxidized products are entrapped in the small pores of the sensing element. If the desorption of these products does not occur fully, the base resistance of the sensing element recovers partially. Therefore, a marginal drift in the baseline resistance is observed. The mathematical formulations of irreversible and reversible sensing are described in the subsequent sections.

4.2.2.1 Irreversible sensing model

It is assumed that the site fraction θ on the sensor surface is occupied by adsorbed oxygen ions O_{ad}^- and test gas (R_{gas}) adsorption takes place over remaining available sites. For an "n"-type metal oxide semiconductor gas sensor, reducing gas reacts with chemiadsorbed oxygen to form an oxidized product as represented by Eq. (26).

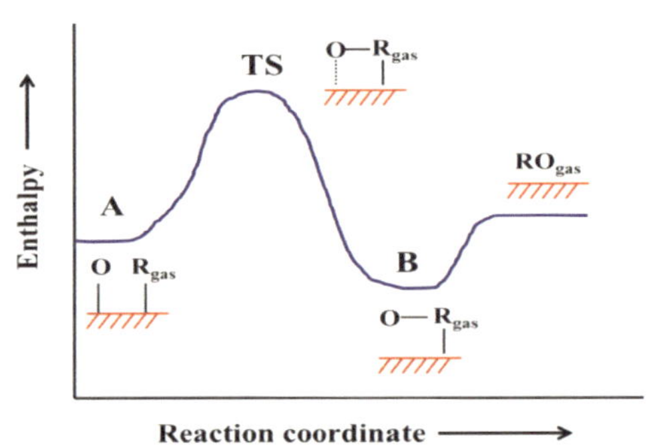

Figure 10.
Schematic representation of Langmuir-Hinshelwood mechanism.

$$R_{gas} + O_{ad^-} \leftrightarrow RO_{ad} \tag{26}$$

At constant operating temperature, it was assumed that the total number of active sites (Fθ) can be expressed by the following relation:

$$\theta\left(\sim \left[O^{ad-}\right]\right) + RO_{ad}\left(occupied\ by\ R_{gas}\right) = F\theta \tag{27}$$

It has been assumed that adsorption of the reaction product RO_{ad} takes place on the same sites over which reducing test gas (R_{gas}) adsorbs initially. Since the formation of RO_{ad} is the rate-determining step for response process, sensor response is directly proportional to the formation of an oxidized reaction product RO_{ad}. Therefore, the sensor response during irreversible sensing can be expressed mathematically by the following relation:

$$\frac{d[RO_{ad}]}{dt} = k[O_{ad}{}^{-}]C_g \tag{28}$$

From Eq. (27), Eq. (28) can be written as

$$\frac{d[RO_{ad}]}{dt} = k[F\theta - RO_{ad}] * C_g \tag{29}$$

Integrating Eq. (29), we obtain

$$\int \frac{d[RO_{ad}]}{dt} = \int k[F\theta - RO_{ad}] * C_g \tag{30}$$

$$\int \frac{d[RO_{ad}]}{[F\theta - RO_{ad}]} = \int k * C_g * dt \tag{31}$$

Simplifying Eq. (31), we can write

$$-ln[F\theta - RO_{ad}] = k * C_g * t + C \tag{32}$$

where C is the constant of integration.

$$At\ t \rightarrow 0, RO_{ad} \rightarrow 0\ and\ C = -\ln F\theta$$

Therefore Eq. (32) can be written as

$$-ln[F\theta - RO_{ad}] = k * C_g * t - \ln F\theta \tag{33}$$

Simplifying Eq. (33), we can write

$$ln\left[1 - \frac{RO_{ad}}{F\theta}\right]A = -k * C_g * t \tag{34}$$

Concentration of RO_{ad} at time t can be expressed as

$$[RO_{ad}](t) = F\theta\left[1 - exp\left(-kC_g t\right)\right] \tag{35}$$

When all active sites (Fθ) are adsorbed by the oxidized reaction product (RO_{ad}), maximum sensor response is obtained. Therefore, the sensor response at a time t can be expressed by the following relation:

$$S(t) = S_{max}\left[1 - exp\left(1 - kC_g t\right)\right] \tag{36}$$

$$S(t) = S_{max}\left[1 - exp(-(t/\tau_{irrev}))\right] \tag{37}$$

where $\tau_{irrev} = \frac{1}{kC_g}$ is termed as the characteristic response time (in seconds) during irreversible sensing.

4.2.2.2 Reversible sensing model

In this case, the equilibrium constant for the sensing reaction (Eq. (38)) is termed as K, and the forward and reverse rate constants are k and k/K. The rate of change of RO_{ad} at time t can be written as

$$\frac{d[RO_{ad}]}{dt} = k\left[O_{ad}{}^{-}\right]C_g - k/K[RO_{ad}] \tag{38}$$

Rearranging Eq. (38),

$$\frac{d[RO_{ad}]}{k[F\theta]C_g - k[RO_{ad}]\left(C_g + \frac{1}{K}\right)} = dt \tag{39}$$

Simplifying Eq. (39), we can write

$$\frac{d[RO_{ad}]}{\left\{\frac{[F\theta]C_g}{C_g + \frac{1}{K}} - [RO_{ad}]\right\}} = k\left(C_g + \frac{1}{K}\right)dt \tag{40}$$

Integrating Eq. (40), one can write

$$\int \frac{d[RO_{ad}]}{\left\{\frac{[F\theta]C_g}{C_g + \frac{1}{K}} - [RO_{ad}]\right\}} = k\left(C_g + \frac{1}{K}\right)\int dt \tag{41}$$

Simplifying Eq. (41), one can write

$$-ln\left\{\frac{(F\theta)C_g}{\left(C_g + \frac{1}{K}\right)}\right\} - [RO_{ad}] = k\left(C_g + \frac{1}{K}\right)t + C \tag{42}$$

$$When\; t \to 0, RO_{ad} \to 0, and\; C = -ln\left\{\frac{(F\theta)C_g}{\left(C_g + \frac{1}{K}\right)}\right\} \tag{43}$$

Therefore, Eq. (43) can be simplified as

$$-ln\left\{\frac{(F\theta)C_g}{\left(C_g + \frac{1}{K}\right)}\right\} - [RO_{ad}] = k\left(C_g + \frac{1}{K}\right)t - ln\left\{\frac{(F\theta)C_g}{\left(C_g + \frac{1}{K}\right)}\right\} \tag{44}$$

Eq. (44) can be simplified to

$$ln\left[\frac{\frac{(F\theta)C_g}{\left(C_g + \frac{1}{K}\right)} - RO_{ad}}{\frac{(F\theta)C_g}{\left(C_g + \frac{1}{K}\right)}}\right]A = -k\left(C_g + \frac{1}{K}\right)t \tag{45}$$

Eq. (45) can be further simplified to

$$1-\frac{[RO_{ad}]}{\frac{(F\theta)C_g}{\left(C_g+\frac{1}{K}\right)}}=exp\left(-k\left(C_g+\frac{1}{K}\right)t\right) \quad (46)$$

The rate of change of RO_{ad} at time t can be written as

$$[RO_{ad}](t)=\left\{\frac{(F\theta)C_g}{\left(C_g+\frac{1}{K}\right)}\right\}\left\{1-exp\left(-k\left(C_g+\frac{1}{K}\right)t\right)\right\} \quad (47)$$

$$[RO_{ad}](t)=\left\{\frac{(F\theta)C_gK}{(1+C_gK)}\right\}\{1-exp-\{(1+C_gK)/K\}kt\} \quad (48)$$

Sensor response (S(t)) can be written as

$$S(t)=S_{max}\left\{\frac{C_gK}{(1+C_gK)}\right\}\{1-exp-\{(1+C_gK)/K\}kt\} \quad (49)$$

Assuming $1+C_gK\approx C_gK$, *one can write*

$$S(t)=S_{max}[1-exp(-t/\tau_{rev})] \quad (50)$$

where τ_{rev} is the characteristic response time during reversible sensing.

Figure 11a, b shows the concentration-dependent resistance transients for H_2 and CO sensing for ZnO thin film. These measurements were taken at 300°C by varying the test gas concentration. During these measurements, drift in baseline resistance was observed as shown by the dashed line in **Figure 11a, b**. Therefore, time-dependent response % for H_2 and CO sensing was fitted using irreversible sensing model as given in Eq. (37). **Figure 11c** shows the typical response % vs. time fitting for 500 ppm H_2 and CO sensing. Inset of **Figure 11c** represents variation of τ_{irrev} (in seconds) with concentration (in ppm) in which τ_{irrev} decreases systematically with increase in test gas concentration which is a typical behavior in case of irreversible gas sensing as reported by [8].

We have also investigated the reversible sensing behavior of plasma sprayed SnO_2 thick film sensor [7]. The SnO_2 thick film sensor shows marginal drift in baseline resistance when exposed to 20–500 ppm CO and H_2. This is shown in **Figure 12**.

Inset of **Figure 12c** shows the variation of characteristic response time τ_{rev} (s) with concentration (ppm). τ_{rev} was almost invariant of the test gas concentration in the range 500–200 ppm. Interestingly, τ_{rev} was also invariant of test gas concentration in the range 100–20 ppm. However, the value of τ_{rev} was found to be higher in the lower concentration range (100–20 ppm) than that of the higher concentration range (500–200 ppm).

Figure 13 shows the surface micrograph for (a) ZnO thin film and (b) SnO_2 thick film sensing elements. ZnO thin film exhibits denser surface morphology than SnO_2 thick film as revealed from **Figure 13a**. In line with this discussion, irreversible sensing behavior for ZnO thin film can be attributed to denser surface morphology. It may be noted here that our results for ZnO thin film are in accordance with [8]. On the other hand, reversible sensing behavior in case of SnO_2 thick film can be attributed to its porous surface morphology as revealed from **Figure 13b**.

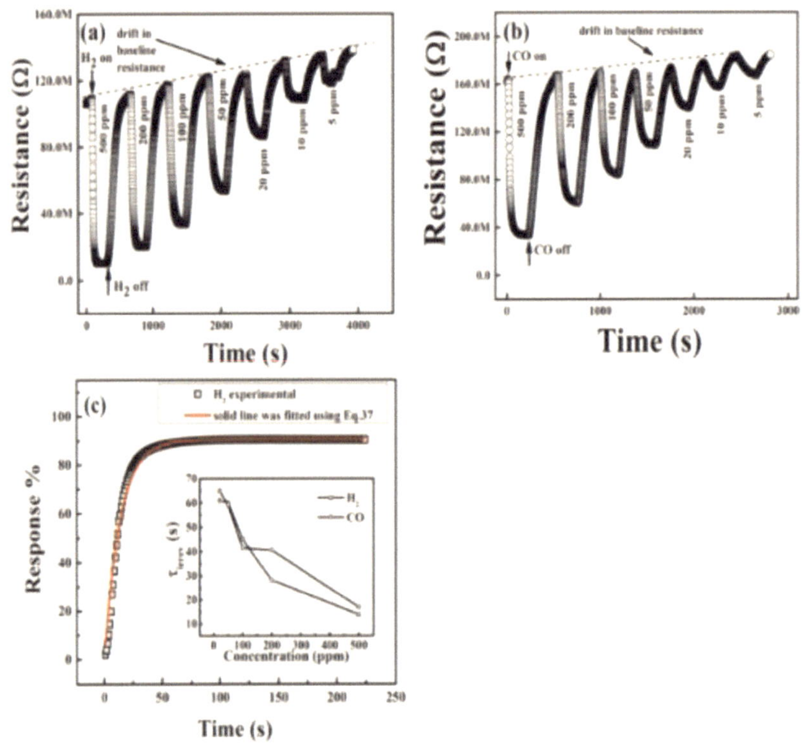

Figure 11.
Concentration-dependent resistance transients recorded at 300°C for (a) H_2, (b) CO, (c) time-dependent response % fitting for 500 ppm H_2 sensing for ZnO thin film. Experimental points (symbols) and the solid line were fitted using Eq. (37). Inset shows the variation of τ_{irrev} with concentration.

4.3 CO_2 sensing characteristics

Carbon dioxide concentrations have increased enormously in the atmosphere in the past hundred years or so. In the preindustrial era, CO_2 concentration was ~280 ppm. However, current global standards put CO_2 concentrations at an alarming level of 400 ppm mainly arising from deforestation and the burning of fossil fuels. India alone emitted 2299 million tons of CO_2 in 2018 according to an International Energy Agency (IEA) report [50]. Most of India's emissions come from the burning of coal. Increased CO_2 emissions have led to global warming—a climate phenomenon that is responsible for the increasing number of storms as well as droughts [51]. In addition, increased levels of CO_2 in the atmosphere affect human health severely. **Table 4** summarizes the effects on human health with increasing CO_2 concentrations. It is thus incumbent that we take necessary steps to detect CO_2 concentrations in the atmosphere.

Research on CO_2 sensing gained prominence in the late 1980s when researchers such as Toshio Maruyama [52] and Yasuhiro Shimizu [53] reported a NASICON ($Na_3Zr_2Si_2PO_{12}$)-based solid electrolyte gas sensor and an electrochemical K_2CO_3-polyethylene glycol CO_2 sensor, respectively. However, it was soon realized that handling these materials was cumbersome and incorporating these into actual devices posed a serious challenge. Research was thus shifted to resistive-type sensors [54, 55] which were far better in terms of stability and could be easily miniaturized to accommodate in sensing devices. However, such sensors were plagued with poor selectivity and low response to CO_2. This prompted the use of mixed

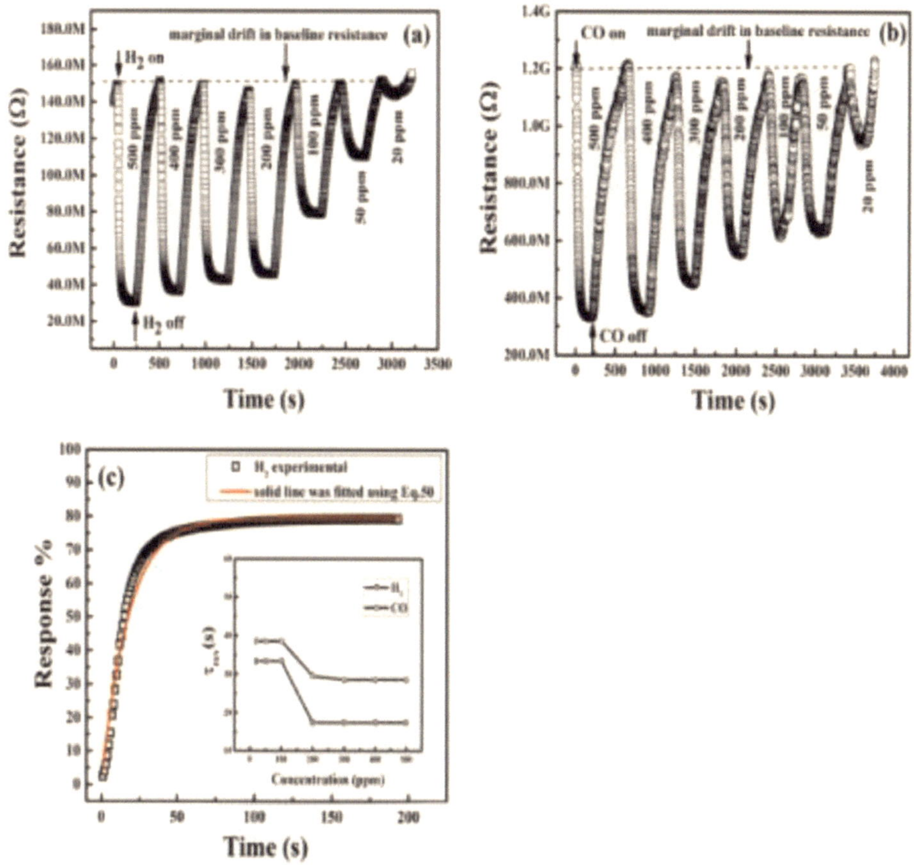

Figure 12.
*Concentration-dependent resistance transients recorded at 250°C for (a) H2, (b) CO, and (c) time-dependent response % fitting for 500 ppm H2 sensing. Experimental points (symbols) and the solid line were fitted using Eq. (50). Inset shows the variation of τ_{rev} with concentration (***Figure 12a–c** *has been reprinted from [7] with permission from Elsevier).*

oxide or composite sensors [56–58], the research potential of which is still being investigated [59, 60].

A comprehensive study of literature (elaborated in **Table 5**) related to CO_2 sensing leads us to the fact that formation of p-n junctions is an important criterion for CO_2 detection. We have reported CO_2 sensing of ~36% achieved using a LFCO-ZnO composite thin film.

CO_2 sensing characteristics of LFCO-ZnO composite thin film was measured in the temperature range 350–150°C. The highest response of ~36% was obtained at 300°C to 2500 ppm CO_2, and the response decreases to 12% at 150°C. The response and recovery times recorded at 300°C are 300 and 630 s, respectively. The composite shows n-type sensing behavior which confirms the dominance of ZnO crystallites [67]. The LFCO-ZnO thin film composite is cross-sensitive to CO_2. A response of 65% was recorded for 500 ppm CO at 300°C. The composite shows higher response to both CO and CO_2 than pure LFCO and ZnO thin films. To address the cross-sensitivity of the composite thin film, principal component analysis of the resistance transients was performed. The relevant features of the resistance transients of the test gases are extracted using fast Fourier transform technique. These features are then used as input parameters for PCA.

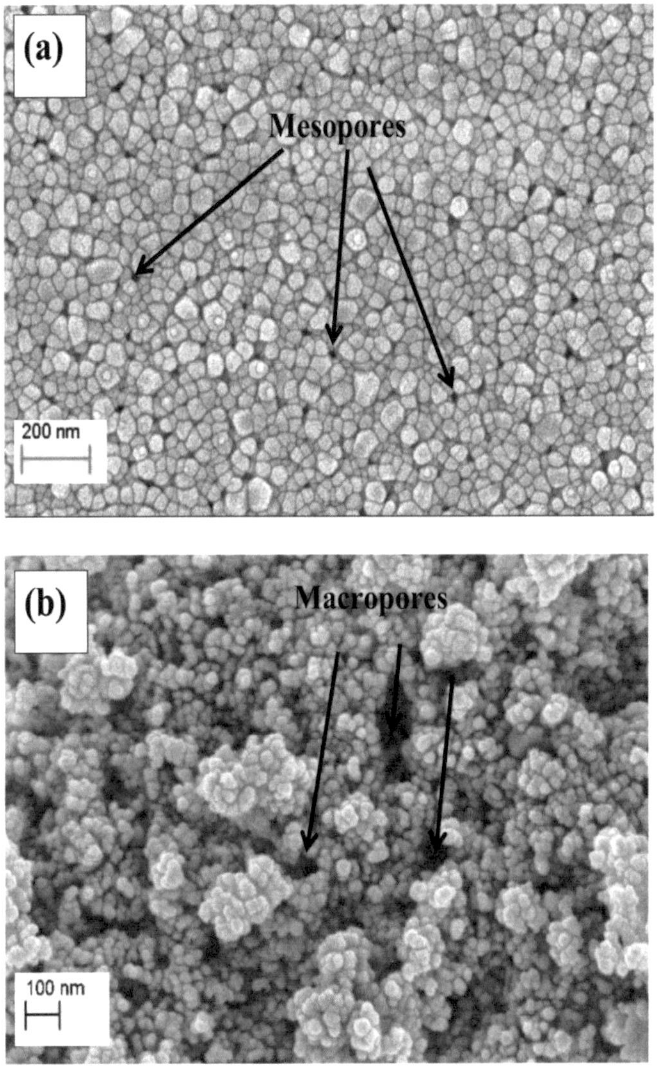

Figure 13.
Surface micrograph: (a) ZnO thin film and (b) SnO_2 thick film.

Concentration of CO_2 (ppm)	Effect on human health
250–300	Normal outdoor air
350–1000	Indoor air
100–2000	Drowsiness (poor air)
2000–5000	Increased heart rate, nausea, headaches
>5000	Oxygen deprivation begins
>40,000	Serious oxygen deprivation leading to brain damage, coma, and death

Table 4.
Effect on human health with increasing concentrations of CO_2.

Figure 14d shows low dispersion of CO and CO_2 for the LFCO-ZnO composite thin film after performing PCA. Thus, FFT-PCA is an efficient tool for the differentiation of CO and CO_2 gases for LFCO-ZnO thin film.

Sensor material	Form	Gas	T_{opt} (°C)	S%	Selectivity	τ_{res}, τ_{rec} (s)	Conc. (ppm)	Ref.
ZHS microcubes on ZnO nanorods	Thin film	CO_2	150	354	No	10, 24	2500	[61]
$LaBaCo_2O_{5+\delta}$	Bulk	CO_2	300	1.04 (R_g/R_a)	No	—	400	[62]
Graphene sheet	Thin film	CO_2	60	26	No	8, 10	100	[63]
$BaCO_3$-Co_3O_4	Bulk	CO_2	150	15	No	227, 245	10,000	[59]
Ag/$BaTiO_3$-CuO	Thin film	CO_2	250	28	No	900, 600	5000	[64]
Ca/Al-doped ZnO	Bulk	CO_2	300	15	No	—	2500	[65]
$LaFeO_3$-SnO_2	Bulk	CO_2	250	3 (R_g/R_a)	No	<20	4000	[60]
Ca-ZnO	Thin film	CO_2	350	53	Yes	—	50,000	[66]
LFCO-ZnO	Thin film	CO_2	300	36	Yes	300, 630	2500	This work

Table 5.
Overview of the carbon dioxide sensing characteristics of various thin and bulk materials from literature.

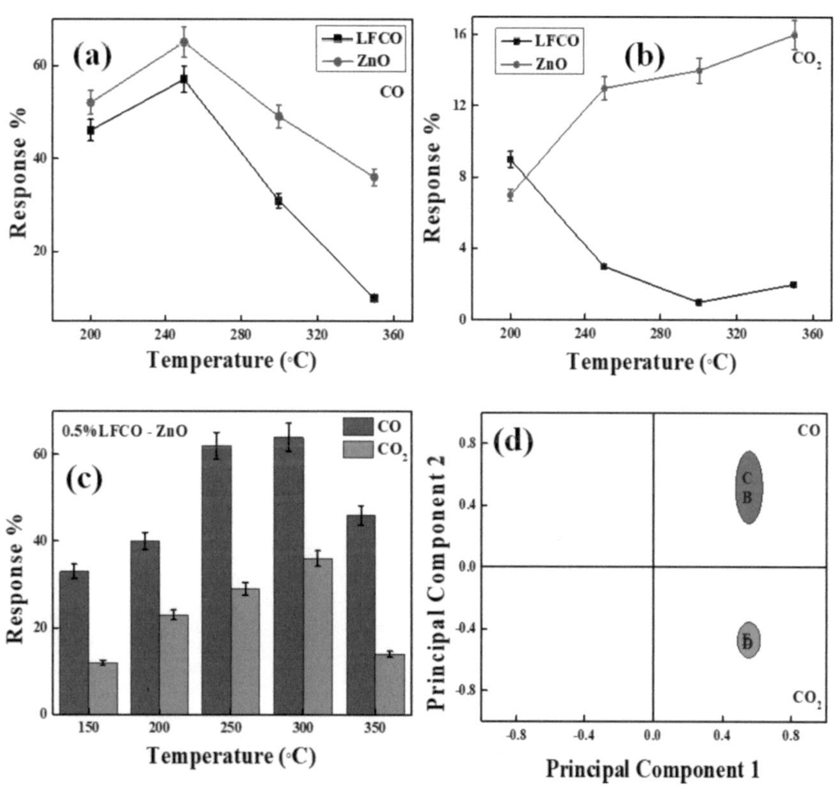

Figure 14.
Temperature variation of response % for pure LFCO and pure ZnO to (a) 500 ppm CO and (b) 2500 ppm CO_2. (c) Temperature variation of response % of 0.5% LFCO-ZnO for 500 ppm CO and 2500 ppm CO_2. (d) Principal component analyses for CO and CO_2 at 300°C.

The LFCO-ZnO composite follows the gas sensing behavior of an n-type semiconductor due to the presence of a large number of ZnO crystallites. Therefore, analyses of the conductance transients of CO and CO_2 gases follow Langmuir adsorption mechanism similar to ZnO thin film sensor. Activation energy, E_a, and heat of adsorption, Q, have been calculated in accordance with the theory described in the previous section. **Figure 15d** represents the variation of E_a with concentration for CO and CO_2 gases. It should be noted that each gas possesses distinct activation energy values. The activation energy decreases from 153.83 kJ/mol to 149.67 kJ/mol for 500 ppm CO to 50 ppm CO, whereas the activation energy decreases from 155.93 kJ/mol to 155.28 kJ/mol for 2500 ppm CO_2 to 1000 ppm CO_2. The variation in E_a is much less in CO_2 than in CO which could pertain to the efficiency of CO_2 sensing of 0.5% LFCO-ZnO thin film.

As discussed earlier, when the sensor surface is exposed to air, oxygen ions adsorb on the sensor surface which leads to the formation of an electron-depleted layer and hence a potential barrier between two grains. The conductance G at temperature T is given by Eq. (12).

Now, assuming Langmuir adsorption (ideal gas), when sensor surface is exposed to gas, it is covered with a monolayer of gas molecules/atoms. The conductance transients are fitted according to Eq. (18) for the response transient. The recovery transient, however, represents decay of conductance, and it is fitted by the following equation:

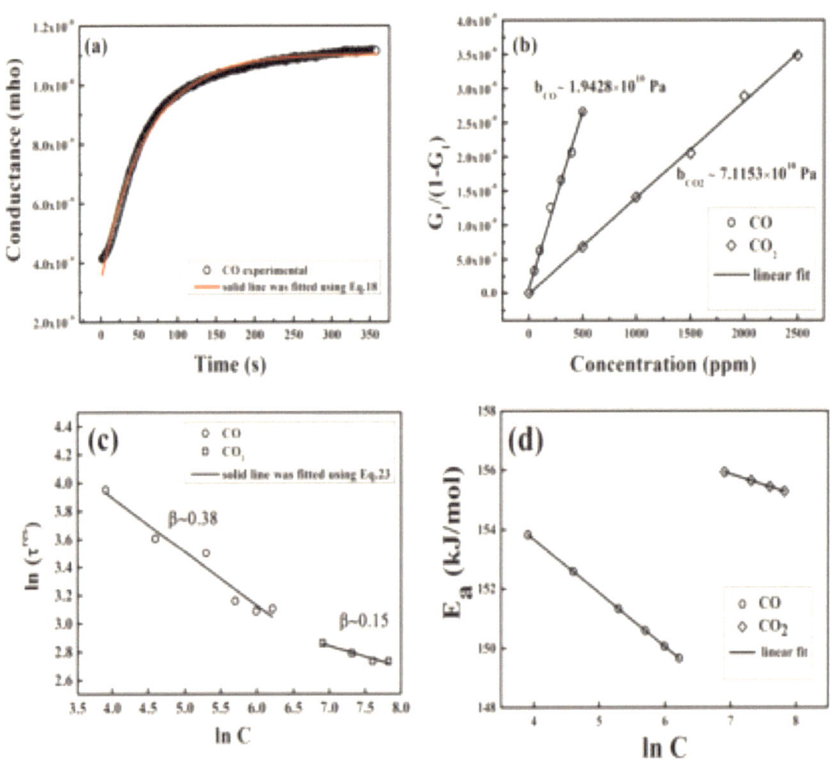

Figure 15.
(a) Conductance transient plot for 500 ppm CO sensing. Experimental points are shown by symbols, and the solid line was fitted using Eq. (18), (b) linear variation of G1/(1-G1) with concentration, (c) log–log plot of τ^{res} with C. experimental points are shown by symbols, and the solid line was fitted using Eq. (23), (d) variation of E_a with ln C for CO and CO_2 sensing. For all these measurements, operating temperature was kept fixed at 300°C.

$$G(t)_{recovery} = G_0' + G\left[exp\left(\frac{-t}{\tau_{recovery}}\right)\right] \tag{51}$$

Figure 16 shows the conductance transients of CO and CO_2 fitted with Eqs. (18) and (51). **Table 6** summarizes the fitted parameters. Each parameter is unique to both the gases. It may be noted that the response times to both CO and CO_2 are almost similar (~60 and 68 s, respectively) although the response % is markedly different (64 and 36%, respectively).

Assuming nonlinear variation of sheet conductance with test gas concentration, the gas sensing response varies with absolute temperature T according to the following relation:

$$\frac{R_a - R_g}{R_a} = \frac{a_0}{6} \times exp\left(-\frac{E_a}{RT}\right)\left[6 + n.m_0^2\, exp\left(-\frac{E_k}{RT}\right)T^{-0.5}\right] \times \left[\frac{C_{AS}}{\cosh\left(m_0.exp\left(-\frac{E}{2RT}\right)T^{-0.25}\right)}\right]^n \tag{52}$$

where R_a is the resistance in air, R_g is the resistance in gas, a_0*is the* pre-exponential constant, n is the sensitivity of the gas, m_0 is the Hatta number, E_k is the activation energy of the first order reaction, and C_{AS} is the test gas concentration at the film surface [6].

The response (S) versus absolute temperature (T) follows a "bell-shaped" dynamic and can be fitted in accordance with Eq. (52) to yield parameters specific to CO and CO_2. **Figure 17** shows the S versus T fitting for CO and CO_2. Sensitivity of the thin film, "n," is higher for CO_2 which shows that the sensor surface is more active toward CO_2 adsorption. Addition of 0.5% LFCO to ZnO structure enhances

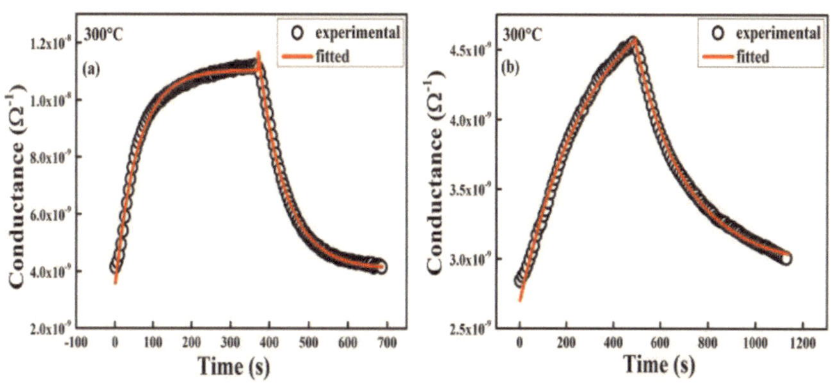

Figure 16.
Fitted conductance transients of (a) CO and (b) CO_2 at 300°C in accordance with Eqs. (18) and (51).

Fitted response parameters	T_{opt}	G_0 (Ω^{-1})	G (Ω^{-1})	τ_{res} (s)
CO	300 °C	3.45165E-9	7.60875E-9	59.74
CO_2	300 °C	2.69149E-9	2.34245E-9	303.00
Fitted recovery parameters	T_{opt}	G_0 (Ω^{-1})	G (Ω^{-1})	τ_{rec} (s)
CO	300°C	4.09001E-9	7.6711E-9	68.56
CO_2	300°C	2.95542E-9	1.57362E-9	220.29

Table 6.
Fitted parameters for the conductance transients of CO and CO_2 at 300°C.

oxygen adsorption on the ZnO surface due to the formation of additional defects. Also, the formation of multiple junctions in the composite leads to the phenomenon of "band bending" which enhances the interaction of CO_2 with oxygen ions. It may be noted that CO interferes with CO_2 sensing in the composite due to the formation of multiple junctions. This problem is eliminated in the ZnO/LFCO bilayer (not shown) which consists of only a single p-n junction and, thus, is selective to CO_2.

Simulated parameters from **Figure 17** are summarized in **Table 7** where E_a and E_k have units of J mol^{-1}. Significant differences may be noted for the simulated parameters of each gas. **Table 7** and **Figure 15d** corroborate the fact that the activation energy is lower for CO which explains the higher response to CO. It may also be emphasized that the parameters are valid for the entire range of operations of the sensor. The simulation may provide useful insights regarding the maximum and minimum temperature at which the sensor is operative. Further, a logarithmic plot of sensor response S and concentration C yields a value of ~0.67 for CO and ~1.56 for CO_2 (**Figure 18**) which is well in agreement with the simulated values. Additionally, it is shown that sensor response decreases sequentially with decreasing concentration as can be seen from **Figure 18**. The limiting concentration for CO is 50 ppm and that for CO_2 is 500 ppm. The detection limit for CO_2 is well above the permissible exposure limit (PEL) which is >5000 ppm. Thus, we may conclude that 0.5% LFCO-ZnO composite thin film is a fair CO_2 sensor and is well suited for practical applications.

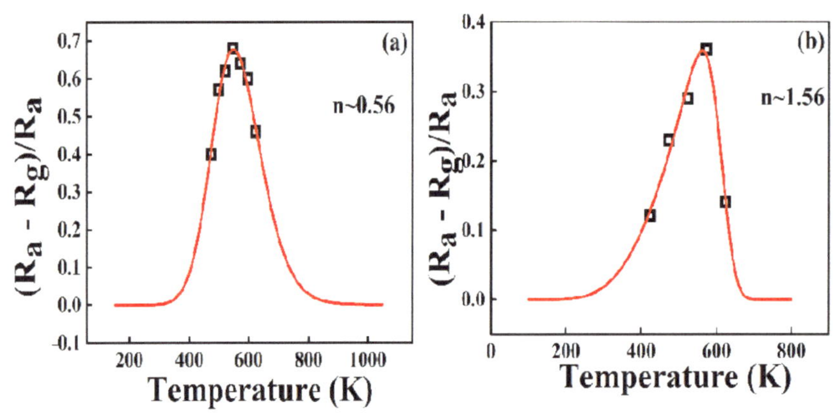

Figure 17.
Estimated value of sensitivity "n" for (a) CO and (b) CO_2 from the fitting of response vs. temperature using Eq. (52).

Fitting parameters	Carbon monoxide	Carbon dioxide
E_k/R	3611.87	15674.17511
$a_0/6$	3278.17	1.22219E-5
E_a/R	6062.32	2007.15851
$m_0 = L(\sqrt{(k/D_k)}$	1240.78	2.9675E6
n (sensitivity)	0.5654	1.56
C_{AS} (ppm)	500	2500

Table 7.
Fitted parameters from the fitting of response versus temperature of CO and CO_2.

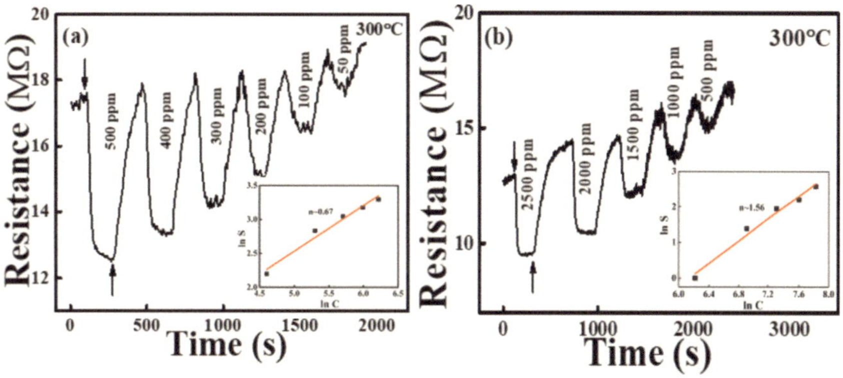

Figure 18.
Concentration variation in 0.5% LFCO-ZnO for (a) CO (500–50 ppm) and (b) CO_2 (2500–500 ppm); inset in both figures shows logarithmic variation of response with concentration which can be fitted linearly to obtain "n," the sensitivity.

4.3.1 CO_2 sensing mechanism of LFCO-ZnO thin film gas sensor

The electrical resistance change in ZnO, which is an n-type conducting material, is attributed to adsorption and desorption kinetics of oxygen on ZnO surface. Chemisorption of molecular oxygen of the type O_2^- occurs above room temperature on ZnO surface, leading to the formation of electron depletion layer (EDL). As the temperature increases up to ~300°C, the molecular oxygen dissociates into atomic form and forms a thicker EDL which pushes inside the ZnO core. Homogeneous particles observed in ZnO thin film microstructure (**Figure 13a**) contribute to the increased rate of oxygen adsorption on ZnO surface. This increases the sensitivity toward CO_2 gas as can be observed from **Figure 14b**. Reaction of carbon dioxide molecules with adsorbed oxygen is given below [68]:

$$CO_2 + e^- \rightarrow CO^-_{2(ads)} \tag{53}$$

$$CO^-_{2(ads)} + O^-_{(ads)} + 2e^- \rightarrow 2CO + 2O_2^- \tag{54}$$

These equations are consistent with the reduction in resistance of ZnO sensor on exposure to CO_2. In principle, ZnO sensor should register an increase in resistance on exposure to CO_2 as it is an oxidizing gas. However, it was found that ZnO demonstrates typical n-type behavior when brought in contact with CO_2. This anomaly can be explained from the defect chemistry and catalytic properties of ZnO. This demands a more detailed study and is currently beyond the scope of this book. Thus, we have confined ourselves to the current discussion of LFCO-ZnO composite thin film sensor.

When n-type ZnO is combined with p-type LFCO ($LaFe_{0.8}Co_{0.2}O_3$) sensor, a p-n junction is created at the interface between the two sensors. On exposure to CO_2, the junction pushes further in to the ZnO core. This leads to an enhanced CO_2 sensing response as compared to pristine LFCO and ZnO sensors.

The general reaction mechanism of $LaFeO_3$ (LFO) was reported by [69] in which LFO reacts with CO_2 in the presence of moisture to form carbonates. The reaction is given below.

$$H_2O + \frac{1}{2}O_2 + 2e^- \rightarrow 2OH^- \tag{55}$$

$$O_2 + e^- \rightarrow O_2^- \quad (56)$$

$$O_2^- + e^- \rightarrow 2O^- \quad (57)$$

$$La^{3+} + 3OH^- + 3CO_2 \rightarrow La(CO_2OH)_3 \quad (58)$$

The carbonates thus formed aid in CO_2 sensing by reducing the activation energy required for the reaction to occur. Further, equalization of Fermi levels takes place via migration of carriers leading to the formation of p-n junction. On exposure to CO_2, hydroxycarbonates form at the junction interface leading to the increase of potential barrier. This increases the concentration of holes at LFCO interface. However, as the concentration of LFCO is only 0.5%, hole accumulation at LFCO surface is impeded, and fall in resistance occurs, and the composite as a whole shows n-type sensing behavior [60].

SnO_2 thick film gas sensor did not show any appreciable CO_2 sensing. This is due to the absence of a Schottky contact in between SnO_2 thick film and silver electrodes. Investigations on CO_2 sensing were not performed for any composite thick film sensor.

4.4 VOC sensing characteristics of thin and thick film

Volatile organic compounds such as ethanol, acetone, and isopropanol pollute auto cabins and indoors of buildings. Long time spent indoors in such an environment ultimately leads to sick building syndrome and also has been found to be linked to leukemia and lymphoma. VOCs are also biomarkers of lung cancer and type 2 diabetes. Ethanol, acetone, and isopropanol have a permissible exposure limit of 1000 ppm over an 8 h work shift as documented by OSHA.

Figure 19 demonstrates the necessity of VOC detection. Traditionally, n-type semiconducting metal oxides such as SnO_2, ZnO, WO_3, TiO_2, In_2O_3, Fe_2O_3, etc. have been used for VOC sensing. However, recent research activities show that p-type semiconducting metal oxides such as copper oxide have been gaining popularity in gas sensing mainly due to its ease of fabrication, low cost, high stability, and high response to volatile organic compounds. CuO architectures such as nanorods [70], nanosheets [71], nanoparticles [72], and thin film [19] play an important role in gas sensing behavior. Thin films, in particular, are preferred over powdered CuO variants due to high stability and ease of device fabrication (**Table 8**). Additionally, composites have also gained prominence in selective VOC sensing. We have synthesized CuO thin films using solgel technique. Quartz substrate was used to deposit

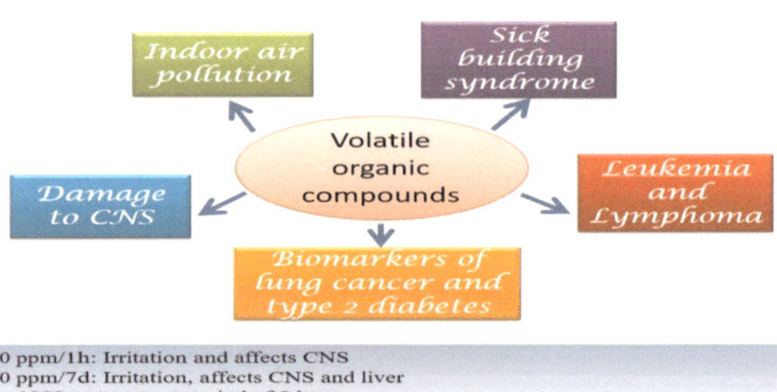

Figure 19.
Permissible exposure limit and harmful effects of volatile organic compounds.

Sensor material	Geometry	Test gas	T_{opt} (°C)	S%	τ_{res}, τ_{rec} (s)	Mixed gas sensing	Conc. (ppm)	Ref.
CuO	Nanowire	n-Propanol	190	6.2	1.2, 6.6	No	100	[76]
CuO	Nanosheets	Ethanol	320	4.22	2–15, 5–24	No	500	[73]
CuO	Thin film	2-Propanol	225	282	12, 279	No	300	[19]
CuO	Nanorods	Ethanol	350	90	—	No	600	[70]
CuO	Nanoparticles	Ethanol	200	5.6	10, 8	No	500	[74]
CuO	Powder	Ethanol	230	2.37 (R_g/R_a)	—	No	210	[75, 76]
CuO	Nanoparticles	Ethanol	220	2.7 (R_g/R_a)	11, 7	No	0.1	[72]
CuO	Thin film	Ethanol Acetone	300	144 168	3, 216; 7, 248	Yes	300	This work

Table 8.
Comparison of VOC sensing performances of CuO powders and thin films reported in recent literature.

the thin films using spin coating. After heat treatment, gold electrodes were deposited on these films using sputter coating following which the gas sensing measurements were performed in a static gas sensing setup. The VOCs were found to be cross-sensitive. Cross-sensitivity was addressed with principal component analysis. Additionally, we have also investigated the VOC sensing characteristics of WO_3-SnO_2 thick film. Investigations into various thick films for VOC sensing has been tabulated in **Table 9**. The gas–solid interaction in this case was modeled with the Freundlich adsorption isotherm which is nonlinear in nature.

The CuO sensors, thus synthesized, were used for detection of ethanol and acetone by varying the operating temperature (200–300°C) and gas concentration (25–300 ppm). It was found that CuO shows high response and excellent stability with respect to both these gases. The response to 300 ppm acetone and ethanol was found to be 168 and 144%, respectively (**Figure 12**). The response and recovery times were found to be 7 and 248 s for acetone and 3 and 216 s for ethanol, respectively. The static gas sensing setup is similar to a real environment response which explains the delayed recovery times. The response times mentioned are not exact as in a static gas sensing setup, where response to the test gas does not saturate before recovery begins [19]. Mixed gas sensing was also performed with ethanol and acetone (300 ppm/300 ppm) at 300°C. It was found that the response was not affected by the order in which the gases were injected into the static chamber. Principal component analysis was used as a discriminatory method to quantify the gases for mixed gas sensing data (**Table 9**).

4.4.1 Ethanol and acetone sensing mechanism in CuO sensor

CuO is a p-type sensor, and as such the sensing mechanism of a p-type sensor differs slightly to that of an n-type sensor. The major carriers in this case are holes. The concentration of holes changes at the sensor surface due to reaction with the target gas. When the CuO sensor is exposed to air at the operating temperature range of 200–300°C, oxygen takes up electrons from the sensor surface, and the oxygen ions formed chemisorb onto the surface. This increases the density of holes resulting in the decrease of the surface Fermi level. A hole accumulation layer (HAL) at the surface increases the conduction of holes which reduces the baseline

Sensor material	Geometry	Gas	T_{opt} (°C)	S %	τ_{res}, τ_{rec} (s)	Selectivity	Conc. (ppm)	Ref.
CeO_x - SnO_2	Thick film	Ethanol	350	95	1.1	Yes	200	[77]
$Mg_{0.5}Zn_{0.5}Fe_2O_4$	Nanoparticles	Ethanol	325	50	—	No	200	[78]
TiO_2	Nanoparticles	Ethanol	375	65	—	No	200	[79]
Barium hexaferrite	Nanoparticles	Ethanol	375	90	—	Yes	200	[80]
Sr-SnO_2	Thick film	Ethanol	300	65	2, 7		100	[81]
SnO_2	Thick film	Ethanol	300	84	8, 340	Yes	300	[20]
WO_3-SnO_2	Thick film	Ethanol	250	73	10, 312	Yes	300	[15]

Table 9.
Comparison of ethanol sensing performances of nanopowders and thick films reported in recent literatures.

resistance of CuO (R_a = resistance in air ambient). The equations involved are listed below [19].

$$O_2\ (gas) \rightarrow O_2(ads) \tag{59}$$

$$O_2\ (ads) + e^- \rightarrow O^-_{ads} + h \tag{60}$$

$$O^-_{ads} + e^- \rightarrow O^{2-}_{ads} + h \tag{61}$$

On exposure to acetone and ethanol, the oxygen ions react with the test gases to produce carbon dioxide and water with the release of electrons. This process reduces the thickness of the hole accumulation layer which increases the sensor resistance (R_g = resistance in gas ambient) [75].

The probable reaction mechanisms for acetone and ethanol are

$$8O^{2-}_{ads} + CH_3COCH_3 \rightarrow 16e^- + 3CO_2 + 3H_2O \tag{62}$$

$$6O^{2-}_{ads} + C_2H_5OH \rightarrow CH_3CHO + H_2O + e^- \tag{63}$$

$$e^- + h \rightarrow null \tag{64}$$

In the end, electron hole recombination takes place and the HAL diminishes.

4.4.2 Mixed gas sensing in CuO thin film sensor and principal component analyses

We have also investigated mixed gas sensing for 300 ppm acetone and 300 ppm ethanol. We found that the mixed gas sensing response (205%) is much higher than individual sensing response of acetone (168%) and ethanol (144%), but it was not an addition of the individual responses (**Figures 20** and **21**). Further, it was found that the order in which the gases were injected into the static chamber had no bearing on the mixed gas sensing response.

This can be attributed to the fact that the gas with the higher reactivity will first interact with the adsorbed O^- ions, leaving the next gas to react with less number of the same ions leading to reduced sensing [77].

The gases can be discriminated by principal component analyses, which quantify the gases in a multi coordinate system. Fast Fourier transform was utilized to extract the important features of the resistance transients of both acetone and ethanol measured at 300°C. These parameters were then fed into a linear unsupervised principal component analyses pattern recognition technique. **Figure 22a** shows the clustering of ethanol and acetone when injected individually.

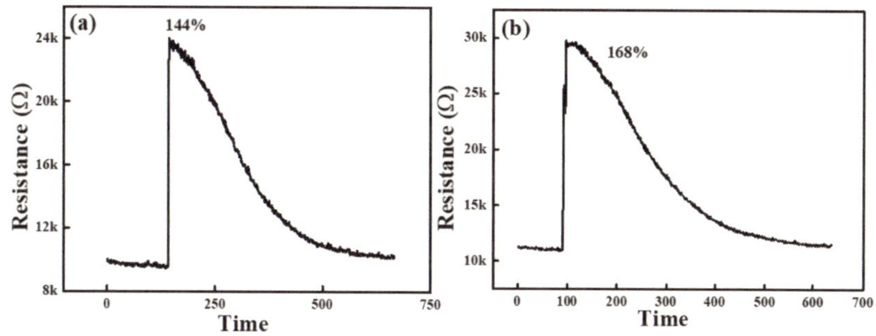

Figure 20.
Resistance transients showing the response % of (a) ethanol and (b) acetone.

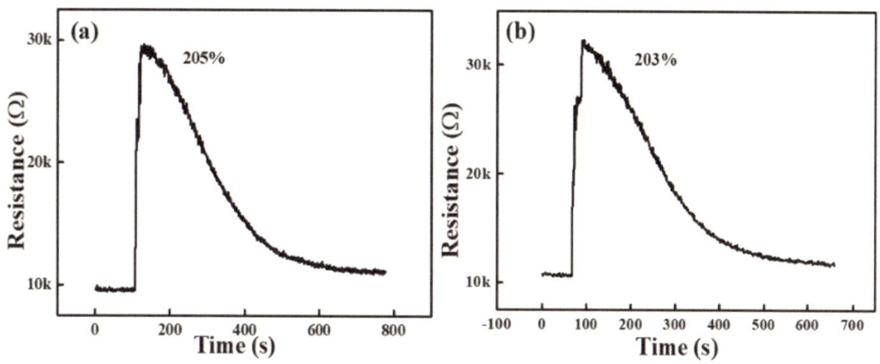

Figure 21.
Resistance transients showing the response % of (a) 300 ppm ethanol and 300 ppm acetone and (b) 300 ppm acetone and 300 ppm ethanol.

The clustering does not change when all the resistance transients of 300 ppm each of acetone and ethanol are analyzed (**Figure 22b**). Thus, PCA can be successfully used to identify the gases in a mixture of acetone and ethanol.

Additionally, we have explored the VOC sensing characteristics of WO_3-SnO_2 composite thick film sensor [15]. Sensing measurements were made in the range 300–150°C for 300 ppm ethanol, acetone, and isopropanol. The composite thick film was found to be cross-sensitive to each of the gases. To address the cross-sensitivity, the conductance transients of the gases were further analyzed. These

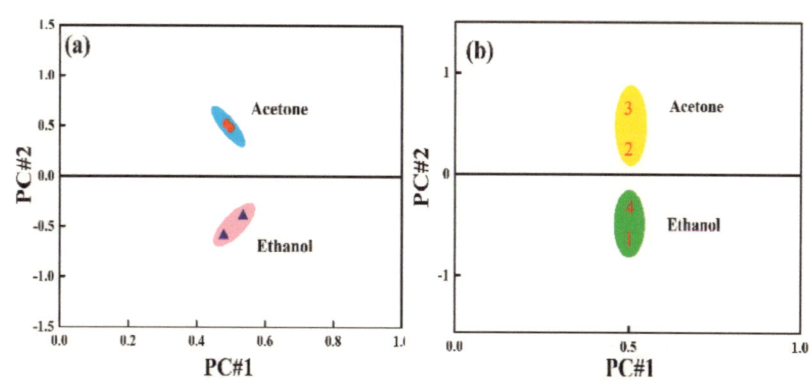

Figure 22.
Principal component analyses for (a) individual gases and (b) mixed gases; note the separate clustering enabling us to distinguish the gases.

conductance transients were obtained at 250°C, and VOC concentration was kept fixed at 300 ppm. Conductance transients for the three VOCs at respective concentration level were fitted using Eq. (18) to estimate the values of parameters such as G_0 (mho), G_1 (mho), and τ^{res} (s). **Figure 23b** envisages the nonlinear variation of G1/(1-G1) with VOC concentration (C) (symbols). In this case, Langmuir adsorption isotherm is not validated since the variation of G1/(1-G1) with concentration is nonlinear for all the gases and does not pass through the origin. In fact, the results are much closer to the Freundlich adsorption isotherm. The relation between G_1 and C in this case obeys the following relation:

$$\frac{G_1}{1-G_1} = \alpha C^{\delta} \tag{65}$$

From the nonlinear fit, we estimated the values of α and δ for all VOCs under consideration. These results indicate that the sensor surface is not homogeneous and heat of adsorption (Q) is not uniform unlike the Langmuir model.

In case of Langmuir adsorption isotherm, assuming $Z = G_1/(1\text{-}G_1)$, differentiating Eq. (19) we can write

$$\frac{dZ}{dC} = \frac{1}{b} \tag{66}$$

Differentiating Eq. (66), we can write

$$\frac{dZ}{dC} = (\alpha\delta)C^{\delta-1} \tag{67}$$

Assuming Z vs. C slope at a particular point ~1/b, we can write

$$\frac{1}{b} = (\alpha\delta)C^{\delta-1} \tag{68}$$

$$b = \frac{1}{(\alpha\delta)C^{\delta-1}} = b_0 \exp\left(-\frac{Q}{RT}\right) \tag{69}$$

Simplification of Eq. (69) yields

$$Q = RT\left[\ln b_o - (1-\delta)\ln C + \ln \alpha\delta\right] \tag{70}$$

We have estimated b_o from the expression given in Eq. (16). Assuming $\upsilon = 10^{13}$ Hz [49], R = 8.314 JK^{-1}, and T = 523 K, Eq. (16) can be simplified to.

$$b_0 = 2.7443 \times 10^{-9}\left(\frac{1}{K_0.S_{area}}\right) \times M^{1/2}\ \text{Pa} \tag{71}$$

As described earlier, considering the surface area, $S_{area} \sim 16.2$ Å^2 for nitrogen (molecular weight = 28 g mol^{-1}), we can easily calculate the surface area for ethanol which is 26.61 Å^2 (molecular weight = 46.0 g mol^{-1}). Assuming condensation coefficient $K_o = 1$ [47], for all test gases, b_o (at T ~ 523 K) can be estimated using Eq. (71). From Eq. (70) we have estimated Q as a function of VOC concentration at 250°C.

Figure 24a shows the log–log plot of characteristic response time (τ^{res}) vs. VOC concentration. As shown in **Figure 24a**, variation of τ^{res} with C follows the following relation:

$$\tau^{res} = \tau_0 \times C^{-\beta} \tag{72}$$

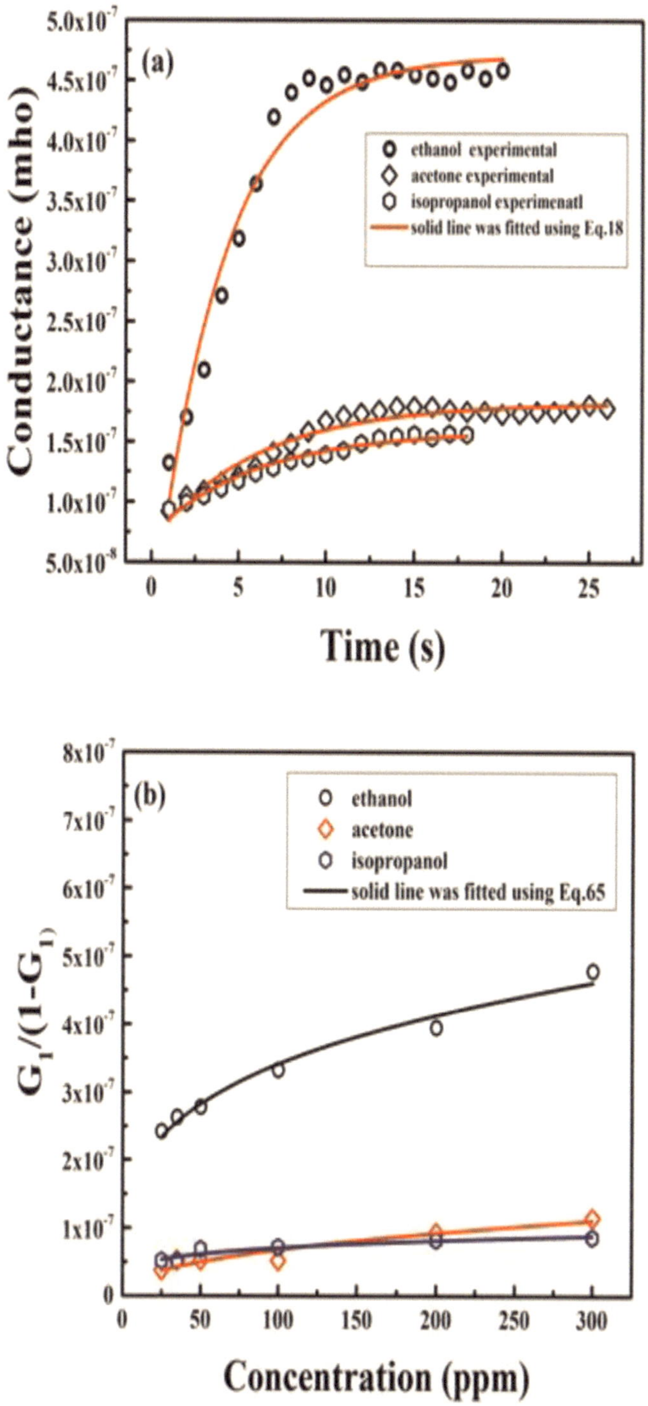

Figure 23.
(a) Conductance transient plots measured at 250°C for 300 ppm ethanol, acetone, and isopropanol sensing. Experimental points are shown by symbols, and the solid line was fitted using Eq. (18), and (b) nonlinear variation of G1/(1-G1) with concentration (ppm). Experimental points are shown by symbols, and the solid line was fitted using Eq. (65) (**Figure 23a, b** *reprinted from [15] with permission from Elsevier).*

The ln (τ^{res}) vs. ln (C) plot was fitted linearly in order to estimate value of β for three VOCs under consideration.

In case of Freundlich adsorption isotherm, temperature-dependent variation of τ^{res} can be written as per the following relation:

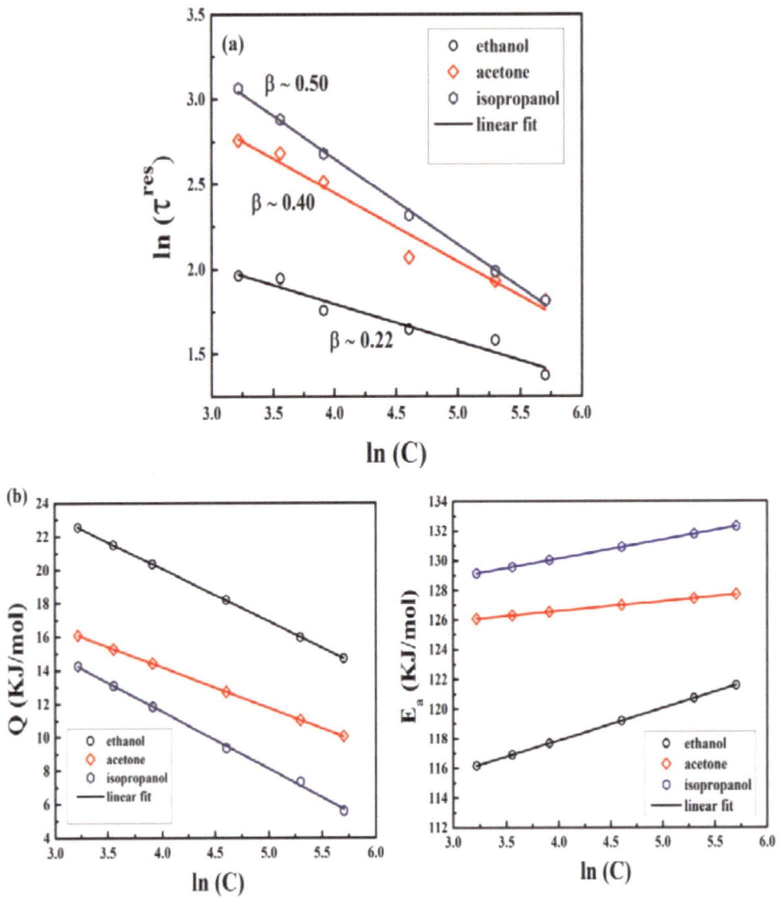

Figure 24.
*(a) Log-log plot of τ^{res} with C. The sensing measurements were carried out at 250°C, and (b) variation of E_a and Q with concentration for ethanol, acetone, and isopropanol sensing (**Figure 24a, b** reprinted from [15] with permission from Elsevier).*

$$\tau^{res} = \upsilon^{-1} \times \exp\left(\frac{E_a + Q}{RT}\right) \tag{73}$$

Using Eqs. (70) and (72), we can easily simplify Eq. (73). Activation energy, E_a, is expressed via the following relation:

$$\mathrm{E_a} = \mathrm{RT} \times \ln\left(\frac{\tau_0.\upsilon.\mathrm{C}^{(1-\delta-\beta)}}{\alpha.\delta.\mathrm{b}_0}\right) \tag{74}$$

From Eqs. (70) and (74), we can determine the linearity of Q versus ln C plot as well as the E_a versus ln C plot. Also, we can see that Q decreases with gas concentration, whereas E_a shows the exact opposite behavior. **Figure 24b** shows the variation of E_a and Q for each VOC concentration (T = 250°C) [15]. From our analysis, we have estimated that E_a for ethanol sensing is lower than that of acetone and isopropanol which justify the larger yield of the composite sensor in terms of ethanol sensing ($S\% \rightarrow$ ethanol>acetone>isopropanol).

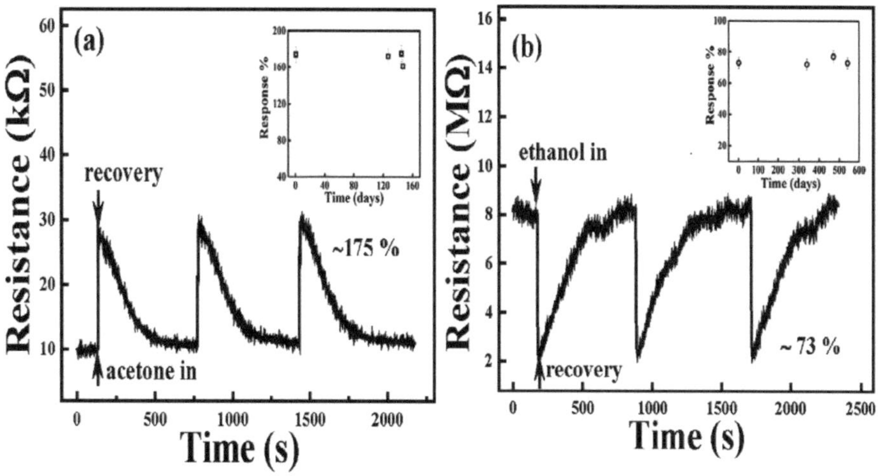

Figure 25.
*Typical resistance transient plots showing reproducibility toward 300 ppm (a) acetone sensing in CuO thin film and (b) ethanol sensing in WO_3-SnO_2 thick film; inset of each figure shows invariance of response % with time (**Figure 25b** reprinted from [15] with permission from Elsevier).*

4.4.3 Reproducibility of CuO thin film and WO_3-SnO_2 thick film

The stability and reproducibility of CuO thin film and WO_3-SnO_2 thick film composite sensors were studied with respect to acetone and ethanol gases (**Figure 25**). Excellent baseline recovery was achieved for both the sensors for an extended period of time with marginal change of initial response %. The response % for CuO sensor was recorded for a period of ~150 days, whereas that for WO_3-SnO_2 sensor was recorded for ~500 days. Stable baseline was recorded for CuO thin film sensor, whereas WO_3-SnO_2 thick film sensor was prone to drift in baseline resistance.

5. Conclusions

This chapter throws light on detection of common air pollutants including carbon monoxide, carbon dioxide, nitrogen dioxide, volatile organic compounds, and also hydrogen. The permissible exposure limits for these gases act as benchmark for minimum concentration of gas detection. Gas sensing terminologies such as response %, response time, recovery time, selectivity, and stability have been defined. The interaction of semiconducting metal oxide sensor with reducing gas has been illustrated. Factors that affect the performance of a gas sensor, namely, receptor, transducer, and utility functions, have been explained. It was reported that the receptor function is influenced by the grain size of the sensor. The transducer function controls the flow of charge carriers between the electrodes. It is highly dependent on purity of the sensor material. The utility function controls the diffusion processes of gas during response and recovery processes.

A modified wet chemical route was chosen as the method to synthesize sols required for thin film deposition process. The sols were deposited on quartz and glass substrates and annealed at a high temperature. The materials (ZnO-G, ZnO, SnO_2, LFCO-ZnO, CuO, and WO_3-SnO_2) for gas sensors were chosen after extensive literature review and earlier experiments conducted by us in our laboratory.

High response (894%) to 5 ppm NO_2 sensing was reported for ZnO-G multilayer as compared to pure ZnO (26%) at 150°C. Also, NO_2 detection is selective at 150°C with respect to 500 ppm each of CO, H_2, and i-C_4H_{10} gases. The high response and selectivity have been attributed to the presence of p-n junction between ZnO and G grains. The formation of a p-n junction increases the amount of available electrons increasing the gas response. In addition, mixed gas sensing for NO_2/CO, NO_2/H_2, and NO_2/i-C_4H_{10} has also been reported. Principal component analyses have been demonstrated as an effective tool for the quantification of these mixed gases. ZnO thin film and SnO_2 thick film were investigated for CO and H_2 detection. Both materials demonstrated cross-sensitivity to the gases. Cross-sensitivity was addressed through modeling based on Langmuir adsorption mechanism. Activation energy, E_a, and heat of adsorption, Q, were calculated based on the model. E_a and Q values were found to be distinct for each gas and varied as a function of gas concentration. Further, the irreversible nature of ZnO thin film and reversible nature of SnO_2 thick film were demonstrated when exposed to a fixed concentration of CO and H_2 gases. CO_2 sensing characteristics of LFCO-ZnO thin film sensor were investigated. The thin film sensor yields 36% response to 2500 ppm CO_2 at 300°C. LFCO-ZnO thin film is cross-sensitive to CO at this temperature. Principal component analyses have been performed to eliminate cross-sensitivity. Further, Langmuir adsorption mechanism is validated in LFCO-ZnO thin film sensor as the variation of conductance and concentration is a straight line passing through origin (not shown). A model has also been developed for the thin film gas sensor relating response with temperature. Modeling of the response with temperature for CO and CO_2 sensing yielded parameters unique to both gases. Volatile organic compounds have been detected with a CuO thin film sensor. Both acetone and ethanol show high response with a response time of only a few seconds. The gases have been quantified using principal component analyses. Mixed gas (ethanol/acetone) sensing has also been reported for CuO thin film. The mixed gases have been identified on the basis of the PCA cluster for individual gases. WO_3-SnO_2 thick film sensor was investigated for ethanol, acetone, and isopropanol sensing characteristics. The thick film sensor yielded highest response to 300 ppm ethanol. Freundlich adsorption mechanism was dominant as the variation of conductance and concentration is nonlinear. Based on this adsorption mechanism, E_a and Q values were estimated for each gas. Both E_a and Q values were found to vary linearly with gas concentration.

Acknowledgements

The research work was partially supported by research grant obtained from CSIR, Government of India, vide sanction letter no. 03/(1371)/16/EMR-II, dated 10 May 2016 and NNetRa grant vide sanction 5 (1)/2017-NANO dated 28 March 2018, DST/NM/NNetRa/2018 (G)-IIT-KGP dated 21 March 2018.

Conflict of interest

The authors declare no conflict of interest.

Nomenclature

C	concentration
CO	carbon monoxide
CO_2	carbon dioxide

CuO	copper oxide
C_g	reducing gas concentration
EDL	electron depletion layer
eV_s	Schottky barrier
E_a	activation energy for gas adsorption during response process
E_d	activation energy for gas desorption during response process
FFT	fast Fourier transform
$F\theta$	total number of active sites
G	graphene
$G(t)$	conductance transient of the sensor
$G\ (t)^{response}$	conductance transient of the sensor during response process
G_0	conductance of the sensor in air
G_1	conductance of the sensor at site 1
HAL	hole accumulation layer
H_2	hydrogen
k	Boltzmann constant
K	equilibrium constant for response process
K_0	fraction of molecules with energy $> E_a$
LFCO	cobalt doped lanthanum ferrite
M	molar mass of gas
MEA	methanolamine
MO	metal oxide
MOE	methoxyethanol
NMP	N-methyl-2-pyrrolidine
NO_2	nitrogen dioxide
N_A	Avogadro number
$N(t)$	number of gas molecules adsorbed at time t
N^*	total number of available adsorption sites
OSHA	Occupational Safety and Health Administration
O_{ads}^-	chemiadsorbed oxygen
$O_{ad-surface}$	physiadsorbed oxygen
P	gas pressure
PCA	principal component analysis
PEL	permissible exposure limit
Q	heat of adsorption
R	universal gas constant
R_a	sensor resistance in air
R_{ad}	physiadsorbed reducing gas
R_g	sensor resistance in reducing gas
RO_{ad}	adsorbed oxidizing product of reducing gas
RO_{gas}	desorbed oxidizing product of reducing gas
S	sensor response (%)
S_{area}	surface area of a single adsorbed molecule
S_p	sensor response (%) of p-type semiconductor
S_n	sensor response (%) of n-type semiconductor
SMO	semiconducting metal oxide
SnO_2	tin oxide
T_{opt}	operating temperature of sensor
VOC	volatile inorganic compound
WO_3	tungsten oxide
ZnO	zinc oxide
κ	selectivity factor

ν	frequency of oscillation of adsorbed gas
τ_{irrev}	characteristic response time in irreversible response process
τ_{rev}	characteristic response time in reversible response process
τ_{res}	response time of sensor
τ_{rec}	recovery time of sensor

Author details

Tynee Bhowmick[1], Vibhav Ambardekar[2], Abhishek Ghosh[3], Moumita Dewan[1], Partha Pratim Bandyopadhyay[2], Sudip Nag[4] and Subhasish Basu Majumder[1*]

1 Materials Science Centre, Indian Institute of Technology, Kharagpur, India

2 Mechanical Engineering, Indian Institute of Technology, Kharagpur, India

3 Department of Biomedical Engineering, University of Michigan, USA

4 Electronics and Electrical Communication Engineering, Indian Institute of Technology, Kharagpur, India

*Address all correspondence to: subhasish@matsc.iitkgp.ac.in

References

[1] Zampolli S, Elmi I, Ahmed F, Passini M, Cardinali GC, Nicoletti S, et al. An electronic nose based on solid state sensor arrays for low-cost indoor air quality monitoring applications. Sensors and Actuators B: Chemical. 2004;**101**:39-46

[2] World Health Organization. Household Air Pollution and Health. Weblog. Available from: https://www.who.int/news-room/fact-sheets/detailhousehold-air-pollution-and-health. 2018

[3] United States Department of Labour. OSHA Annotated PELs—Occupational Safety and Health Administration. Weblog. Available from: https://www.osha.gov/dsg/annotated-pels/tablez-1.html. 2019

[4] Bhargav KK, Ram S, Labhsetwar N, Majumder SB. Correlation of carbon monoxide sensing and catalytic activity of pure and cation doped lanthanum iron oxide nano-crystals. Sensors and Actuators B: Chemical. 2015;**206**:389-398

[5] Mukherjee K, Majumder SB. Analyses of response and recovery kinetics of zinc ferrite as hydrogen gas sensor analyses of response and recovery kinetics of zinc ferrite as hydrogen gas sensor. Journal of Applied Physics. 2009;**106**:064912

[6] Ghosh A, Majumder SB. Modeling the sensing characteristics of chemi-resistive thin film semi-conducting gas sensors. Physical Chemistry Chemical Physics. 2017;**19**:23431-23443

[7] Ambardekar V, Bandyopadhyay PP, Majumder SB. Hydrogen sensing performance of atmospheric plasma sprayed tin dioxide coating. International Journal of Hydrogen Energy. 2019;**44**(26):14092-14104

[8] Mukherjee K, Gaur APS, Majumder SB. Investigations on irreversible and reversible-type gas sensing for ZnO and $Mg_{0.5}Zn_{0.5}Fe_2O_4$ chemi-resistive sensors. Journal of Physics D: Applied Physics. 2012;**45**(50):505306

[9] Bhowmick T, Banerjee A, Nag S, Majumder SB. Gas sensing characteristics in ZnO thin film explicated through the analysis of conductance transients and the concept of activation energy. Proceedings of IEEE Sensors. INSPEC Accession Number 18329455. 2018. DOI: 10.1109/ICSENS.2018.8589612

[10] Barsan N, Weimar U. Conduction model of metal oxide gas sensors. Journal of Electroceramics. 2001;**7**(3):143-167

[11] Kim H, Lee J. Highly sensitive and selective gas sensors using p-type oxide semiconductors: Overview. Sensors and Actuators B: Chemical. 2014;**192**:607-627

[12] Kalisadhan M. Gas sensing characteristics of wet chemical synthesized spinel ferrites [PhD dissertation]. Kharagpur: Materials Science Centre, Indian Institute of Technology, Kharagpur; 2011

[13] Bochenkov VE, Sergeev GB. Chapter 2: Sensitivity, selectivity, and stability of gas-sensitive metal-oxide nanostructures. In: Metal Oxide Nanostructures and Their Applications. Vol. 3. Valencia, California, USA: American Scientific Publishers; 2010. pp. 31-52. Available from: http://www.chem.msu.su/rus/books/2011/sergeev/all.pdf

[14] Ghosh A, Bhowmick T, Majumder SB. Multi-layered zinc oxide-graphene composite thin films for selective nitrogen dioxide sensing.

Journal of Applied Physics. 2018;**123**: 084501

[15] Ambardekar V, Bandyopadhyay PP, Majumder SB. Understanding on the ethanol sensing mechanism of atmospheric plasma sprayed 25 wt.% WO_3-75 wt.% SnO_2 coating. Sensors and Actuators B: Chemical. 2019;**290**: 414-425

[16] Ghosh A, Schneller T, Waser R, Majumder SB. Understanding on the selective carbon monoxide sensing characteristics of copper oxide-zinc oxide composite thin films. Sensors and Actuators B: Chemical. 2017;**253**: 685-696

[17] Korotcenkov G. Chapter 5: Conductance transient analyses of metal oxide gas sensors for the example of spinel ferrite gas sensors. In: Mukherjee, Kalisadhan, Majumder SB, editors. Chemical Sensors: Simulation and Modeling. Volume 1: Microstructural Characterization and Modeling of Metal Oxides. New York, USA: Momentum Press; 2012. Available from: http://www.momentum press.net/books/chemical-sensors-simulation-and-modeling-volume-1-microstructural-characterization-and-modelin

[18] Maity A. Development of nano-structured ceramic oxide gas sensing system for air quality monitoring [PhD dissertation]. Kharagpur: Materials Science Centre, Indian Institute of Technology, Kharagpur; 2015

[19] Ghosh A, Maity A, Banerjee R, Majumder SB. Volatile organic compound sensing using copper oxide thin films: Addressing the cross sensitivity issue. Journal of Alloys and Compounds. 2017;**692**:108-118

[20] Ambardekar V, Bandyopadhyay PP, Majumder SB. Atmospheric plasma sprayed SnO_2 coating for ethanol detection. Journal of Alloys and Compounds. 2018;**752**(2):440-447

[21] Sriskandan K, Pettingale KW. Numismatist's pneumonitis. A case of acute nitrogen dioxide poisoning. Postgraduate Medical Journal. 1985;**61**: 819-821

[22] CBC. 'Inadequate Ventilation' at Super 8 led to Carbon Monoxide Poisoning, Officials say. CBC News [Newspaper on the Internet]. July 10, 2019. Available from: https://www.cbc.ca/news/canada//manitoba/carbon-monoxide-poisoning-hotel-winnipeg-super8-1.5206755 [Cited: July 30, 2019]

[23] Permentier K, Vercammen S, Soetaert S, Schellemans C. Carbon dioxide poisoning: A literature review of an often forgotten cause of intoxication in the emergency department. International Journal of Emergency Medicine. 2017;**10**(14):17-20

[24] Kumarvel V, Da Fonseca J. Acetone poisoning—A diagnostic dilemma. EJA. 2007;**24**:803-816

[25] Dilisi GA. Disaster: Combining physics and history in the laboratory. Physics Teacher. 2017;**55**:268-273

[26] Jacob K. Nitrogen Emissions Going Up: Study. The Hindu [Newspaper on the Internet]. June 6, 2018. Available from: https://www.thehindu.com/sci-tech/energy-and-environment/nitrogen-emissions-going-up-study/article 24090131.ece [Cited: July 30, 2019]

[27] Zhang J, Wang S, Wang Y, Wang Y, Zhu B, Xia H, et al. NO_2 sensing performance of SnO_2 hollow-sphere sensor. Sensors and Actuators B: Chemical. 2009;**135**(2):610-617

[28] Lee D, Nam K, Lee D. Effect of substrate on NO_2-sensing properties of WO_3 thin film gas sensors. Thin Solid Films. 2000;**2**:142-146

[29] PCKKJ L. NO_2 sensing characteristics of ZnO nanorods

prepared by hydrothermal method. Journal of Electroceramics. 2006;**2**: 975-978

[30] Neri G, Bonavita A, Galvagno S, Siciliano P, Capone S. CO and NO_2 sensing properties of doped-Fe_2O_3 thin films prepared by LPD. Sensors and Actuators B: Chemical. 2002;**82**(2): 40-47

[31] Srivastava S, Jain K, Singh VN, Singh S, Vijayan N, Dilawar N, et al. Faster response of NO_2 sensing in graphene–WO_3 nanocomposites. Nanotechnology. 2012;**23**:205501

[32] Zhang H, Feng J, Fei T, Liu S, Zhang T. SnO_2 nanoparticles-reduced graphene oxide nanocomposites for NO_2 sensing at low operating temperature. Sensors and Actuators B: Chemical. 2014;**190**(2):472-478

[33] Maity A, Majumder SB. NO_2 sensing and selectivity characteristics of tungsten oxide thin films. Sensors and Actuators B: Chemical. 2015;**206**(2): 423-429

[34] Hübert T, Boon-Brett L, Black G, Banach U. Hydrogen sensors—A review. Sensors and Actuators B: Chemical. 2011;**157**(2):329-352

[35] Gal M. Enhanced optical detection of hydrogen using the excitation of surface plasmons in palladium. Applied Surface Science. 1993;**68**:135-138

[36] Jamnani SR, Moghaddam HM, Leonardi SG, Donato N, Neri G. Applied surface science synthesis and characterization of Sm_2O_3 nanorods for application as a novel CO gas sensor. Applied Surface Science. 2019;**487**: 793-800

[37] Hou L, Zhang C, Li L, Du C, Li X, Kang X, et al. CO gas sensors based on p-type CuO nanotubes and CuO nanotubes: Morphology and surface structure effects on the sensing performance. Talanta. 2018;**188**:41-49

[38] Chung W, Sakai G, Shimanoe K, Miura N. Preparation of indium oxide thin film by spin-coating method and its gas-sensing properties. Sensors and Actuators B: Chemical. 1998;**46**:139-145

[39] Singh A, Sharma A, Tomar M, Gupta V. Growth of highly porous ZnO nanostructures for carbon monoxide gas sensing. Surface and Coating Technology. 2018;**343**:49-56

[40] Pati S, Maity A, Banerji P, Majumder SB. Qualitative and quantitative differentiation of gases using ZnO thin film gas sensors and pattern recognition analysis. The Analyst. 2014;**139**(7):1796

[41] Pati S, Banerji P, Majumder SB. n- to p- type carrier reversal in nanocrystalline indium doped ZnO thin film gas sensors. International Journal of Hydrogen Energy. 2014;**39**(27): 15134-15141

[42] Brunet E, Maier T, Mutinati GC, Steinhauer S, Köck A, Gspan C, et al. Comparison of the gas sensing performance of SnO_2 thin film and SnO_2 nanowire sensors. Sensors and Actuators B: Chemical. 2012;**165**:110-118

[43] Drmosh QA, Yamani ZH, Hossain MK. Hydrogen gas sensing performance of low partial oxygen-mediated nanostructured zinc oxide thin film. Sensors and Actuators B: Chemical. 2017;**248**:868-877

[44] Chang S. Oxygen chemisorption on tin oxide: Correlation between electrical conductivity and EPR measurements. Journal of Vacuum Science and Technology. 2002;**17**(1):366-369

[45] Lenaerts S, Roggen J, Maes G. FT-IR characterization of tin dioxide gas sensor materials under working conditions. Spectrochimica Acta Part A:

Molecular and Biomolecular Spectroscopy. 1995;**51**(5):883-894

[46] Mirzaei A, Leonardi SG, Neri G. Detection of hazardous volatile organic compounds (VOCs) by metal oxide nanostructures-based gas sensors: A review. Ceramics International. 2016; **42**(14):15119-15141

[47] Hu H, Trejo M, Nicho ME, Saniger JM, García-Valenzuela A. Adsorption kinetics of optochemical NH_3 gas sensing with semiconductor polyaniline films. Sensors and Actuators B: Chemical. 2002;**82**(1):14-23

[48] Mukherjee K, Majumder SB. Analyses of conductance transients to address the selectivity issue of zinc ferrite gas sensors. Electrochemical and Solid-State Letters. 2010;**13**:J25

[49] Adamson AW. Physical Chemistry of Surfaces. New York, USA: Wiley; 1982. pp. 519-523

[50] Special Correspondent. India's Carbon Dioxide Emissions up 5%. The Hindu [Newspaper on the Internet]. March 27, 2019. Available from: https://www.thehindu.com/sci-tech/energy-and-environment/indias-carbon-dioxide-emissions-up-5/article 26646376.ece54 [Cited: July 30, 2019]

[51] NASA. Global Patterns of Carbon Dioxide. Weblog. Available from: https://earthobservatory.nasa.gov./images/82142/global-patterns-of-carbon-dioxide. 2013

[52] Maruyama T, Sasaki S, Saito Y. Potentiometric gas sensor for carbon dioxide using solid electrolytes. Solid State Ionics. 1987;**23**:107-112

[53] Shimizu Y, Komori K, Egashira M. Carbon dioxide sensor consisting of K_2CO_3-polyethylene glycol solution supported on porous ceramics. Journal of the Electrochemical Society. 1989; **136**(8):2256-2260

[54] Ishihara T, Kometani K, Mizuhara Y, Takita Y. A new type of CO_2 gas sensor based on capacitance changes. Sensors and Actuators B: Chemical. 1991;**5**:97-102

[55] Hoefer U, Kuhner G, Schweizer W, Sulz G, Steiner K. CO and CO2 thin-film SnO_2, gas sensors on Si substrates. Sensors and Actuators B: Chemical. 1994;**22**:115-119

[56] Ishihara T, Kometani K, Mizuhara Y, Takita Y. Capacitive-type gas sensor for the selective detection of carbon dioxide. Sensors and Actuators B: Chemical. 1993;**14**:470-472

[57] Mizuno N, Yoshioka T, Kato K, Iwamoto M. CO_2 sensing characteristics of SnO_2 element modified by La_2O_3. Sensors and Actuators B: Chemical. 1993;**14**:473-475

[58] Haeusler A, Meyer JS. A novel thick film conductive type CO_2 sensor. Sensors and Actuators B: Chemical. 1996;**34**:388-395

[59] Yeob D, Kang H, Choi N, Hyun K, Lee H. A carbon dioxide gas sensor based on cobalt oxide containing barium carbonate. Sensors and Actuators B: Chemical. 2017;**248**:987-992

[60] Zhang W, Xie C, Zhang G, Zhang J, Zhang S. Porous $LaFeO_3/SnO_2$ nanocomposite fi lm for CO_2 detection with high sensitivity. Materials Chemistry and Physics. 2017;**186**: 228-236

[61] Juang F, Chern W, Chen B. Carbon dioxide gas sensing properties of ZnSn $(OH)_6$-ZnO nanocomposites with ZnO nanorod structures. Thin Solid Films. 2018;**660**(March):771-776

[62] Gómez L, Galeano V, Parra R, Michel CR, Paucar C, Morán O. Carbon dioxide gas sensing properties of ordered oxygen deficient perovskite

$LnBaCo_2O$=(Ln= La,Eu). Sensors and Actuators B: Chemical. 2015;**221**: 1455-1460

[63] Joong H, Han D, Ho J, Zhou Z. Carbon dioxide gas sensor using a graphene sheet. Sensors and Actuators B: Chemical. 2011;**157**(1):310-313

[64] Herrán J, Mandayo GG, Castaño E. Semiconducting $BaTiO_3$-CuO mixed oxide thin films for CO_2 detection. Thin Solid Films. 2009;**517**(22):6192-6197

[65] Dhahri R, Leonardi SG, Hjiri M, El Mir L, Bonavita A, Donato N, et al. Enhanced performance of novel calcium/aluminum co-doped zinc oxide for CO_2 sensors. Sensors and Actuators B: Chemical. 2017;**239**:36-44

[66] Ghosh A, Zhang C, Zhang H, Shi S. CO_2 sensing behavior of calcium-doped ZnO thin film: A study to address the cross-sensitivity of CO_2 in H_2 and CO environment. Langmuir. 2019;**35**: 10267-10275

[67] Kim S, Na CW, Hwang I, Lee J. One-pot hydrothermal synthesis of CuO – ZnO composite hollow spheres for selective H_2S detection. Sensors and Actuators B: Chemical. 2012;**168**:83-89

[68] Martinez L, Holguin-Momaca JT, Karthik TVK, Olive-Mendez SF, Campos-Alvarez J, Agarwal V. Sputtering temperature dependent growth kinetics and CO2 sensing properties of ZnO deposited over porous silicon. Superlattices and Microstructures. 2016;**98**:8-17

[69] Goldsmith J, Ross S. Factors affecting the infra-red spectra of some planar anions with D_{8h} symmetry-III. The spectra of rare-earth carbonates and their thermal decomposition products. Spectrochimica Acta. 1967;**23A**: 1909-1915

[70] Bhuvaneshwari S, Gopalakrishnan N. Room temperature ammonia and VOC sensing properties of CuO nanorods. AIP Conference Proceedings. 2016;**1731**: 050112-1-050112-3

[71] Yan H, Tian X, Ma F, Sun J. CuO nanoparticles fabricated by direct thermo-oxidation of sputtered Cu film for VOCs detection. Sensors and Actuators B: Chemical. 2015;**221**: 599-605

[72] Wang F, Li H, Yuan Z, Sun Y, Chang F, Deng H, et al. A highly sensitive gas sensor based on CuO nanoparticles synthetized via a sol-gel method. RSC Advances. 2016;**6**(83): 79343-79349

[73] Yan H, Tian X, Sun J, Ma F. Enhanced sensing properties of CuO nanosheets for volatile organic compounds detection. Journal of Materials Science: Materials in Electronics. 2014;**26**(1):280-287

[74] Yan H, Tian X, Ma F, Sun J. Chemical CuO nanoparticles fabricated by direct thermo-oxidation of sputtered Cu film for VOCs detection. Sensors and Actuators B. 2015;**221**:599-605

[75] Xia S, Zhu H, Cai H, Zhang J, Yu J, Tang Z. Hydrothermally synthesized CuO based volatile organic compound gas sensor. RSC Advances. 2014;**3**: 57975-57982

[76] Tan J, Dun M, Li L, Zhao J, Li X, Hu Y, et al. Self-template derived CuO nanowires assembled microspheres and its gas sensing properties. Sensors and Actuators B: Chemical. 2017;**252**:1-8

[77] Kotchasak N, Wisitsoraat A, Tuantranont A, Phanichphant S, Yordsri V, Liewhiran C. Highly sensitive and selective detection of ethanol vapor using flame-spray-made CeOx-doped SnO_2 nanoparticulate thick films. Sensors and Actuators B: Chemical. 2018;**255**:8-21

[78] Karmakar M, Das P, Pal M, Mondal B, Majumder SB, Mukherjee K.

Acetone and ethanol sensing characteristics of magnesium zinc ferrite nano-particulate chemi-resistive sensor. Journal of Materials Science. 2014:5766-5771

[79] Das P, Mondal B, Mukherjee K. Simultaneous adsorption-desorption processes in the conductance transient of anatase titania for sensing ethanol: A distinctive feature with kinetic perception. Journal of Physical Chemistry C. 2017;**121**:1146-1152

[80] Karmakar M, Mondal B, Pal M, Mukherjee K. Acetone and ethanol sensing of barium hexaferrite particles: A case study considering the possibilities of non-conventional hexaferrite sensor. Sensors and Actuators B: Chemical. 2014;**190**: 627-633

[81] Shaikh FI, Chikhale LP, Mulla IS, Suryavanshi SS. Synthesis and enhanced ethanol sensing performance of nanostructured Sr doped SnO_2 thick film sensor. Journal of Materials Science: Materials in Electronics. 2017;**28**(4): 3128-3139

Chapter 9

Nano Layers of 2D Graphene Versus Graphene Oxides for Sensing Hydrogen Gas

Anuradha Kashyap, Shikha Sinha, Partha Bir Barman and Surajit Kumar Hazra

Abstract

Hydrogen is one of the most useful but dangerous gases because of its broad combustion range and small ignition temperature. Currently, there is a great need for hydrogen detectors with selectivity, high sensitivity and reliable operations in view of its safe production, storage, transportation and other applications. In this regard, nano thin films of two dimensional materials like graphene, graphene oxide (GO) and reduced graphene oxide (rGO) have immense promise because their material attributes can be exceptionally tuned to achieve the desired characteristics. Also graphene oxide and reduced graphene oxide serve as potential sensing hosts due to the presence of functional groups on their surfaces. In this chapter, an attempt has been made to compare the work done in the field of hydrogen sensors using pure graphene and graphene derivatives such as graphene oxide and reduced graphene oxide. The response parameters like sensitivity, stability, selectivity, response time, recovery time, detection limit, linearity, dynamic range, and working temperatures for various graphene based sensors have been elaborately compared. Finally, a conclusion and future outlook on nano scale thin film of graphene and graphene oxides for gas sensing have been briefly discussed.

Keywords: nano thin film, graphene, graphene oxide, reduced graphene oxide, hydrogen sensor, sensing mechanism

1. Introduction

Dimensionality plays a vital role in determining the structure of a nanomaterial. Nanomaterials fall in the range of nanoscale defined by at least one dimension (D), and categorized using the concept of quantum confinement. Quantum confinement may be defined as a state in which a nanocrystal is comparatively smaller as compared to Bohr exciton radius of that particular material for at least one dimension. When it is quantum confined in one direction, it is termed as a film (i.e., a 2D material). For two dimensional quantum confinement, rod (i.e., a 1D material) is the name given to it, and quantum dot (i.e., a 0D material) has all the three dimensions confined. All these have specific shapes for the density of states which arises due to difference in the degree of confinement of electrons. Bulk materials fall under the category of 3D materials in which none of the domains are confined. Graphene is a 1D confined material and is obtained from its bulk counterpart

IntechOpen

graphite. The advantage of 2D materials is that they possesses "all surface" properties that can be conveniently tuned by chemical functionalization and other surface treatments. It even beholds its magnificent properties when synthesized in the form of nano flakes. This is the reason why the field of 2D materials is growing at such a rapid pace [1–3].

The mechanical exfoliation method to obtain graphene from graphite using scotch tape is not scalable to industrial practice owing to its limited yield. Although the quality yield is lofty, yet the mass scale production is an issue. Since it is much easier to obtain graphene oxide using wet chemical method as compared to graphene, so a careful study of both graphene, graphene oxide and reduced graphene oxide would help in gaining more insights about this family as well as the hydrogen sensing attributes on a comparative scale.

2. About graphene and graphene oxide

Graphene is structured as 2D sheet of six carbon (sp^2 hybridized) ring and confined in a hexagonal lattice. Substituting some of the hexagons with pentagons, imparts a spherical curvature culminating into 0D structure known as fullerene. Rolling up graphene sheets form carbon nanotubes (1D), and racked up graphene sheets into layers results into 3D graphite [4, 5].

Graphene lattice holds two sub lattices involving atomic sites A and B which are arranged in a planar trigonal fashion. Putting together the A and B sites along parallel layers (similar to a multilayer graphene) can generate the unit cell. The normal separation between two carbon atoms is 1.42 Å. However, for a hexagonal carbon lattice, the parameters are a = 2.46 and c = 6.7 Å. Graphene forms two types of bonds viz. sigma (σ) bond (three in number) and pi (π) bond (one in number). The σ bond is a consequence of interaction of its hybridized s, p(x) and p(y) orbitals with the neighboring carbon atoms; likewise π bond being formed through its interaction with the p(z) orbital [5].

Owing to arduous extraction and production of graphene, several graphene derivatives took the limelight namely graphene oxide (GO) and reduced graphene oxide (rGO). Where graphene involves only oxygen and hydrogen bonds, and functional groups attached to them like CO, COOH, OH groups, it is known as graphene oxide. This possesses a partial aromatic nature as compared to graphene which is completely aromatic in nature. Due to addition of these functional groups, a band gap could be achieved. The reduced form of graphene oxide bears much resemblance with graphene rather than graphene oxide in respect that all the functional groups have been removed with some oxygen traces [6].

3. Device configurations used for graphene based sensors

There are various configurations present by which graphene based gas sensors can be fabricated. Some of these configurations are chemiresistor, field effect transistor (FET), capacitance sensor, surface acoustic wave (SAW), surface work function change transistor, optical fiber sensor, etc. (**Figure 1**).

In some transistor configuration, the surface work function of the sensing material can be tuned, while in optical fiber sensors the optical properties such as transmission/absorption are affected due to solid-gas interaction.

There are few technically uncommon sensor devices. For instance, Schedin and group fabricated a graphene device for selective gas sensing. Electrical contacts were made by using electron beam lithography on the mechanically exfoliated

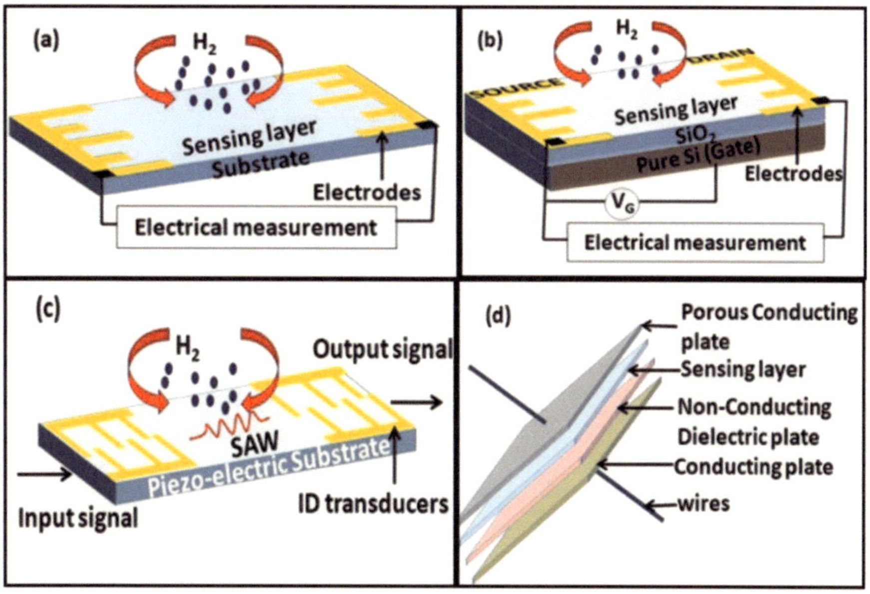

Figure 1.
Schematic representation of gas sensors: (a) chemiresistor (b) field effect transistor, (c) surface acoustic wave and (d) capacitance sensor.

graphene sheets on the Si substrates. The as prepared sensing device was placed in a superconducting magnet and characterized using field effect measurements within a temperature range in order to calculate the value of mobility of charge carriers and no. of graphene layers [7]. Dutta and group fabricated a heterojunction by putting a TiO_2 layer on graphene synthesized by chemical vapor deposition method [8]. Prezoiso and group dispersed GO in water and deposited on the interdigitated Pt electrodes patterned on Si_3N_4 substrate by drop casting technique and dried at 50°C. Wang and group chemically synthesized GO and fabricated a hydrogen sensor by dielectrophoresis technique [9]. In 2014, Niu and group proposed a very simple method for the fabrication of phosphorus modified graphene based gas sensor. Doping was done by annealing graphene oxide with triphenylphosphine within a temperature range from 400 to 800°C under inert atmosphere. The sensing device was fabricated by spreading the phosphorus doped graphene powder on the ceramic substrate with Pd electrodes [10].

4. Tuning of the sensing layer

Pristine graphene have a poor response towards gas sensing, whereas graphene oxide and reduced graphene oxide shows good response towards gas sensing. A small gas molecule adsorbed on the GO or rGO surface brings change in the conductivity. When compared, rGO shows better sensing response than GO. Peng and Li [11] demonstrated adsorption of NH_3 on graphene and graphene oxide surfaces and reported that although the calculated adsorption energy of graphene is lower than GO, the presence of surface hydroxyl and epoxy groups in GO promotes the NH_3 adsorption on the electron deficient sites (or hole site). This leads to easy electron transfer from nitrogen to the hole site in GO matrix. Similarly a comparative study showed graphene with functional groups as better sensing host as compared to pristine graphene (Gr) (**Table 1**). This can be attributed to the fact that graphene

Sensing layer	Gr	RGO	RGO-Gr
Device resistance	Low	High	Low
Response (%)	0.48%	3.32%	4.7%
SNR	Moderate	Low	High

Table 1.
Comparison of various types graphene based material performance.

Sensing layer	t (°C)	C (k)	Response	τ_{res}(s)
Pd/Gr	RT	1	$\Delta G/G_{air}$ = 26%	40
Pt/Gr	30	10	$\Delta R/R_{air}$ = 1.6%	1
Pd/Ag/Gr	105	0.5	$\Delta R/R_{air}$ = 9.96%	102
MoO_3/Gr	RT	1	R_{air}/R_{gas} = 20.5	10
CuO/rGO/CuO	RT	0.1	$\Delta R/R_{air}$ = 4.2%	<80
Pd/WO_3/Gr	RT	1	ΔI = 12 μA	17

t = temperature; C = concentration (μL/L); τ_{res}/τ_{rec} = response time/recovery time; RT = room temperature; k = 1000; Ref = references; G = conductance; R = resistance; I = current.

Table 2.
Graphene/nano-particle based gas sensors.

is perfect lattice with no functional groups hence no adsorption site for the gas molecules, while graphene oxide or reduced graphene oxide offer larger number of sites for the gases to adsorb on the surface. This efficiency was also confirmed from the high signal to noise ratio (SNR) [12].

Similar graphene based material tuning was done with metal or oxide nanoparticles to obtain interesting response with hydrogen (**Table 2**) [11, 13–18].

Gadipelli and Guo have clearly analyzed the graphene system for the uptake of carbon dioxide and hydrogen. They stated that more hydrogen can be adsorbed on modified porous structured graphene in comparison with normal chemically exfoliated graphene. Also, porous or doped graphene has been stated as a better host to adsorb carbon dioxide [15]. Hence porosity tuning is an important synthesis concern.

5. Electrical characteristics of graphene based materials in air

Due to the presence of oxygen functional group graphene oxide is insulating in nature which becomes electrically conducting on removal of oxygen by controlled reduction process. This results in the formation of reduced graphene oxide which on reduction still contains some of the oxygen groups. The remaining oxygen functional group is responsible for limiting the electron transport in rGO. The electrical characteristics can be analyzed by comparing parameters such as conductivity, barrier, etc., with respect to material doping, junction formation, material porosity, composite formation (with different concentrations), etc. The conductivity range for rGO is from 0.05 to 500 S/cm. This range depends on the reduction degree as this degree is related to the ratio of graphitic regions to that of oxidized regions. In one of the works, Jung et al. reported thermal and chemical combined reduction gave conductivity five times greater as compared to the samples that was reduced via thermal and chemical processes alone [16].

Gao et al. [19] synthesized sulfur doped graphene to study the electrical behavior of doped/undoped pristine graphene. The sheet resistance value S-doped graphene is 6.28×10^3 Ω/□, which is greater than that of pristine graphene. Hence doped system has lower conductivity. The presence of scattering centers (defects) in the doped sample is responsible for the lower conductivity (**Figure 2a**).

Jimenez and Dartora [17] gave a theoretical model which was successful in explaining the I-V characteristics of graphene based p-n tunnel junction. The study considered the effect of potential barrier of the junction and material doping (n- or p-type). It was concluded that the sign of potential barrier made no difference on the I-V characteristic and the conductivity increased with the increase in potential barrier (**Figure 2b**). This interesting fact is typical of graphene material which depicts unusual electronic properties in doped and undoped states. However, it needs further critical analysis.

Awasthi et al. [18] reported the I-V characteristics of multilayer graphene with polyaniline (mixed composite) at low temperature and room temperature with varying concentration. When the concentration of graphene nanosheets (GNS) was 3 wt%, the resistance came out to be 1.034 GΩ which decreased up to 4.106 MΩ on increasing the concentration to 6 wt% (**Figure 2c**). This increase in conductivity is attributed to the change in concentration of graphene only as all the other parameters were kept constant.

Haditale et al. [20] compared the I-V characteristics of graphene (G)/porous silicon (PS) hybrid structures with that of graphene/silicon (Si) by varying the graphene solution volume. The junctions were fabricated by taking 1 and 2 ml of graphene solutions, which were deposited on the substrates (Si or PS) by thermal spray pyrolysis. The G/Si junctions showed rectifying behavior and 1 ml devices showed higher values of current in comparison to 2 ml devices. Low volume (1 ml) resulted into low layered ordered structure of graphene which came out to be more conductive in nature. On increasing the volume to 2 ml, the structure became less ordered with lower conductivity value. On analyzing G/PS junctions following conclusions were drawn: (1) Reverse biased devices exhibited higher current values as compared to that of forward biased devices. (2) In reverse bias condition, G/Si have

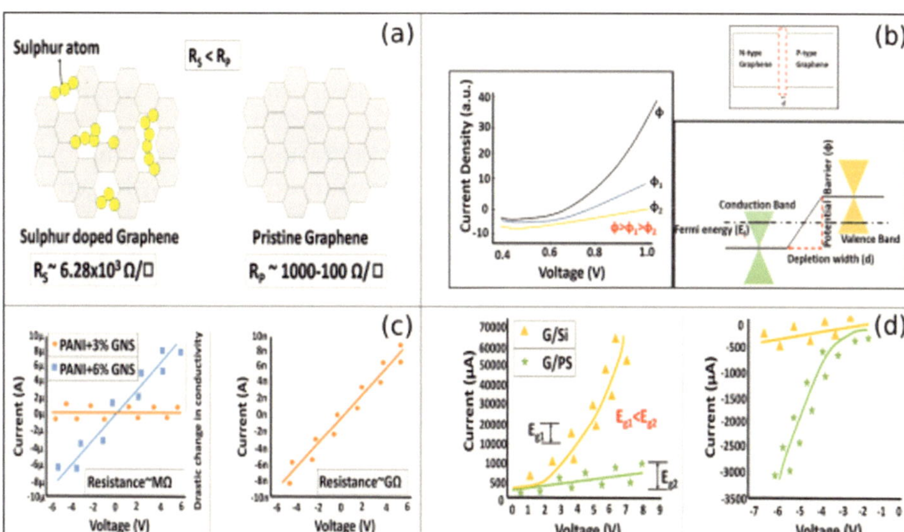

Figure 2.
Electrical characteristics of graphene based material in air to highlight the (a) effect of doping, (b) effect of barrier, (c) I-V change due to incorporation of other components like polyaniline (PANI), and (d) I-V change due to porosity.

zero current values (for low bias) whereas G/PS devices have non zero and high current values (for low bias). (3) As the graphene volume increases, the current value increases in G/PS devices. (4) The forward current is high in G/Si and low in G/PS.

These differences are due to the porosity. PS surface have large surface which can easily accommodate tiny graphene flakes in large quantities. Hence, higher current is obtained with increase in graphene concentration. However the defects in PS is responsible for the enhancement of reverse leakage characteristics. Moreover, there are two effective junctions (G/PS and PS/Si) in G/PS devices, and only one in G/Si. The unique existence of multiple junctions is due to quantum confinement effect that increases the bandgap of PS which is responsible for creating the effective potential barrier of G/PS devices. So, probably the forward barrier is high in G/PS and low in G/Si. Hence the forward I-V characteristic of G/Si is superior in comparison to G/PS (**Figure 2d**).

6. Electrical characteristics of graphene based materials in hydrogen

Any solid-gas (oxidizing or reducing) interaction is known to change the conductance of the solid. When the sensor is exposed to some oxidizing gas, a reaction will be triggered which is given as Eq. (1):

$$O_2 + e^{-1} \rightarrow 2O^{-1} \quad (1)$$

where O_2 is the charge accepting molecule and takes up the electron from sensor surface. This will reduce the conductance due to reduction in electron concentration. Or else when exposed to reducing gas like hydrogen, the reaction given below will be triggered as shown in Eq. (2).

$$H_2 + O^{-1} \rightarrow H_2O + e^{-1} \quad (2)$$

Here hydrogen will interact with oxygen ions adsorbed on the surface and result in increase in charge concentration. This will increase the device current [21]. The general experimental trend is to observe the device current/resistance variation at fixed bias. The I-V data in presence/absence of hydrogen can also be considered to evaluate various forward and reverse bias characteristics, and the voltage shift upon changing the ambient from air to hydrogen.

Temperature plays a crucial role in solid gas interaction and hence the electrical output can be tuned. In one report the concentration of hydrogen was fixed up to 10 ppm, and change in device current (from its value in air) was found to vary (initial increase and then decrease) with the increase in temperature [22]. The initial increase in the enhancement of adsorption while the later decrease is due to slow dominance of desorption over adsorption. The role of temperature was also reported by Dutta et al. in which the current of Pd/graphene junctions increased in hydrogen up to 100°C, beyond which the junction current decreased [23].

Composite morphology modulated the electrical output in hydrogen. As reported for graphene, the change in device current (from its value in air) was appreciably large for graphene/Pd nano-composite and negligibly small for the pure graphene matrix [24].

Similarly via other research effort it is well established that the electrical characteristics in presence of hydrogen not only depends on the gas concentration, but on parameters such humidity [25], catalytic modification [26], etc.

7. Parametric response analysis

Some of basic parameters that are considered for gas sensors are sensitivity (or response percent), selectivity, detection limit, dynamic range, linearity, response time, recovery time and working temperature. **Figure 3** presents pictorially some of the important parameters. All these parameters are very important to describe the properties of a sensing device or material. Sensitivity of a device is the ratio (per unit analyte concentration) of measured signal magnitude at a particular concentration limit of analyte and in absence of analyte. The response percent is the ratio of change in signal magnitude due to a particular concentration limit of analyte and the initial signal magnitude without analyte, which is expressed as a percentage Eq. (3):

$$\text{Response percent} = \frac{|I_g - I_a|}{I_a} \times 100 \tag{3}$$

Selectivity is the ability of a gas sensing device to act in response to a particular group of analytes or specifically to a single analyte. The lowest concentration of gas that can be detected by a sensing device at a specific temperature is known as detection limit. Detection range is the concentration range (minimum to maximum) that the sensor can detect. Linearity is nothing but a deviation from an ideal straight line to experimentally obtained calibration graph. Response time is the time needed by sensor to reach 90% of the final saturation value while responding to a concentration change, i.e., from zero to a certain value of concentration. Recovery time is the time the sensor signal takes to return to its 90% its initial baseline value. Working temperature is the temperature at which maximum sensitivity (or percent response) or minimum response time is achieved. High values of sensitivity, selectivity, and detection range while low values of detection limit, response time and recovery time are some characteristics of an ideal sensing

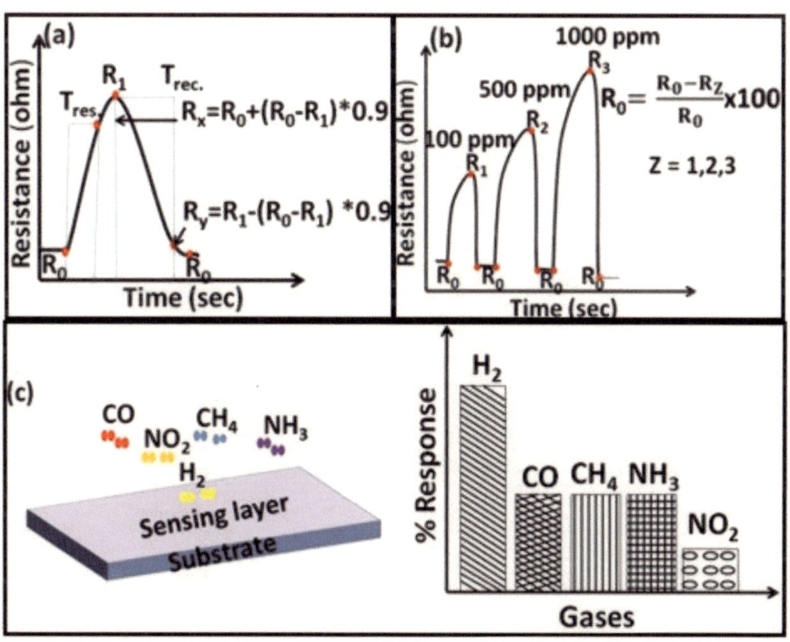

Figure 3.
Schematic representation of (a) response time, and recovery time (b), sensitivity (c) selectivity of gas sensors.

device [27]. These parameters have been improved by various ways. For instance surface decoration improved the sensing parameters in an H_2 gas sensor based on reduced graphene oxide sheets decorated with nano-structured platinum [28]. The surface decorated graphene sensors showed higher sensitivity (of 16% at room temperature) than surface decorated carbon nanotubes. An extraordinary response to extremely low concentration of hydrogen was obtained with sensor based on rGO loaded ZnO nano-fibers [29]. They concluded that the properties

SL	OM	t	D	S	t_{res}	t_{rec}	Ref
Gr	—	50	—	69%	19 s	612 s	[34]
rGO	CuO	RT	10–1500 ppm	—	80 s	60 s	[35]
rGO	Pt	50	0.5%	8%	63 s	104 s	[36]
GO	—	RT	150 ppm	81%	—	—	[37]
Gr	Pd	RT	—	—	140 s	250 s	[38]
GO	Pd	RT	1%	~0.5%	—	—	[39]
rGO	PANI	RT	0.06–1%	16.57%	—	—	[40]
rGO	Pt-SnO_2	50	0.5–3%	—	2–6 s	2–6 s	[41]
GA	Pt	—	10,000 ppm	1.6%	0.97 s	0.72 s	[42]
Gr	Pd	—	1000 ppm	32.5%	—	—	[43]
rGO	Ni-Pd	—	0.1%	11%	3 s	12 s	[44]
Gr	Pd	—	0.1%	7%	4 s	20 s	[45]
Gr	Pd	—	0.1%	33%	15 s	25 s	[46]
Gr	Pd	—	1%	50%	1 s	2 s	[39]
Gr	Pt-Pd	—	2%	4%	2 s	60 s	[47]
Gr	Pd-Ag	245	1 ppm	—	16 s	14 s	[48]
rGO	Pd -Pt	RT	6 ppm	—	3 min	1.2 min	[49]
rGO	Pd	RT	1%	—	1 s	9 s	[50]
GO	—	RT	200 ppm	6%	11 s	36 s	[51]
rGO	NiO	80	1%	1.58%	15 s	61 s	[52]
rGO	Pd	RT	100 ppm	1.6%	—	—	[53]
rGO	Pt	RT	4000 ppm	17%	—	—	[28]
rGO	Pt	RT	200 ppm	40%	—	—	[54]
rGO	Pd	RT	1000 ppm	13%	—	—	[55]
rGO	Pd	RT	1000 ppm	5%	—	—	[56]
rGO	SnO_2	RT	10,000 ppm	1.8%	—	—	[41]
rGO	ZnO	150	200 ppm	3.5%	—	—	[57]
rGO	NiO	100	2000 ppm	24%	—	—	[58]
Gr	TiO_2	125	5000 ppm	8%	—	—	[59]
rGO	MoS_2	60	200 ppm	15.6%	—	—	[60]

SL = sensing layer; OM = other material (used); t = sensing temperature (°C); D = detection range; S = sensitivity; t_{res} = response time; t_{rec} = recovery time; RT = room temperature; G = graphene; GA = graphene aerogel; Ref = reference.

Table 3.
Detailed data of various graphene based hydrogen sensors.

of p type rGO sheets and semiconductor to metal transition in ZnO jointly are responsible for excellent sensing performance towards such a low concentration (100 ppb) of hydrogen. The selectivity of graphene based sensors was analyzed theoretically by Maity et al. by considering surface functional groups such as carboxyl, carbonyl, hydroxyl, epoxy and hydroxyl [30]. They reported that with GO nanoflakes, COOH group helps in selective NH_3 detection, while carbonyl group favors selective NO_2 detection. Other ways to improve the selectivity of graphene based sensors is to use filtration polymer based membranes, which was experimentally demonstrated by Hong et al. [31]. The PMMA (poly(methylmethacrylate)) polymeric coating on palladium/single layer graphene allowed only hydrogen molecules to pass through and reach the palladium surface. As a result, molecules like CH_4, CO, and NO_2 showed no response.

Another group in 2016 studied the properties of bimetallic-rGO based sensor for hydrogen detection under different working circumstances. Pt-Pd nano-particles were decorated on rGO surface via one step chemical reduction method and the sensor study was done by observing the resistance change upon exposure to different concentrations of hydrogen at room temperature. They concluded that the response percent was dependent on concentration and operating temperature but was independent of flow rate [32]. Moreover, the sensor recovery was faster in air than nitrogen, which were used as carrier gases. In 2009, the sensing properties of Pt decorated rGO sheets deposited on SiC substrates were studied by Shafiei et al. [33]. By studying the reverse I-V characteristics (while maintaining constant device current) in 1% of hydrogen in air, they observed a significant 100 mV shift at 100°C.

Similar other reports have also been tabulated in **Table 3**, which gives an idea about the variation of the response parameters in the field of graphene based hydrogen sensors.

8. Suggested mechanism of hydrogen sensing

A general overview or in particular few important hydrogen sensing highlights as proposed by research groups are compiled to explain the sensing mechanism of graphene based hydrogen sensors [13, 61]. Basically four factors critically influence the sensing mechanism: (a) relevance of material, (b) development of junction, (c) surface adsorbed species and (d) external perturbation during sensing.

8.1 Relevance of materials

Introduction of nanoparticles in the base matrix helps in preventing agglomeration of layered materials (such as graphene), and increase the surface area by forming three dimensional networks; an essential part to improve the base matrix performance during sensing [41]. Also, it is reported that catalytic metals such as palladium normally improves the adsorption/desorption kinetics [62]. Furthermore, synergistic effect (cumulative response) due to the presence of different sensing hosts is very important from gas sensor perspectives. Synergistic effect between copper oxide and graphene was the proposed mechanism by Zhang et al. [35]. Copper oxide (CuO) acts as good p-type semiconductor while graphene (rGO) acts as excellent conducting platform. Number of holes in the accumulation layer between CuO-rGO account for the electrical conductance change during hydrogen sensing (**Figure 4**).

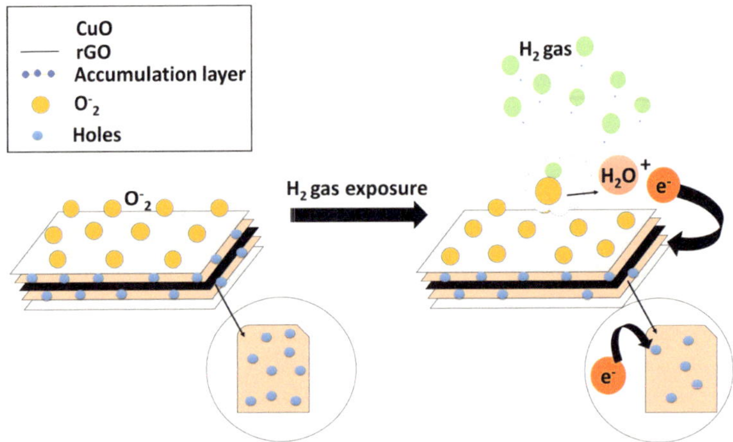

Figure 4.
Sensing mechanism graphene based material highlighting the role of accumulation layer.

8.2 Development of junction

The gas sensing properties can be enhanced by forming p-n heterojunctions between metal or metal oxide and graphene. Dutta and co-workers [59] have suggested a mechanism for p-TiO_2/n-graphene with palladium contacts. When this device was exposed to hydrogen gas due to interaction of palladium with the gas, dissociative adsorption took place as shown in Eq. (4):

$$H_2(g) \rightarrow 2H(ads) \quad (4)$$

Thereafter diffusion of atomic hydrogen into bulk palladium occur. A non-stoichiometric hydride PdH_x is formed which serves as dipole because of charge separation. Here hydrogen acts as positive center and palladium as negative center. An electron from diffusing hydrogen is now shared between the semiconductor and the palladium. Due to this an increase in concentration of electron at the interface leads to the reduction of work function of palladium. Shift in Fermi level is suggested due to which the heterojunction barrier changes. Change in this barrier during gas exposure is responsible for the change in device current. They observed that the current increased when exposed to hydrogen gas, implying reduction in barrier height. During recovery, the hydrogen gas desorb from the surface. This shows decrease in current in the circuit. Operating temperature has a major role in determining the adsorption and desorption of the gas molecules.

8.3 Surface adsorbed species

Adsorbed oxygen in hybrid materials that contain metal oxides such as SnO_2, CuO, ZnO, etc., tune the device resistance in presence and absence of test gas [13]. Sensing mechanism is based on the phenomenon of adsorption and dissociation of gas molecules on the host surface. Due to these interactions either a carrier is generated or a carrier is annihilated, which affects the resistance of the base matrix. Oxygen adsorbed on the graphene surface aids in good sensing. This oxygen accommodates an electron during its surface adsorption and stays in charged state on the surface of the sensing layer. During sensing there is interaction between the charged oxygen radical and the gas molecules and an electron is released. Hence, during gas exposure the surface oxygen gets removed from the surface, which is again recovered in sensor recovery process.

8.4 External perturbation during sensing

Tang and group in a sensing study with palladium and graphene discussed the formation of two phases, "α" and "β" in hydrogen adsorbed palladium [61] (**Figure 5**). At room temperature, pressure as low as 0.01 bar leads to the saturation of "α" phase. It results in 10% ratio of H/Pd. Their mechanism is based on the change in resistance of graphene which in turn is dependent on the Pd work function. Since Pd has higher work function (5.2 eV) than graphene's (4.7 eV), therefore on Pd decoration, resistance of graphene tend to decrease. So considering p-type graphene, it is apparent that there is increase in hole density of graphene upon Pd decoration because the electron transfer occurs from graphene to palladium. During hydrogen exposure PdH_x is formed and its work function (3.2 eV) is less than graphene. Therefore in accordance to the previous argument, the resistance will increase and hole density will decrease. Also, the electron transfer is from PdH_x to graphene. This mechanism was more like a shark-fin (signal increase and decrease were slow and response percent low) curve, which were observed only in dark condition (without external light energy perturbation). Under light illumination (photon energy ~3.1 eV) square type (signal increase and decrease were fast and response percent high) response was observed (**Figure 5**). This difference is due to the extra energy provided by the external light source. During response, illumination energy is increasing the adsorption amount as well as accelerating fast dissociation of hydrogen molecules on the surface of palladium. During recovery, the low dissociation energy (2.96 eV) of PdH_x in comparison to incident photon energy helps in easy release of the adsorbed hydrogen, and the Pd work function slowly reverts back to its initial value. So the response parameters improved due to external perturbation [61]. Novoselov et al. also showed that it is possible to regenerate the sensor to its original state

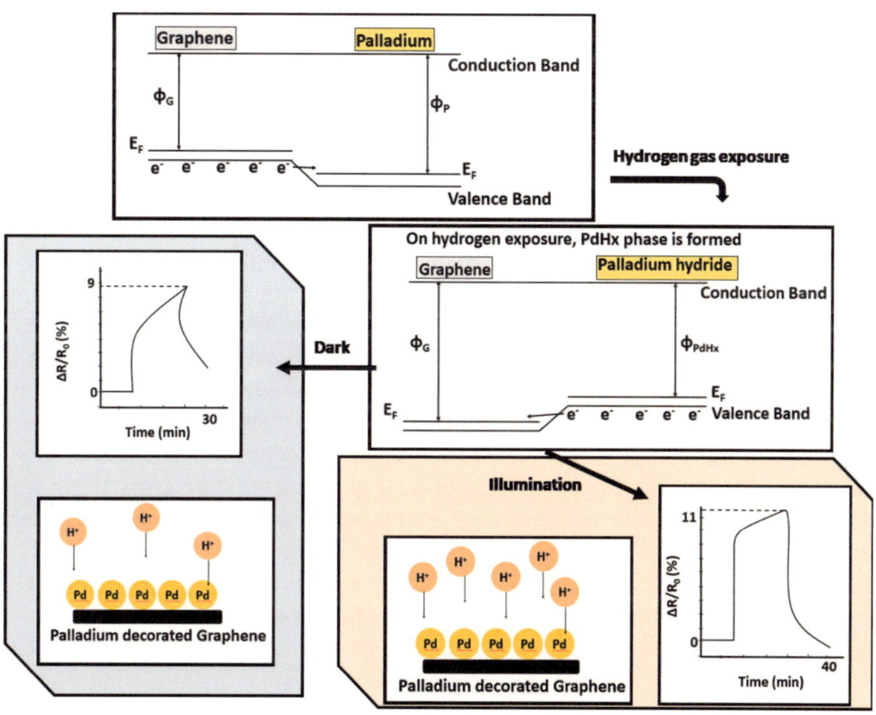

Figure 5.
Effect of external perturbation (light illumination) during sensing.

by annealing at 150°C in vacuum or by short UV radiation exposure [7]. This can only be achieved within 100–200 s.

9. Performance rating

Graphene and graphene oxide based devices in sensing must have the following qualities such as (a) reliability, (b) reproducibility and (c) long term stability.

The accuracy and precision of a sensing device leads to its reliability in practical applications. The sensing system's capability to give same results when subjected to same or slightly distinct conditions is known as the reproducibility of the system. When same parameter is measured after a period of time, the ability of a sensing sensor to give the same output is known as its long term stability.

In order to test these qualities, the device is tested with respect to (a) the magnitude of response parameters within error limits under perturbed ambient conditions (such as high humidity or domestic/commercial/industrial locations), (b) the nature of response of a single device or multiple devices prepared under similar conditions and (c) time of aging (by continuous or repeated operation). Few notable experimental demonstrations in this regard are discussed below:

Dutta et al. have produced same results (with p-TiO_2/n-graphene heterojunction) in consecutive 3 days of sensor studies with 1% concentration of hydrogen in air [59]. The testing period can be extended from days to weeks or months for monitoring the response parameters. The reliability and reproducibility of graphene sensors were tested by Noll et al. by monitoring input characteristics continuously of a single device and then with large number of devices, respectively [63].

10. Conclusion and future outlook

Graphene and graphene oxides can be abundantly used as sensing materials due to their amazing properties, which are easy to tune. Reports regarding graphene and graphene oxide based gas sensors for hydrogen detection have proved that these materials can be used to selectively detect hydrogen. Hence there is possibility of fabrication of gas sensors with desired sensitivity, selectivity and stability with these materials. Furthermore, the possibility of doping or modification with other materials widens the future scope of graphene and graphene oxide based gas sensors. Reduced graphene oxide is also proven to be a tremendous sensing material for hydrogen along with graphene and graphene oxide. In fact the performance of reduced graphene oxide as a sensing material is better as compared to that of pristine graphene and oxidized graphene. Also, modification of graphene based layer with metal nano-particles have improved the response parameters. In addition to monometallic metal nanoparticles, bi-metallic or tri-metallic nano-particles also exhibit better catalytic properties due to synergistic effect between them. So it is apparent that graphene based layers modified with bi-metallic nano-particles or tri metallic nano-particles to form ternary or quaternary composites can be widely used as sensing hosts in near future. Hence within the purview of this review, it can be concluded that the potentiality of these materials is tremendous in the field of hydrogen sensing, which can further be cultivated to obtain desired selective response characteristics with other gases.

DOI: http://dx.doi.org/10.5772/intechopen.88538

Author details

Anuradha Kashyap, Shikha Sinha, Partha Bir Barman and Surajit Kumar Hazra*
Department of Physics and Materials Science, Jaypee University of Information, Technology, Solan, Himachal Pradesh, India

*Address all correspondence to: surajithazra@yahoo.co.in

References

[1] Butler SZ, Hollen SM, Cao L, Cui Y, Gupta JA, Gutiérrez HR, et al. Progress, challenges, and opportunities in two-dimensional materials beyond graphene. ACS Nano. 2013;7:2898-2926. DOI: 10.1021/nn400280c

[2] Zhou SY, Gweon GH, Graf J, Fedorov AV, Spataru CD, Diehl RD, et al. First direct observation of Dirac fermions in graphite. Nature Physics. 2006;**2**:595. DOI: 10.1038/nphys393

[3] Geim AK, Novoselov KS. The rise of graphene. Nature Materials. 2007;**6**: 183-191. DOI: 10.1038/nmat1849

[4] Zhu Y, Murali S, Cai W, Li X, Suk JW, Potts JR, et al. Graphene and graphene oxide: Synthesis, properties, and applications. Advanced Materials. 2010;**22**:3906-3924. DOI: 10.1002/adma.201001068

[5] Edwards RS, Coleman KS. Graphene synthesis: Relationship to applications. Nanoscale. 2013;**5**:38-51. DOI: 10.1039/C2NR32629A

[6] Fitzer E, Kochling KH, Boehm HP, Marsh H. Recommended terminology for the description of carbon as a solid (IUPAC recommendations 1995). Pure and Applied Chemistry. 1995;**67**:473-506. DOI: 10.1351/pac199567030473

[7] Schedin F, Geim AK, Morozov SV, Hill EW, Blake P, Katsnelson MI, et al. Detection of individual gas molecules adsorbed on graphene. Nature Materials. 2007;**6**:652. DOI: 10.1038/nmat1967

[8] Dutta D, Das J, Hazra SK, Sarkar CK, Basu S. Influence of graphene growth temperature by chemical vapour deposition on the hydrogen response of palladium-graphene junction. Journal of Materials Science: Materials in Electronics. 2017;**28**:13217-13228. DOI: 10.1007/s10854-017-7157-2

[9] Wang J, Singh B, Park JH, Rathi S, Lee IY, Maeng S, et al. Dielectrophoresis of graphene oxide nanostructures for hydrogen gas sensor at room temperature. Sensors and Actuators B: Chemical. 2014;**194**:296-302. DOI: 10.1016/j.snb.2013.12.009

[10] Niu F, Tao LM, Deng YC, Wang QH, Song WG. Phosphorus doped graphene nanosheets for room temperature NH_3 sensing. New Journal of Chemistry. 2014;**38**:2269-2272. DOI: 10.1039/C4NJ00162A

[11] Peng Y, Li J. Ammonia adsorption on graphene and graphene oxide: A first-principles study. Frontiers of Environmental Science & Engineering. 2013;**7**:403-411. DOI: 10.1007/s11783-013-0491-6

[12] Wang C, Lei S, Li X, Guo S, Cui P, Wei X, et al. A reduced GO-graphene hybrid gas sensor for ultra-low concentration ammonia detection. Sensors. 2018;**18**:3147. DOI: 10.3390/s18093147

[13] Jin L, Chen W, Zhang Y. Application of graphene hybrid materials in fault characteristic gas detection of oil-immersed equipment. Frontiers in Chemistry. 2018;**6**(399):1-8. DOI: 10.3389/fchem.2018.00399

[14] Chen M, Zou L, Zhang Z, Shen J, Li D, Zong Q, et al. Tandem gasochromic-Pd-WO_3/graphene/Si device for room-temperature high-performance optoelectronic hydrogen sensors. Carbon. 2018;**130**:281-287. DOI: 10.1016/j.carbon.2018.01.013

[15] Gadipelli S, Guo ZX. Graphene-based materials: Synthesis and gas sorption, storage and separation. Progress in Materials Science. 2015;**69**: 1-60. DOI: 10.1016/j.pmatsci.2014.10.004

[16] Singh V, Joung D, Zhai L, Das S, Khondaker SI, Seal S. Graphene based

materials: Past, present and future. Progress in Materials Science. 2011;**56**:1178-1271. DOI: 10.1016/j.pmatsci.2011.03.003

[17] Jimenez MS, Dartora CA. The I-V characteristics of a graphene tunnel diode. Physica E: Low-dimensional Systems and Nanostructures. 2014;**59**: 1-5. DOI: 10.1016/j.physe.2013.12.014

[18] Awasthi S, Gopinathan PS, Rajanikanth A, Bansal C. Current-voltage characteristics of electrochemically synthesized multi-layer graphene with polyaniline. Journal of Science: Advanced Materials and Devices. 2018;**3**:37-43. DOI: 10.1016/j.jsamd.2018.01.003

[19] Gao H, Liu Z, Song L, Guo W, Gao W, Ci L, et al. Synthesis of S-doped graphene by liquid precursor. Nanotechnology. 2012;**23**(27):275605

[20] Haditale M, Zabihipour A, Koppelaar H. A comparison of IV characteristics of graphene silicon and graphene-porous silicon hybrid structures. Superlattices and Microstructures. 2018;**122**:387-393. DOI: 10.1016/j.spmi.2018.07.005

[21] Kim Y, Choi YS, Park SY, Kim T, Hong S, Moon CW, et al. Hydrogen gas detection of gold nanoparticles on self-activated graphene layers. In: 17th International Meeting on Chemical Sensors—IMCS; 2018. DOI 10.5162/IMCS2018/P2GS.7

[22] Zhang Z, Zou X, Xu L, Liao L, Liu W, Ho J, et al. Hydrogen gas sensor based on metal oxide nanoparticles decorated graphene transistor. Nanoscale. 2015;**7**:10078-10084. DOI: 10.1039/C5NR01924A

[23] Dutta D, Hazra SK, Das J, Sarkar CK, Basu S. Temperature- and hydrogen-gas-dependent reversible inversion of n-/p-type conductivity in CVD-grown multilayer graphene (MLG) film. Journal of Electronic Materials. 2016;**45**(6):2861-2869

[24] Gutés A, Hsia B, Sussman A, Mickelson W, Zettl A, Carraro C, et al. Graphene decoration with metal nanoparticles: Towards easy integration for sensing applications. Nanoscale. 2012;**4**:438-440. DOI: 10.1039/C1NR11537E

[25] Alfano B, Massera E, Polichetti T, Miglietta ML, Di Francia G. Effect of humidity on the hydrogen sensing in graphene based devices. In: Convegno Nazionale Sensori. Vol. 539. Springer, Cham; 2018. pp. 11-16. DOI: 10.1007/978-3-030-04324-7_2

[26] Wang T, Huang D, Yang Z, Xu S, He G, Li X, et al. A review on graphene-based gas/vapor sensors with unique properties and potential applications. Nano-Micro Letters. 2016;**8**:95-119. DOI: 10.1007/s40820-015-0073-1

[27] Bochenkov VE, Sergeev GB. Sensitivity, selectivity, and stability of gas-sensitive metal-oxide nanostructures. In: Metal Oxide Nanostructures and their Applications. Vol. 3. American Scientific Publishers; 2010. pp. 31-52

[28] Kaniyoor A, Jafri RI, Arockiadoss T, Ramaprabhu S. Nanostructured Pt decorated graphene and multi walled carbon nanotube based room temperature hydrogen gas sensor. Nanoscale. 2009;**1**:382-386. DOI: 10.1039/B9NR00015A

[29] Abideen ZU, Kim HW, Kim SS. An ultra-sensitive hydrogen gas sensor using reduced graphene oxide-loaded ZnO nanofibers. Chemical Communications. 2015;**51**:15418-15421. DOI: 10.1039/C5CC05370F

[30] Maity I, Ghosh K, Rahaman H, Partha B. Selectivity tuning of graphene oxide based reliable gas sensor devices by tailoring the oxygen functional

groups: A DFT study based approach. IEEE Transactions on Device and Materials Reliability. 2017;**17**(4):738-745

[31] Hong J, Lee S, Seo J, Pyo S, Kim J, Lee TA. Highly sensitive hydrogen sensor with gas selectivity using a PMMA membrane-coated Pd nanoparticle/single-layer graphene hybrid. ACS Applied Materials & Interfaces. 2015;7(6):3554-3561

[32] Peng Y, Ye J, Zheng L, Zou K. The hydrogen sensing properties of Pt-Pd/reduced graphene oxide based sensor under different operating conditions. RSC Advances. 2016;**6**:24880-24888. DOI: 10.1039/C5RA26618A

[33] Shafiei M, Arsat R, Yu J, Kalantar-Zadeh K, Wlodarski W, Dubin S, et al. Pt/graphene nano-sheet based hydrogen gas sensor. In: SENSORS, 2009 IEEE; 2009. pp. 295-298. DOI: 10.1109/ICSENS.2009.5398157

[34] Dutta D, Bontempi E, You Y, Sinha S, Das J, Hazra SK, et al. Surface topography and hydrogen sensor response of APCVD grown multilayer graphene thin films. Journal of Materials Science: Materials in Electronics. 2017;**28**:157-166. DOI: 10.1007/s10854-016-5506-1

[35] Zhang D, Yin N, Jiang C, Xia B. Characterization of CuO-reduced graphene oxide sandwiched nanostructure and its hydrogen sensing characteristics. Journal of Materials Science: Materials in Electronics. 2017;**28**:2763-2768. DOI: 10.1007/s10854-016-5856-8

[36] Lu X, Song X, Gu C, Ren H, Sun Y, Huang J. Freeze drying-assisted synthesis of Pt@ reduced graphene oxide nanocomposites as excellent hydrogen sensor. Journal of Physics and Chemistry of Solids. 2018;**116**:324-330. DOI: 10.1016/j.jpcs.2018.02.006

[37] Pavithra A, Rakkesh RA, Durgalakshmi D, Balakumar S. Room temperature detection of hydrogen gas using graphene based conductometric gas sensor. Journal of Nanoscience and Nanotechnology. 2017;**17**:3449-3453. DOI: 10.1166/jnn.2017.13054

[38] Kim KS, Chung GS. Fabrication and characterization of hydrogen sensors based on transferred graphene synthesized by annealing of Ni/3C-SiC thin films. Surface Review and Letters. 2014;**21**:1450050. DOI: 10.1142/S0218625X14500504

[39] Lange U, Hirsch T, Mirsky VM, Wolfbeis OS. Hydrogen sensor based on a graphene–palladium nanocomposite. Electrochimica Acta. 2011;**56**:3707-3712. DOI: 10.1016/j.electacta.2010.10.078

[40] Al-Mashat L, Shin K, Kalantar-zadeh K, Plessis JD, Han SH, Kojima RW, et al. Graphene/polyaniline nanocomposite for hydrogen sensing. The Journal of Physical Chemistry C. 2010;**114**:16168-16173. DOI: 10.1021/jp103134u

[41] Russo PA, Donato N, Leonardi SG, Baek S, Conte DE, Neri G, et al. Room-temperature hydrogen sensing with hetero nanostructures based on reduced graphene oxide and tin oxide. Angewandte Chemie (International Ed. in English). 2012;**51**:11053-11057. DOI: 10.1002/anie.201204373

[42] Harley-Trochimczyk A, Chang J, Zhou Q, Dong J, Pham T, Worsley MA, et al. Catalytic hydrogen sensing using microheated platinum nanoparticle-loaded graphene aerogel. Sensors and Actuators B: Chemical. 2015;**206**:399-406. DOI: 10.1016/j.snb.2014.09.057

[43] Goto A, Takeuchi G, Yamachi R, Tanaka T, Takahashi T, Uchida K. Impact of hydrogen on carrier mobility and concentration in graphene decorated with Pd nanoparticle. ECS

Transactions. 2016;**72**:7-12. DOI: 10.1149/07214.0007ecst

[44] Phan DT, Chung GS. Reliability of hydrogen sensing based on bimetallic Ni–Pd/graphene composites. International Journal of Hydrogen Energy. 2014;**39**:20294-20304. DOI: 10.1016/j.ijhydene.2014.10.006

[45] Phan DT, Chung GS. Characteristics of resistivity-type hydrogen sensing based on palladium-graphene nanocomposites. International Journal of Hydrogen Energy. 2014;**39**:620-629. DOI: 10.1016/j.ijhydene.2013.08.107

[46] Chung MG, Kim DH, Seo DK, Kim T, Im HU, Lee HM, et al. Flexible hydrogen sensors using graphene with palladium nanoparticle decoration. Sensors and Actuators B: Chemical. 2012;**169**:387-392. DOI: 10.1016/j.snb.2012.05.031

[47] Kumar R, Varandani D, Mehta BR, Singh VN, Wen Z, Feng X, et al. Fast response and recovery of hydrogen sensing in Pd-Pt nanoparticle-graphene composite layers. Nanotechnology. 2011;**22**:275719. DOI: 10.1088/0957-4484/22/27/275719

[48] Sharma B, Kim JS. MEMS based highly sensitive dual FET gas sensor using graphene decorated Pd-Ag alloy nanoparticles for H_2 detection. Scientific Reports. 2018;**8**:5902. DOI: 10.1038/s41598-018-24324-z

[49] Phan DT, Chung GS. Effects of palladium nanocrystal morphologies on hydrogen sensors based on palladium-graphene hydrid. In: 2015 IEEE SENSORS; 2015. pp. 1-4. DOI: 10.1109/ICSENS.2015.7370203

[50] Ha NH, Nam NH, Dung DD, Phuong NH, Thach PD, Hong HS. Hydrogen gas sensing using palladium-graphene nanocomposite material based on surface acoustic wave. Journal of Nanomaterials. 2017;**9057250**:1-6. DOI: 10.1155/2017/9057250

[51] Wang J, Singh B, Maeng S, Joh HI, Kim GH. Assembly of thermally reduced graphene oxide nanostructures by alternating current dielectrophoresis as hydrogen-gas sensors. Applied Physics Letters. 2013;**103**:083112. DOI: 10.1063/1.4819378

[52] Ren H, Gu C, Joo SW, Cui J, Sun Y, Huang J. Preparation of SnO_2 nanorods on reduced graphene oxide and sensing properties of as-grown nanocomposites towards hydrogen at low working temperature. Materials Express. 2018;**8**:263-271. DOI: 10.1166/mex.2018.1428

[53] Yatskiv R, Grym J. Hydrogen sensing using reduced graphene oxide sheets supported by Pd nanoparticles. Journal of Physics: Conference Series. 2013;**450**:012020. DOI: 10.1088/1742-6596/450/1/012020

[54] Vedala H, Sorescu DC, Kotchey GP, Star A. Chemical sensitivity of graphene edges decorated with metal nanoparticles. Nano Letters. 2011;**11**:2342-2347. DOI: 10.1021/nl2006438

[55] Pandey PA, Wilson NR, Covington JA. Pd-doped reduced graphene oxide sensing films for H_2 detection. Sensors and Actuators B: Chemical. 2013;**183**:478-487. DOI: 10.1016/j.snb.2013.03.089

[56] Pak Y, Kim SM, Jeong H, Kang CG, Park JS, Song H, et al. Palladium-decorated hydrogen-gas sensors using periodically aligned graphene nanoribbons. ACS Applied Materials & Interfaces. 2014;**6**:13293-13298. DOI: 10.1021/am503105s

[57] Anand K, Singh O, Singh MP, Kaur J, Singh RC. Hydrogen sensor based on graphene/ZnO nanocomposite. Sensors

and Actuators B: Chemical. 2014;**195**:409-415. DOI: 10.1016/j.snb.2014.01.029

[58] Kamal T. High performance NiO decorated graphene as a potential H_2 gas sensor. Journal of Alloys and Compounds. 2017;**729**:1058-1063. DOI: 10.1016/j.jallcom.2017.09.124

[59] Dutta D, Hazra SK, Das J, Sarkar CK, Basu S. Studies on p-TiO_2/n-graphene heterojunction for hydrogen detection. Sensors and Actuators B: Chemical. 2015;**212**:84-92. DOI: 10.1016/j.snb.2015.02.009

[60] Venkatesan A, Rathi S, Lee IY, Park J, Lim D, Kang M, et al. Molybdenum disulfide nanoparticles decorated reduced graphene oxide: Highly sensitive and selective hydrogen sensor. Nanotechnology. 2017;**28**:365501. DOI: 10.1088/1361-6528/aa7d66

[61] Tang X, Haddad PA, Mager N, Geng X, Reckinger N, Hermans S, et al. Chemically deposited palladium nanoparticles on graphene for hydrogen sensor applications. Scientific Reports. 2019;**9**:3653. DOI: 10.1038/s41598-019-40257-7

[62] Hazra S, Basu S. Graphene-oxide nano composites for chemical sensor applications. C—Journal of Carbon Research. 2016;**2**:12. DOI: 10.3390/c2020012

[63] Noll D, Schwalke U. Yield and reliability of nanocrystalline graphene field-effect gas sensors. ECS Transactions. 2018;**86**(9):41-49. DOI: 10.1149/08609.0041ecst

Chapter 10

Multilayered Nanostructures Integrated with Emerging Technologies

Maria L. Braunger, Rafael C. Hensel, Gabriel Gaál, Mawin J.M. Jimenez, Varlei Rodrigues and Antonio Riul Jr

Abstract

Surface and interface functionalization are crucial steps to introduce new functionalities in numerous applications, as faster dynamics occur on surfaces rather than bulk. Within this context, the layer-by-layer (LbL) technique is a versatile methodology to controllably form organized nanostructures from the spontaneous adsorption of charged molecules. It enables the assembly of multilayered LbL films on virtually any surface using non-covalent molecular interactions, allowing the nanoengineering of interfaces and creation of multifunctional systems with distinct building blocks (polymers, clays, metal nanoparticles, enzymes, organic macromolecules, etc.). Several applications require thin films on electrodes for sensing/biosensing, and here we explore LbL films deposited on interdigitated electrodes (IDEs) that were 3D-printed using the fusing deposition modeling (FDM) technique. IDEs covered with LbL films can be used to form multisensory systems employed in the analysis of complex liquids transforming raw data into specific patterns easily recognized by computational and statistical methods. We extend the FDM 3D-printing methodology to simplify the manufacturing of electrodes and microchannels, thus integrating an e-tongue system in a microfluidic device. Moreover, the continuous flow within microchannels contributes to faster and more accurate analysis, reducing the amount of sample, waste, and costs.

Keywords: LbL assembly, interdigitated electrodes, microchannel, 3D printing, electrical measurements, e-tongue

1. Introduction

In 1966, Iler proposed the deposition of multilayered films through the consecutive adsorption of charged inorganic colloids [1]. In 1997, Decher and coworkers extended Iler's method in an even more general approach using the spontaneous physical adsorption of materials from aqueous solutions [2]. It enables the assembly of layer-by-layer (LbL) nanostructured thin films on solid supports using a wide variety of building blocks (DNA, enzymes, proteins, polymers, inorganic particles, clays, colloids, etc.). The widespread use of the LbL technique in numerous applications has led to the development of conventional procedures that were organized in five distinct categories [3, 4]: immersive, spin, spray, electromagnetic, and fluidic assemblies. **Figure 1** shows a simplified schematic diagram of the LbL immersive

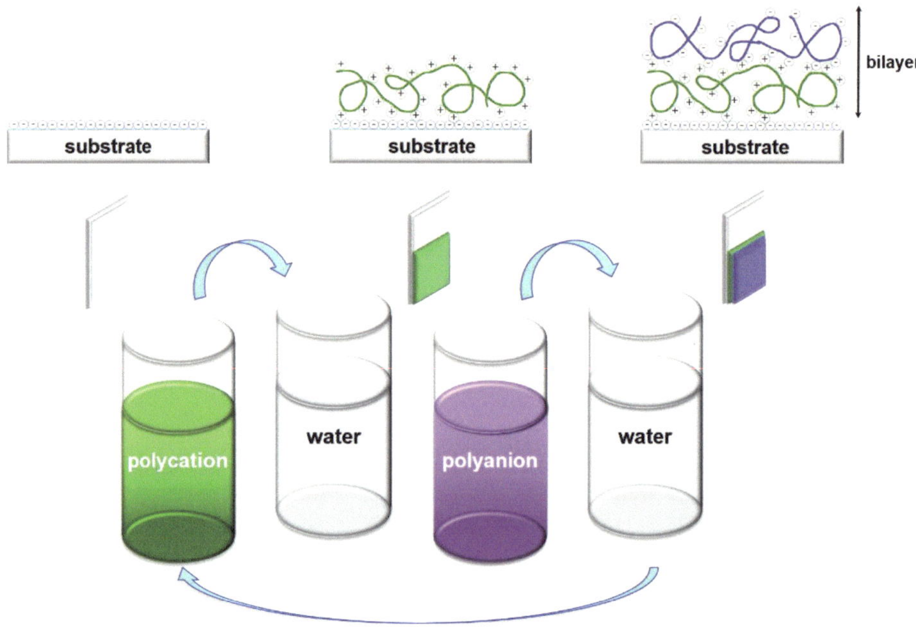

Figure 1.
Schematic representation of the LbL immersive assembly and the formation of the first cationic/anionic bilayer.

assembly and the consecutive bilayer formation. Briefly, a solid substrate is dipped in an aqueous dispersion containing the material to be adsorbed. Subsequently, the substrate is rinsed to remove loosely bound material, allowed to dry, and then immersed in another polyelectrolyte solution of opposite charge. Then, multilayered thin films are formed by spontaneously alternating the adsorption between cationic and anionic layers.

There are a myriad of LbL applications in the literature [5], being electronic tongues (e-tongue) a quite interesting example of nanostructures used in the analysis of liquid samples through specific pattern recognition by computational and statistical methods. In 2016, Alessio et al. applied a microfluidic e-tongue based on LbL films to evaluate different coffee brands [6]. The results demonstrated the device's ability to distinguish selected gourmet, organic, and premium coffees from unselected grains. In 2017, Daikuzono et al. also used LbL films in a microfluidic e-tongue setup to detect gliadin (a gluten protein) in order to evaluate gluten-containing foodstuff [7]. Authors were able to detect trace amounts of gliadin in ethanol solutions (up to 0.005 mg kg^{-1}) using computational analysis (Interactive Document Map, IDMAP), detecting also the contamination of gluten-free foodstuff with gliadin, paving the way for the detection in real-life situations.

Salvo-Comino et al. have used LbL films fabricated by spin-coating process and electrochemical characterization in milk analysis [8]. They have used LbL architectures based on chitosan (CHI), 1-butyl-3-methylimidazolium tetrafluoroborate (IL), copper(II) tetrasulfonated phthalocyanines ($CuPc^S$), and galactose oxidase (GAO). Two identical LbL CHI+IL/$CuPc^S$ architectures were made, with the GAO enzyme deposited on top of one of them. The formed biosensor enabled the distinction of milk samples having different lactose contents. Authors also compared Principal Component Analysis (PCA) results with Partial Least Squares (PLS-1) models, obtaining extra information from the milk samples.

LbL films have also been employed in environmental monitoring, aiming the detection of water, air, or soil contamination. In 2017, Barros et al. used

Square-Wave Anodic Stripping Voltammetry (SWASV) using indium tin oxide (ITO) electrodes functionalized with LbL films for the detection of trace levels of cadmium (Cd^{2+}), lead (Pb^{2+}), and copper (Cu^{2+}) [9]. The incorporation of AuNPs in LbL structures composed by emeraldine salt polyaniline (PANi-ES) and sodium montmorillonite clay mineral (Na^{+}MMT) enhanced the electrocatalytic response of the sensors to heavy metal ions, achieving a lower detection limit (from mgL^{-1} to μgL^{-1}). Also in 2017 Andre et al. tested hybrid LbL films as gas sensor to detect ammonia (NH_3) in hazardous pollutant monitoring [10]. LbL films composed by polyaniline (PANi), graphene oxide (GO), and zinc oxide (ZnO) were combined in a tetralayered structure (PANi/GO/PANi/ZnO) onto gold interdigitated electrodes (IDEs). The best performance in NH_3 detection using impedance spectroscopy was achieved with three tetralayers. Two years later, Rodrigues et al. have used LbL films of graphene nanoplatelets and AuNPs in the detection of methyl parathion [11], a highly toxic organophosphorus ester insecticide according to the World Health Organization [12]. The electrochemical results showed the successful detection of methyl parathion in real samples, such as tap water, soil, and cabbage.

LbL films have also been applied for diagnosis in medical and pharmaceutical areas. In 2018, de Lucena et al. tested composite LbL films on the electrochemical detection of dopamine in the presence of natural interferents such as ascorbic acid and uric acid [13]. The limit of detection (LOD) obtained was consistent with those found in the literature for other LbL architectures, with a higher range of detection (5–150 μmol L^{-1}). In 2019, Camilo et al. proposed the synthesis of AuNP conjugated to anti-PSA (prostate-specific antigen) antibody to produce a label-free impedimetric immunosensor based on LbL architectures [14]. The film was assembled onto gold electrodes previously modified with a thiol monolayer, followed by the deposition of an LbL framework incorporating biological materials for specific PSA detection. A LOD of 0.17 ng mL^{-1} was achieved without demanding secondary antibodies, reducing costs, and pointing out to a high potential in the actual detection of prostate cancer. In 2009, Silva et al. applied sensors based on LbL films in the detection of cholesterol in egg yolks, with high potential for food analysis and diet management [15]. Cholesterol oxidase (ChOx) was immobilized on the top of the LbL films composed by conventional polyelectrolytes, gold nanoparticles, and carbon nanotubes, with the sensor first tested to detect and evaluate the cholesterol concentration in commercial solutions and later in food samples by monitoring amperometric signals.

Tunable properties of the LbL films were reported by Fontinele et al. using Norbixin (NBx), a carotenoid extracted from urucum [16]. They have fabricated (NBx/PANi) LbL composite films onto ITO substrates, enabling an electrochemically stable charge transfer. That was possible as the presence of NBx in the film architecture did not inhibit PANi conductive characteristics. Cortez et al. showed highly organized LbL structures of lipid-like surfactants and polyelectrolytes, a distinct feature of the normally obtained "fuzzy" formation from polycation-polyanion multilayer assemblies [17]. LbL films fabricated with poly(allylamine) hydrochloride (PAH) and sodium dodecyl phosphate (DP) displayed a well-stratified layer formation and highly oriented lamellar structures, with X-ray data revealing the presence of Bragg peaks up to fourth order, an interesting configuration for applications requiring naturally ordered spacers in a nacre-like structure.

Here, we will focus on multilayered thin films to form multifunctional coatings and the application of the 3D-printing technique as an alternative methodology for fast prototyping and design of sensors. The emerging 3D-printing technique also enabled an easy integration with other research areas such as microfluidics.

2. Alternative techniques and materials for sensor development

In this section we will describe some subjects and methodologies involved in the development of multilayered nanostructures used in sensory analysis.

2.1 Alternative materials for electrodes

2.1.1 3D-printed interdigitated electrodes

IDEs are a category of planar parallel plate capacitor formed by two metallic combs facing each other. It is a powerful tool for sensing and biosensing owing to the high sensitivity of getting small signal variations close to the IDEs' surface, being widely explored in impedimetric e-tongues used in the analysis of complex liquid systems. Traditional photolithography and microfabrication processes are usually employed in the manufacture of IDEs, enabling the fabrication at the micrometer range. However, it demands multistep processes, specialized facilities, trained personnel, and the use of toxic and expensive reagents. However, in the last decade ink-jet and 3D-printing technologies have emerged as alternative methods for simpler fabrication of electrodes using nonconventional conductive materials [18–20].

3D-printed electrodes are built up typically at the submillimeter range due to limitations imposed by nozzles; nonetheless, the technique has attracted considerable attention due to the fast prototyping, use of alternative materials, and reduction steps in the fabrication of complex geometries within a few minutes. It generally comes straight from the computer screen, with no need of photomasks and intermediate microfabrication steps that usually take a couple of hours using conventional techniques. The materials used as conductive tracks are also attractive because they can be readily deposited onto flexible substrates and integrated with organic and soft electronics [18]. Within this milieu of fast and automated prototyping, the Fused Deposition Modeling 3D printing (FDM) technique shows great potential to fabricate a conductive electrode pattern in a one-step process within minutes. FDM uses thermoplastic filaments to build up the construct, with many commercial materials available to be exploited in the fabrication of devices [21, 22]. To illustrate, conductive filaments enable the manufacture of electrodes, while conventional filaments can be used to board the body of the device. It is possible to prototype the whole electrode in a single printing process if someone uses a multifilament FDM printer, thus increasing the flexibility in the buildup of complex structures with a reduced fabrication time [23]. Recently, a homemade dual-extruder FDM printer was used to fabricate macro planar IDEs and sensor body in e-tongue applications. Each sensing unit was printed within 8 minutes having good reproducibility in the fabrication process of conventional e-tongue devices [24].

2.1.2 Transparent conductive electrodes

Transparent conductive electrodes are indispensable components for potential applications in optoelectronic devices, such as solar cells, flexible displays, light emitting diodes, and touch screens [25–28]. Materials with high transmittance and adequate electrical conductivity are desired in such applications [29]; therefore, the most commonly used material is indium tin oxide (ITO) due to its good optical and electrical properties (e.g., transmittance~80% at visible wavelength and low sheet resistance <40 Ω/sq) [30]. However, ITO is brittle by nature limiting flexible applications [31] and presents~75% indium in its composition, which might hamper display and optoelectronic uses due to its scarcity and increasing cost [32]. Many new transparent conductive materials have been developed to circumvent this problem, including hybrid/composites and multilayered thin films [33–36]. However, some transparent electrodes have complicated structures or require complex processing, demanding the search for new strategies to produce transparent conductive materials at low-cost and with a compatible upscale production.

Many expectations were placed on graphene as a prominent material for the next generation of electronics due to its unique properties derived from the honeycomb configuration of carbon atoms placed in a one-atom thick sheet having sp^2 hybridizations [37]. However, pristine graphene is a semimetallic material having a null bandgap that limits its use in some devices. Graphene has been mostly obtained from chemical vapor deposition (CVD), however, with the cost of a nontrivial separation/transfer process of the produced graphene layers, as the CVD growth is generally made on copper substrates. The separation process can seriously compromise the quality of the graphene samples, and the chemical synthesis of reduced graphene oxide (rGO) appears as a promising alternative. It enables a large scale production, and rGO has properties resembling pristine graphene, with an adjustable bandgap from structural defects and the presence of oxygenated groups in its structure [38]. The presence of oxygen at the basal plane allows the functionalization of the material with polymers and biomolecules, expanding its use with other techniques and enhancing potential applications [39, 40]. The control of dimensionality, quantum confinement, and edge effects is an important factor defining the electronic properties in graphene derivatives [41]. Within this context, the LbL assembly is an elegant way to control both thickness and molecular architecture of rGO structures at nanoscale, potentiating the use of graphene-based materials in flexible electronics, sensors, biosensing, transparent electrodes, and energy storage devices [42–44].

In the last decade, several studies were conducted to study the properties of rGO nanoplatelets (two-dimensional material) as well as zero-dimensional graphene quantum dots (GQDs) [45, 46]. However, there are still scarce studies on the electrical properties and behavior of charge carriers in nanostructured LbL films. Recently, we have shown that multilayered rGO films allowed a fine-tuning on both optical and electrical properties (**Figure 2**). The formation of nanocomposites from rGO chemically functionalized with sodium polystyrene sulfonate (PSS), named here as GPSS and also as quantum dots (GPSS:QDs) or nanoplatelets (GPSS:NPLs), with poly(diallyldimethylammonium chloride) (PDDA), enabled an intraplanar 2D conduction mechanism due to the confinement of charge carriers within the

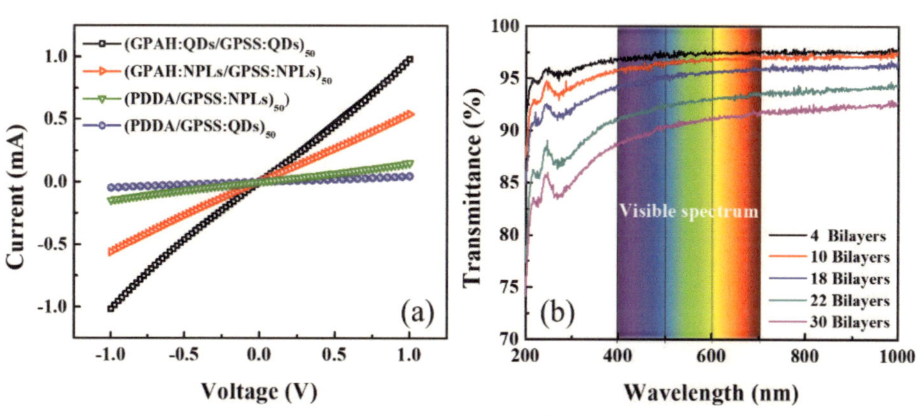

Figure 2
(a) I-V measurements for different LbL films of rGO deposited onto gold interdigitated electrodes.
(b) Transmittance spectra of (GPAH:QDs/GPSS:QD)n LbL film.

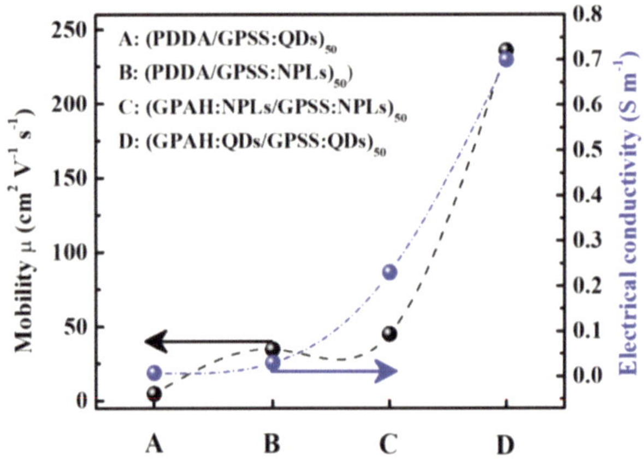

Figure 3.
Mobility of charge carriers and electrical conductivity of different LbL architectures formed mainly by rGO.

conductive layers formed by GPSS in the PDDA/GPSS LbL assembly. **Figure 2a** shows current vs. voltage (I-V) curves from distinct rGO LbL architectures. A strong dependence of the electrical properties with the size of the rGOs is clearly seen [38, 47, 48]. The $(GPAH:QDs/GPSS:QDs)_n$ LbL film, being n the number of deposited bilayers, displayed high transmittance (from 89–92% in the visible region for $n = 30$), as shown in **Figure 2b**.

Notable differences were observed in the mobility of charge carriers depending on the size of the nanoplatelets involved in the LbL film formation being, respectively, 4.7 $cm^2\ V^{-1}\ s^{-1}$ for $(PDDA/GPSS:QDs)_{50}$ and 34.7 $cm^2\ V^{-1}\ s^{-1}$ for $(PDDA/GPSS:NPLs)_{50}$ [47]. Nevertheless, higher mobility values could be achieved exploring synergistic size effects in the LbL assembly formed only by rGOs, either as quantum dots (QDs) or nanoplatelets (NPLs). That was reached changing PDDA in the LbL structures by rGO wrapped with poly(allylamine hydrochloride) (PAH), named here as GPAH. The (GPAH:QDs/GPSS:QD) LbL film exhibited high mobility values ($\mu = 236\ cm^2\ V^{-1}\ s^{-1}$) [48], with changes illustrated in **Figure 3**. The improvement in the electrical characteristics was attributed to rGO size effects, with a 3D driving mechanism assisted mainly by phonons when in QD form. The successful achievement of multilayered rGO:QDs films presenting high transmittance and noticeable mobility values paves the way for future applications in high performance, low-cost optoelectronic devices.

2.2 Film growth monitored by electrical measurements

Electrolyte-insulator-semiconductor (EIS) electrochemical capacitors are charged sensitive devices detecting charge variation at electrolyte-insulator interfaces [49]. They are widely applied in the monitoring of electrical features during the LbL film growth [50, 51]. Briefly, the electrical behavior of (PAH/PSS) LbL films on EIS devices is characterized by a zigzag-like capacitance behavior after each adsorption step, attributed to the outermost charge layer alternation [52]. Furthermore, a similar behavior was also observed in C-V measurements preserving the capacitance constant after each adsorbed layer onto the EIS sensor [53]. However, EIS devices work in salt medium, and such environmental condition certainly modifies the electrical properties in LbL structures. As a simple alternative, IDEs can be used to analyze the LbL films properties in air or in water, with no need of salt addition. However, experimental methods were developed to avoid the influence of aqueous medium on the electrical characterization of LbL films

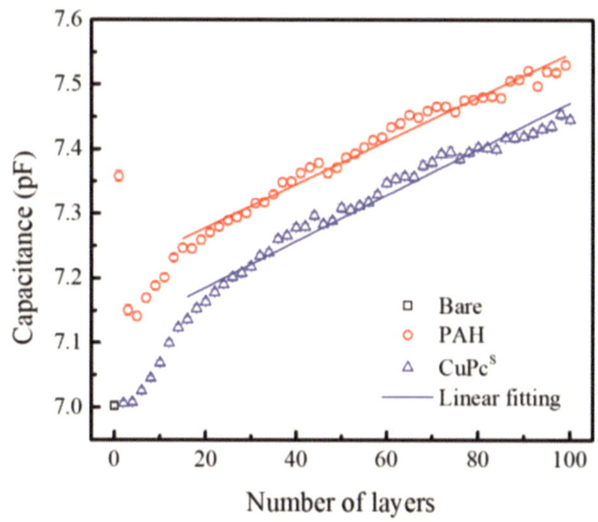

Figure 4.
Capacitance variation as a function of the number of adsorbed layers. Dark square corresponds to the bare IDE, red circles to capacitance measurements after PAH adsorption, and blue triangles after CuPc^S adsorption.

[47, 54]. **Figure 4** presents the capacitance variation after the deposition of each PAH or copper phthalocyanine (CuPc^S) adsorbed layer, followed by spontaneous drying onto gold IDEs. A capacitance alternation according to the outermost deposited layer (PAH or CuPc^S) due to charge overcompensation was observed. For measurements performed in air, the amplitude of the capacitance alternation did not depend on the number of deposited layers, a consequence of the charge density after each deposited PAH layer. In addition, **Figure 4** presents a linear trend after the 16th adsorbed layer of both polyelectrolytes, which was associated with the replacement of an air layer on the top of the IDE by the LbL film, thus increasing the dielectric constant near the IDEs surface. Accordingly, the observed linearity was associated with the deposition of the same amount of material after each adsorption step. Therefore, it allows for the evaluation of both appropriate material and thickness selection for a desired application based on the electrical measurements performed in the LbL film growth.

2.3 Incorporation of nanoparticles by physical methods

The selection of materials having desired functionalities can tune properties in LbL assemblies, such as superhydrophobicity [55], ionic permeability [56], electrical conductivity [57], etc. As an example, the incorporation of gold nanoparticles into LbL films composed by PANi and Na^+MMT enabled the electrochemical detection of trace levels of cadmium, lead, and copper ions [9]. The inclusion of gold nanoparticles increased also the discrimination between skimmed, semi-skimmed, and full-fat milk samples [58]. Although the deposition of nanoparticles between layers in LbL structures might increase both sensitivity and selectivity [9, 58], it is not possible to control an adequate dispersion of them during the spontaneous adsorption process. Consequently, it precludes a better comprehension of the influence of nanoparticles on enhanced sensitivity and selectivity. An alternative is the use of physically prepared metal nanoparticles to assembly samples by independently controlling size, concentration, and distribution of nanoparticles free of surfactants at interfaces. To illustrate, palladium nanoparticles were successfully implanted in poly(methyl methacrylate), effectively modifying

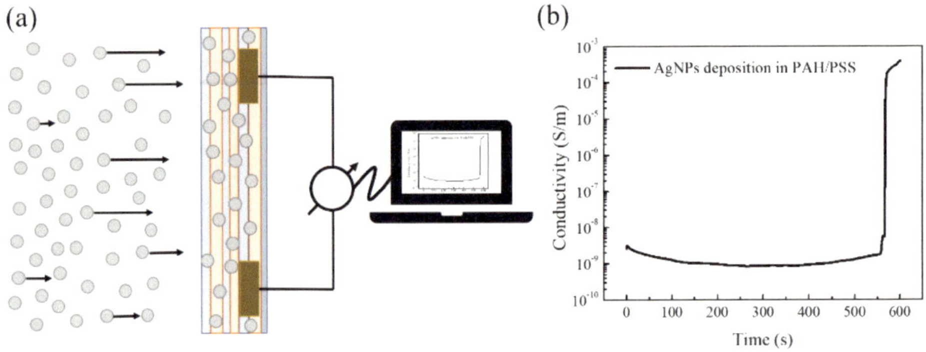

Figure 5.
(a) Scheme to illustrate in situ conductivity monitoring of nanoparticle deposition into LbL films onto IDE. Note the scheme presents just a pair of digits to facilitate viewing. (b) Electrical conductivity variation of PAH/PSS LbL films 200 nm thick onto gold IDEs during the deposition of Ag nanoparticles.

the electrical properties of the polymer in a specific region [59]. In this context, we have incorporated 3 nm silver nanoparticles preformed in the gas phase into (PAH/PSS) LbL films. The silver nanoparticles were formed in an atomic cluster source, mass-selected using ionic optics, with in situ analysis of their size by a time of flight mass spectrometer (TOF) [60]. Then, the mass-selected silver nanoparticles were deposited in $(PAH/PSS)_n$ LbL films previously deposited onto gold IDEs. **Figure 5** illustrates the electrical changes in the LbL film during the nanoparticles' deposition. Initially, tiny variations on the electrical characteristics that were associated with the insertion of charged nanoparticles in the LbL film structure were observed. An increase in the nanoparticle's concentration created conductive pathways in the LbL films, thus increasing the conductivity four orders of magnitude after reaching a percolation limit. Therefore, independent control on nanoparticles' deposition in a surfactant-free process allowed a better comprehension of their influence in the physicochemical properties of hybrid LbL nanostructures.

2.4 Multifunctional materials

Intrinsic self-healing materials have received great attention in the last decade due to the recovery of mechanical properties after damage through reversible dynamic covalent and non-covalent interactions [61]. Distinct building blocks and strategies have been extensively investigated to enable a reliable, functional healing as some polymers involved in the process are vulnerable to abrasion. Again, the LbL technique is as a powerful tool in the buildup of multifunctional interfaces based on spontaneous non-covalent interactions of materials, tuning the interface characteristics [62–64]. Such interactions introduce properties not exhibited by individual materials, which can be easily exemplified with polyethylenimine (PEI) and poly(acrylic acid) (PAA). The materials individually do not present self-regeneration; therefore, when combined in (PEI/PAA) LbL architectures, the intrinsic self-healing capability appears after~30 deposited bilayers, as the first deposited layers are strongly compromised in interactions with the substrate [65, 66]. The healing ability can also be controlled by changing the pH, salt concentration, and temperature during the deposition process due to changes in the polymer configuration [67]. The film properties can also be tuned by post-deposition treatments or by adding fillers to the polyelectrolytes. Chen et al. showed that after the deposition process, one can freeze-dry the PEI/PAA film in order to create microporous structures inside the interface, which were then loaded with drug solutions by wicking or adsorption method [68]. By adding water to the film, the porous structures were healed, and the drug was kept inside the film that was later used to create a controlled drug delivery system.

2.5 Electronic tongue devices

Aiming the integration of the FDM 3D-printing technology in e-tongue devices, microchannels were printed onto gold IDEs having 30 pairs of digits separated 40 μm away from each other, 3 mm in length and 40 μm in width. The electrodes were previously manufactured by photolithography on transparency sheets, and polylactic acid (PLA) was chosen to fabricate the microchannels due to its relative transparency when printed in thin layers, facilitating the flow visualization inside microchannels. The deposition of the IDEs on thin transparency sheets was planned to promote a better sealing with the 3D-printed PLA microchannel (**Figure 6a**). A functional integrated e-tongue device was then developed combining both photolithography and 3D-printing technologies, with LbL films deposited onto the IDEs to form distinct sensing units inside the 3D-printed microchannels (**Figure 6b**). After long usage in repetitive experiments, the detachment of the IDEs inside the microchannels was observed. Although the device was not as robust as conventional ones with IDEs manufactured on glass slides, it can be easily used in fast analysis [22] requiring disposable devices.

Traditional soil chemical investigation is an expensive and time-consuming process, thus motivating the development of alternative approaches in precision agriculture [69–71]. Recently, we have used an e-tongue system based on impedance spectroscopy to recognize different soil samples enriched with macronutrients [24, 72]. Initially, it was used a traditional PDMS microchannel sealed onto gold IDEs fabricated on glass substrates by photolithography (**Figure 7a**). After that, it was used an e-tongue system with 3D-printed IDEs using graphene-based polylactic acid filaments (**Figure 7b**). The sensing units were fabricated using the same materials to facilitate additional comparisons.

The e-tongue systems were tested checking their capability to distinguish soil samples individually enriched with N, P, K, Ca, Mg, and S nutrients, delivered to the soil via chemical fertilizers. The soil samples were simply diluted in water and measured using electrical impedance spectroscopy, with raw data analyzed by PCA. **Figure 8** presents the PCA plot obtained from both devices. Although the results are visually not identical, it is possible to note in both cases a clear distinction of all soil samples tested in PCA plots. A cluster between control and P soil

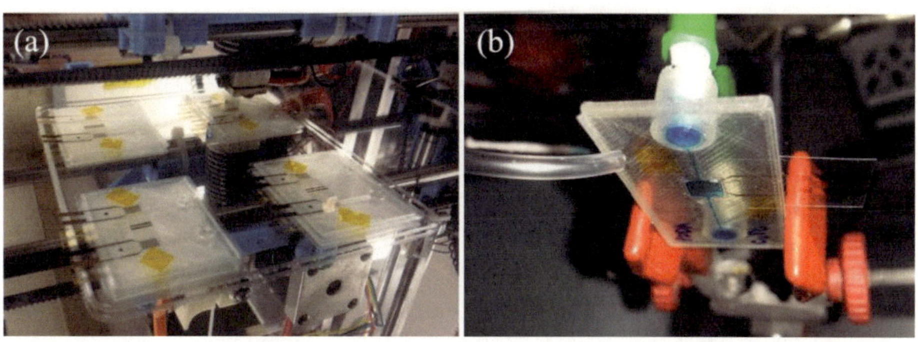

Figure 6.
(a) PLA microchannels while printing by FDM onto transparency sheets and (b) mixed device formed by 3D-printed microchannel and IDE previously manufactured by photolithography. In blue the deposition of the $CuPc^S$ layer in an LbL film structure is evidenced (PDDA/$CuPc^S$).

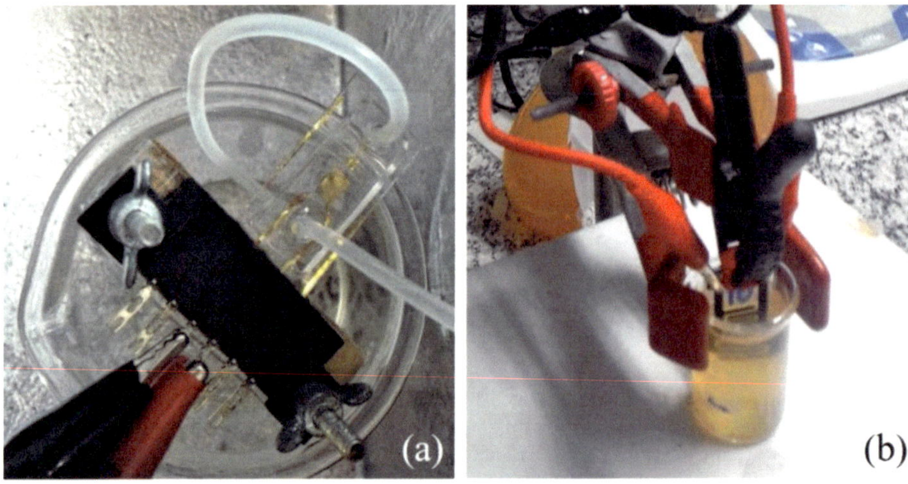

Figure 7.
Different devices used as sensing units for e-tongue system in soil analysis: (a) gold IDE inside a PDMS microchannel and (b) graphene-based 3D-printed IDE.

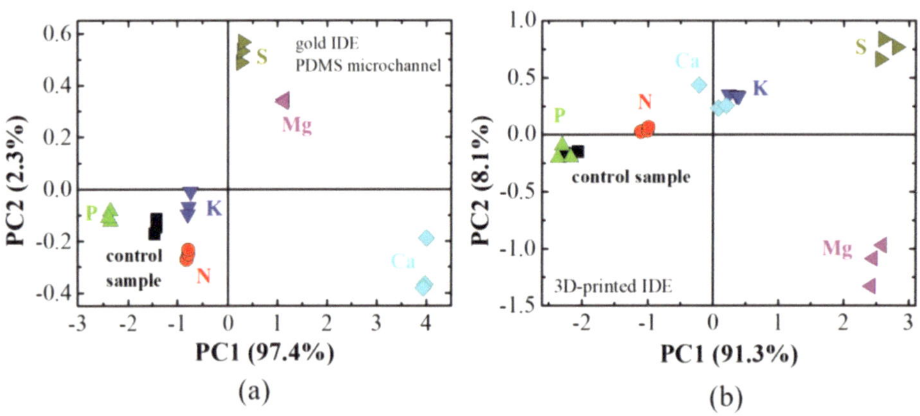

Figure 8.
PCA for the 1 kHz capacitance results of the e-tongue systems on (a) microfluidic and (b) on 3D-printed electrodes, both applied in soil samples diluted in water (control, N, P, K, Ca, Mg, and S).

samples using the 3D-printed electrodes was observed (**Figure 8b**), which does not occur with the conventional device (**Figure 8a**) fabricated with gold IDEs and PDMS microchannels. However, considering the contribution of the third main Principal Component (PC3), all the samples could also be easily separated with the 3D-printed device [24].

3. Conclusions

The LbL technique is a striking research line in materials science based on nanostructured films fabricated from distinct building blocks (polymers, organic macromolecules, clays, and reduced graphene oxides). The molecular level thickness control offered by the LbL assembly enables multilayered nanostructures exhibiting interesting assignments in numerous applications. The use of physical methods to controllably embed metallic nanoparticles into LbL films is an interesting alternative approach to give a better understanding of modifications introduced in hybrid materials. Overall, the use of LbL films in sensing applications requires previous electrode fabrication, and within this context we have compared gold IDEs manufactured by photolithography with 3D-printed electrodes using a graphene-based filament. The 3D-printing technique also enables the successful integration of e-tongue devices into microfluidic systems, reducing processing steps and costs. Devices fabricated by different methods show pros and cons, and the decision of which one to use depends on the application, budget, and time available. Therefore, an e-tongue multisensory system was used to illustrate the application of multilayered thin films in the analysis of soil samples enriched with plant macronutrients, by comparing electrodes manufactured by both conventional methods (photolithography) and an emergent technology (3D-printing). The 3D-printing technology has proved to be an interesting approach in integrating different areas such as microfluidics, sensing, and electrode manufacturing, with a reduction in costs, processing steps, and manufacturing time, expanding at the same time novel opportunities in sensorial analysis.

Acknowledgements

The authors are grateful for the financial support from the Brazilian funding agencies: Sao Paulo Research Foundation (FAPESP, grants no 2015/14836-9 and 2017/19862-3), Coordination for the Improvement of Higher Education Personnel (CAPES), and National Council for Scientific and Technological Development (CNPq). They also thank the Brazilian Nanotechnology National Laboratory (LNNano)/National Center for Research in Energy and Materials (CNPEM) (LMF project n° 24749) for providing photolithography facilities.

Conflict of interest

The authors declare no conflict of interest.

Author details

Maria L. Braunger, Rafael C. Hensel, Gabriel Gaál, Mawin J.M. Jimenez, Varlei Rodrigues and Antonio Riul Jr*
Department of Applied Physics, "Gleb Wataghin" Institute of Physics (IFGW), University of Campinas (UNICAMP), Campinas, SP, Brazil

*Address all correspondence to: riul@ifi.unicamp.br

References

[1] Iler RK. Multilayers of colloidal particles. Journal of Colloid and Interface Science. 1966;**21**:569-594. DOI: 10.1016/0095-8522(66)90018-3

[2] Decher G. Fuzzy nanoassemblies: Toward layered polymeric multicomposites. Science. 1997;**277**: 1232-1237. DOI: 10.1126/science.277.5330.1232

[3] Richardson JJ, Björnmalm M, Caruso F. Technology-driven layer-by-layer assembly of nanofilms. Science. 2015;**348**:2491. DOI: 10.1126/science.aaa2491

[4] Richardson JJ, Cui J, Björnmalm M, et al. Innovation in layer-by-layer assembly. Chemical Reviews. 2016;**116**:14828-14867. DOI: 10.1021/acs.chemrev.6b00627

[5] Zhao S, Caruso F, Dähne L, et al. The future of layer-by-layer assembly: A tribute to ACS Nano associate editor Helmuth Möhwald. ACS Nano. 2019;**13**:6151-6169. DOI: 10.1021/acsnano.9b03326

[6] Alessio P, Constantino CJL, Daikuzono CM, et al. Analysis of coffees using electronic tongues. In: Mendez MR, editor. Electronic Noses and Tongues in Food Science. Oxford, UK: Academic Press; 2016. pp. 171-177

[7] Daikuzono CM, Shimizu FM, Manzoli A, et al. Information visualization and feature selection methods applied to detect Gliadin in gluten-containing foodstuff with a microfluidic electronic tongue. ACS Applied Materials & Interfaces. 2017;**9**:19646-19652. DOI: 10.1021/acsami.7b04252

[8] Salvo-Comino C, García-Hernández C, García-Cabezón C, Rodríguez-Méndez ML. Discrimination of milks with a multisensor system based on layer-by-layer films. Sensors. 2018;**18**:1-12. DOI: 10.3390/s18082716

[9] de Barros A, Constantino CJL, da Cruz NC, et al. High performance of electrochemical sensors based on LbL films of gold nanoparticles, polyaniline and sodium montmorillonite clay mineral for simultaneous detection of metal ions. Electrochimica Acta. 2017;**235**:700-708. DOI: 10.1016/j.electacta.2017.03.135

[10] Andre RS, Shimizu FM, Miyazaki CM, et al. Hybrid layer-by-layer (LbL) films of polyaniline, graphene oxide and zinc oxide to detect ammonia. Sensors Actuators, B Chem. 2017;**238**:795-801. DOI: 10.1016/j.snb.2016.07.099

[11] Rodrigues GHS, Miyazaki CM, Rubira RJG, et al. Layer-by-layer films of Graphene Nanoplatelets and gold nanoparticles for methyl parathion sensing. ACS Appl Nano Mater. 2019;**2**:1082-1091. DOI: 10.1021/acsanm.9b00007

[12] World Health Organization. Available from: www.who.int

[13] de Lucena NC, Miyazaki CM, Shimizu FM, et al. Layer-by-layer composite film of nickel phthalocyanine and montmorillonite clay for synergistic effect on electrochemical detection of dopamine. Applied Surface Science. 2018;**436**:957-966. DOI: 10.1016/j.apsusc.2017.12.117

[14] Camilo DE, Miyazaki CM, Shimizu FM, Ferreira M. Improving direct immunoassay response by layer-by-layer films of gold nanoparticles – Antibody conjugate towards label-free detection. Materials Science and Engineering: C. 2019;**102**:315-323. DOI: 10.1016/j.msec.2019.04.055

[15] Silva DPB, Miyazaki CM, Mascagni DBT, Ferreira M. Layer-by-layer films of gold nanoparticles and carbon nanotubes for improved Amperometric detection of cholesterol. Journal of Nanoscience and Nanotechnology. 2019;**19**:5483-5488. DOI: 10.1166/jnn.2019.16534

[16] Fontinele LP, de Sousa RC, Viana VGF, et al. Norbixin extracted from urucum (Bixa orellana L.) for the formation of conductive composites with potential applications in electrochemical sensors. Surfaces and Interfaces. 2018;**13**:92-100. DOI: 10.1016/j.surfin.2018.08.002

[17] Cortez ML, Lorenzo A, Marmisollé WA, et al. Highly-organized stacked multilayers: Via layer-by-layer assembly of lipid-like surfactants and polyelectrolytes. Stratified supramolecular structures for (bio) electrochemical nanoarchitectonics. Soft Matter. 2018;**14**:1939-1952. DOI: 10.1039/c8sm00052b

[18] Cummins G, Desmulliez MPY. Inkjet printing of conductive materials: A review. Circuit World. 2012;**38**:193-213. DOI: 10.1108/03056121211280413

[19] Singh M, Haverinen HM, Dhagat P, Jabbour GE. Inkjet printing-process and its applications. Advanced Materials. 2010;**22**:673-685. DOI: 10.1002/adma.200901141

[20] Paula KT, Gaál G, Almeida GFB, et al. Femtosecond laser micromachining of polylactic acid/graphene composites for designing interdigitated microelectrodes for sensor applications. Optics and Laser Technology. 2018;**101**:74-79. DOI: 10.1016/j.optlastec.2017.11.006

[21] Canessa E, Fonda C, Zennaro M. Low-Cost 3D Printing for Science, Education & Sustainable Development. Trieste, It: ICTP Science Dissemination Unit; 2013

[22] Gaal G, Mendes M, de Almeida TP, et al. Simplified fabrication of integrated microfluidic devices using fused deposition modeling 3D printing. Sensors and Actuators B: Chemical. 2017;**242**:35-40. DOI: 10.1016/j.snb.2016.10.110

[23] Foster CW, Down MP, Yan Z, et al. 3D printed Graphene based energy storage devices. Scientific Reports. 2017;7:1-11. DOI: 10.1038/srep42233

[24] Gaál G, Silva TA, Gaál V, et al. 3D printed e-tongue. Frontiers in Chemistry. 2018;**6**:151. DOI: 10.3389/fchem.2018.00151

[25] Sharma S, Shriwastava S, Kumar S, et al. Alternative transparent conducting electrode materials for flexible optoelectronic devices. Opto-electronics Rev. 2018;**26**:223-235. DOI: 10.1016/j.opelre.2018.06.004

[26] Lee HB, Jin WY, Ovhal MM, et al. Flexible transparent conducting electrodes based on metal meshes for organic optoelectronic device applications: A review. Journal of Materials Chemistry C. 2019;7:1087-1110. DOI: 10.1039/c8tc04423f

[27] Li Y, Xu G, Cui C, Li Y. Flexible and semitransparent organic solar cells. Advanced Energy Materials. 2018;**8**:1-28. DOI: 10.1002/aenm.201701791

[28] Koo JH, Kim DC, Shim HJ, et al. Flexible and stretchable smart display: Materials, fabrication, device design, and system integration. Advanced Functional Materials. 2018;**28**:1-23. DOI: 10.1002/adfm.201801834

[29] Ellmer K. Past achievements and future challenges in the development of optically transparent electrodes. Nature Photonics. 2012;**6**:809-817. DOI: 10.1038/nphoton.2012.282

[30] Wang DH, Kyaw AKK, Gupta V, et al. Enhanced efficiency parameters

of solution-processable small-molecule solar cells depending on ito sheet resistance. Advanced Energy Materials. 2013;**3**:1161-1165. DOI: 10.1002/aenm.201300277

[31] Sakamoto K, Kuwae H, Kobayashi N, et al. Highly flexible transparent electrodes based on mesh-patterned rigid indium tin oxide. Scientific Reports. 2018;**8**:3-4. DOI: 10.1038/s41598-018-20978-x

[32] Hecht DS, Hu L, Irvin G. Emerging transparent electrodes based on thin films of carbon nanotubes, graphene, and metallic nanostructures. Advanced Materials. 2011;**23**:1482-1513. DOI: 10.1002/adma.201003188

[33] Ok KH, Kim J, Park SR, et al. Ultra-thin and smooth transparent electrode for flexible and leakage-free organic light-emitting diodes. Scientific Reports. 2015;**5**:1-8. DOI: 10.1038/srep09464

[34] Zhang C, Khan A, Cai J, et al. Stretchable transparent electrodes with solution-processed regular metal mesh for an electroluminescent light-emitting film. ACS Applied Materials & Interfaces. 2018;**10**:21009-21017. DOI: 10.1021/acsami.8b06691

[35] Ji Y, Yang J, Luo W, et al. Ultraflexible and high-performance multilayer transparent electrode based on ZnO/Ag/CuSCN. ACS Applied Materials & Interfaces. 2018;**10**:9571-9578. DOI: 10.1021/acsami.7b15902

[36] Altin Y, Tas M, Borazan I, et al. Solution-processed transparent conducting electrodes with graphene, silver nanowires and PEDOT: PSS as alternative to ITO. Surf Coatings Technol. 2016;**302**:75-81. DOI: 10.1016/j.surfcoat.2016.05.058

[37] Singh V, Joung D, Zhai L, et al. Graphene based materials: Past, present and future. Progress in Materials Science. 2011;**56**:1178-1271. DOI: 10.1016/j.pmatsci.2011.03.003

[38] Jimenez MJM, de Oliveira RF, Shimizu FM, et al. Poole-Frenkel emission on functionalized, multilayered-packed reduced graphene oxide nanoplatelets. Nanotechnology. 2018;**29**:505703. DOI: 10.1088/1361-6528/aae18e

[39] Wang Y, Li Z, Wang J, et al. Graphene and graphene oxide : Biofunctionalization and applications in biotechnology. Trends in Biotechnology. 2011;**29**:205-212. DOI: 10.1016/j.tibtech.2011.01.008

[40] Li D, Müller MB, Gilje S, et al. Processable aqueous dispersions of graphene nanosheets. Nature Nanotechnology. 2008;**3**:101-105. DOI: 10.1038/nnano.2007.451

[41] Diao S, Zhang X, Shao Z, et al. 12.35% efficient Graphene quantum dots/silicon Heterojunction solar cells using Graphene transparent electrode. Nano Energy. 2017;**31**:359-366. DOI: 10.1016/j.nanoen.2016.11.051

[42] Lee T, Min SH, Gu M, et al. Layer-by-layer assembly for Graphene-based multilayer Nanocomposites : Synthesis and applications. Chemistry of Materials. 2015;**27**(11):3785-3796. DOI: 10.1021/acs.chemmater.5b00491

[43] Morales-Narváez E, Merkoçi A. Graphene oxide as an optical biosensing platform: A Progress report. Advanced Materials. 2019;**31**:1-12. DOI: 10.1002/adma.201805043

[44] Kruk T, Socha RP, Szyk-Warszyńska L, Warszyński P. Flexible and ultrathin polyelectrolyte conductive coatings formed with reduced graphene oxide as a base for advanced new materials. Applied Surface Science. 2019;**484**:501-510. DOI: 10.1016/j.apsusc.2019.04.051

[45] Tian P, Tang L, Teng KS, Lau SP. Graphene quantum dots from chemistry to applications. Mater Today Chem. 2018;**10**:221-258. DOI: 10.1016/j.mtchem.2018.09.007

[46] Krishnan SK, Singh E, Singh P, et al. A review on graphene-based nanocomposites for electrochemical and fluorescent biosensors. RSC Advances. 2019;**9**:8778-8781. DOI: 10.1039/c8ra09577a

[47] Jimenez MJM, Oliveira RF, Almeida TP, et al. Charge carrier transport in defective reduced graphene oxide as quantum dots and nanoplatelets in multilayer films. Nanotechnology. 2017;**28**:495711. DOI: 10.1088/1361-6528/aa91c2

[48] Jimenez MJM, de Oliveira RF, Bufon CCB, et al. Enhanced mobility and controlled transparency in multilayered reduced graphene oxide quantum dots: A charge transport study. Nanotechnology. 2019;**30**:275701. DOI: 10.1088/1361-6528/ab118e

[49] Schöning MJ, Poghossian A, Yoshinobu T, Lüth H. Semiconductor-based field-effect structures for chemical sensing. Adv Environ Chem Sens Technol. 2001;**4205**:188-198. DOI: 10.1117/12.417449

[50] Poghossian A, Abouzar MH, Amberger F, et al. Field-effect sensors with charged macromolecules: Characterisation by capacitance-voltage, constant-capacitance, impedance spectroscopy and atomic-force microscopy methods. Biosensors & Bioelectronics. 2007;**22**:2100-2107. DOI: 10.1016/j.bios.2006.09.014

[51] Abouzar MH, Poghossian A, Razavi A, et al. Characterisation of capacitive field-effect sensors with a nanocrystalline-diamond film as transducer material for multi-parameter sensing. Biosensors & Bioelectronics. 2009;**24**:1298-1304. DOI: 10.1016/j.bios.2008.07.056

[52] Poghossian A, Abouzar MH, Sakkari M, et al. Field-effect sensors for monitoring the layer-by-layer adsorption of charged macromolecules. Sensors and Actuators B. 2006;**118**:163-170. DOI: 10.1016/j.snb.2006.04.013

[53] Poghossian A, Weil M, Cherstvy AG, Schöning MJ. Electrical monitoring of polyelectrolyte multilayer formation by means of capacitive field-effect devices. Analytical and Bioanalytical Chemistry. 2013;**405**:6425-6436. DOI: 10.1007/s00216-013-6951-9

[54] Hensel RC, Rodrigues KL, Pimentel V do L, et al. Automated self-assembly and electrical characterization of nanostructured films. MRS Commun. 2018;**8**:283-288. DOI: 10.1557/mrc.2018.47

[55] Han JT, Zheng Y, Cho JH, et al. Stable superhydrophobic organic-inorganic hybrid films by electrostatic self-assembly. The Journal of Physical Chemistry. B. 2005;**109**:20773-20778. DOI: 10.1021/jp052691x

[56] Sato K, Takahashi S, Anzai J. Layer-by-layer thin films and microcapsules for biosensors and controlled release. Analytical Sciences. 2012;**28**:929-938. DOI: 10.2116/analsci.28.929

[57] Khillan RK, Su Y, Lvov YM, Varahramyan K. Layer-by-layer nanoarchitecture of ultrathin films of PEDOT-PSS and PPy to act as hole transport layer in polymer light emitting diodes and polymer transistors. IEEE Trans Components Packag Technol. 2005;**28**:748-753. DOI: 10.1109/TCAPT.2005.859754

[58] Mercante LA, Scagion VP, Pavinatto A, et al. Electronic tongue based on nanostructured hybrid films of

gold nanoparticles and Phthalocyanines for Milk analysis. Journal of Nanomaterials. 2015;**2015**:1-7. DOI: 10.1155/2015/890637

[59] Ravagnan L, Divitini G, Rebasti S, et al. Poly(methyl methacrylate)-palladium clusters nanocomposite formation by supersonic cluster beam deposition: A method for microstructured metallization of polymer surfaces. Journal of Physics D: Applied Physics. 2009;**42**:082002. DOI: 10.1088/0022-3727/42/8/082002

[60] de Sá ADT, Abrao Oiko VT, di Domenicantonio G, Rodrigues V. New experimental setup for metallic clusters production based on hollow cylindrical magnetron sputtering. J Vac Sci Technol B, Nanotechnol Microelectron Mater Process Meas Phenom. 2014;**32**:061804. DOI: 10.1116/1.4900847

[61] Blaiszik BJ, Kramer SLB, Olugebefola SC, et al. Self-healing polymers and composites. Annual Review of Materials Research. 2010;**40**:179-211. DOI: 10.1002/9781118082720

[62] Ai H, Jones SA, Lvov YM. Biomedical applications of electrostatic layer-by-layer Nano-assembly of polymers, enzymes, and nanoparticles. Cell Biochemistry and Biophysics. 2003;**39**:23-43

[63] Wijeratne S, Liu W, Dong J, et al. Layer-by-layer deposition with polymers containing Nitrilotriacetate, a convenient route to fabricate metal- and protein-binding films. ACS Applied Materials & Interfaces. 2016;**8**:10164-10173. DOI: 10.1021/acsami.6b00896

[64] Tang Z, Kotov NA, Magonov S, Ozturk B. Nanostructured artificial nacre. Nature Materials. 2003;**2**:413-418. DOI: 10.1038/nmat906

[65] Hager MD, Greil P, Leyens C, et al. Self-healing materials. Advanced Materials. 2010;**22**:5424-5430

[66] Xiang Z, Zhang L, Yuan T, et al. Healability demonstrates enhanced shape-recovery of Graphene-oxide-reinforced shape-memory polymeric films. ACS Applied Materials & Interfaces. 2018;**10**:2897-2906. DOI: 10.1021/acsami.7b14588

[67] Vidyasagar A, Sung C, Gamble R, Lutkenhaus JL. Thermal transitions in dry and hydrated layer-by-layer assemblies exhibiting linear and exponential growth. ACS Nano. 2012;**6**:6174-6184. DOI: 10.1021/nn301526b

[68] Chen XC, Ren KF, Zhang JH, et al. Humidity-triggered self-healing of microporous polyelectrolyte multilayer coatings for hydrophobic drug delivery. Advanced Functional Materials. 2015;**25**:7470-7477. DOI: 10.1002/adfm.201503258

[69] Mishra P, Mapara S, Vyas P. Testing/monitoring of soil chemical level using wireless sensor network technology. Int J Appl or Innov Eng Manag. 2015;**4**:114-117

[70] Rogovska N, Laird DA, Chiou CP, Bond LJ. Development of field mobile soil nitrate sensor technology to facilitate precision fertilizer management. Precision Agriculture. 2019;**20**:40-55. DOI: 10.1007/s11119-018-9579-0

[71] Hartemink AE, McBratney AB. Proximal Soil Sensing. New York, USA: Springer; 2010

[72] Braunger ML, Shimizu FM, Jimenez MJM, et al. Microfluidic electronic tongue applied to soil analysis. Chem. 2017;**5**:1-10. DOI: 10.3390/chemosensors5020014

Chapter 11

Spin Transport in Nanowires Synthesized Using Anodic Nanoporous Alumina Films

Supriyo Bandyopadhyay

Abstract

Spin transport in restricted dimensionality structures (e.g., nanowires) have unusual features not observed in bulk samples. One popular method to synthesize nanowires of different materials is to electrodeposit them selectively within nanometer diameter pores in anodic alumina films. Different materials can be sequentially deposited within the pores to form nanowire "spin valves" consisting of a spacer nanowire sandwiched between two ferromagnetic nanowires. This construct allows one to study spin transport in the spacer nanowire, with the ferromagnetic contacts acting as spin injector and detector. Some of our past work related to the study of spin transport in organic and inorganic nanowire spin valves produced using nanoporous anodic alumina films is reviewed in this chapter.

Keywords: spin transport, nanowires, nanoporous anodic alumina films, self-assembly

1. Introduction

The use of anodic porous alumina films as templates to fabricate nanowires of different materials has a long history [1–15]. Such films have been used to fabricate organic [3], semiconducting [4] and metallic [6] nanowires in our group over the past three decades. While these structures have been used to demonstrate a wide variety of optical [7–9], magnetic [10, 11] and electronic [12–15] phenomena and devices, here we focus on spin transport properties investigated in nanowire "spin valves" fabricated by sequentially electrodepositing a ferromagnetic material, a semiconductor/metal/organic, and finally another ferromagnet selectively within the nanopores of anodic alumina films.

Figure 1 shows the schematic of a vertically standing array of nanowire spin-valves fabricated by electrodepositing different materials sequentially within the nanopores of an anodic alumina film.

To produce the nanoporous anodic film, first, a commercial grade aluminum foil of thickness 0.1 mm is diced into 2 × 2 cm coupons. Each coupon is electropolished in a solution of perchloric acid, ethanol, butyl cellusolve and distilled water to reduce the surface roughness to about 3 nm [16] which gives the surface a shiny appearance as shown in **Figure 2**. The electropolishing is carried out at 40 V dc in six intervals of 5 s each in order to not overheat the electropolishing solution. Next, the electropolished foil is rinsed in distilled water and anodized in an anodizing bath (**Figure 3a**) in 0.3 M oxalic acid using the foil as the anode and a platinum grid

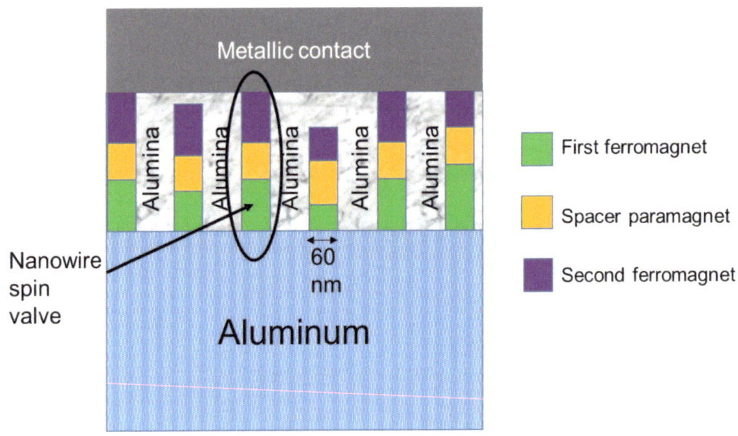

Figure 1.
Schematic of a vertically standing array of spin-valves in an anodic alumina matrix.

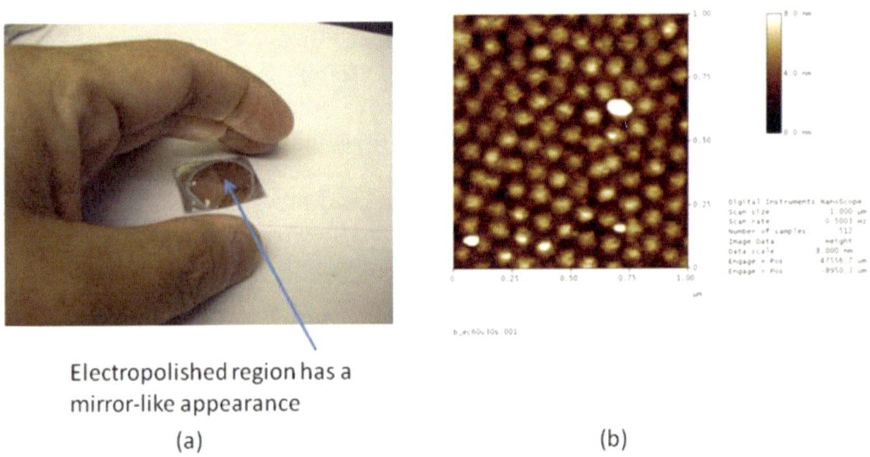

Figure 2.
(a) An electropolished sample and (b) atomic force micrograph of the electropolished sample showing that the rms surface roughness is about 3 nm.

as the cathode. The anodization is carried out at 40 V for 12 min and then the voltage is gradually reduced in steps of 10 V per min for another 3 min. This step anodization weakens the barrier layers that form at the bottom of the pores (**Figure 3d**). Subsequent soaking in 5% phosphoric acid for 45 min removes the barrier layers in most of the pores and at the same time widens the pores slightly. After this step, one obtains ~60 nm diameter pores in an anodic alumina film whose bottoms are exposed to aluminum as shown in **Figure 1**. The thickness of the anodic film that is produced depends on the duration of anodization. For the 15-min anodization process described above, the film thickness is typically ~1 μm.

In order to form the nanowire spin-valves, a ferromagnetic material, a semiconducting/metallic/organic spacer material and another ferromagnetic material are sequentially electrodeposited selectively within the nanopores to produce the structure in **Figure 1**. Electrodeposition is carried out in solutions containing the material to be deposited within the pores. For electrodepositing Co or Ni (ferromagnets), aqueous solutions of $NiSO_4$ and $CoSO_4$ with a slight amount of Boric acid (to aid ionization in the solution) are used. The Co^{2+} or Ni^{2+} ions in the solution will flow selectively into the pores since they offer the least resistance paths for the

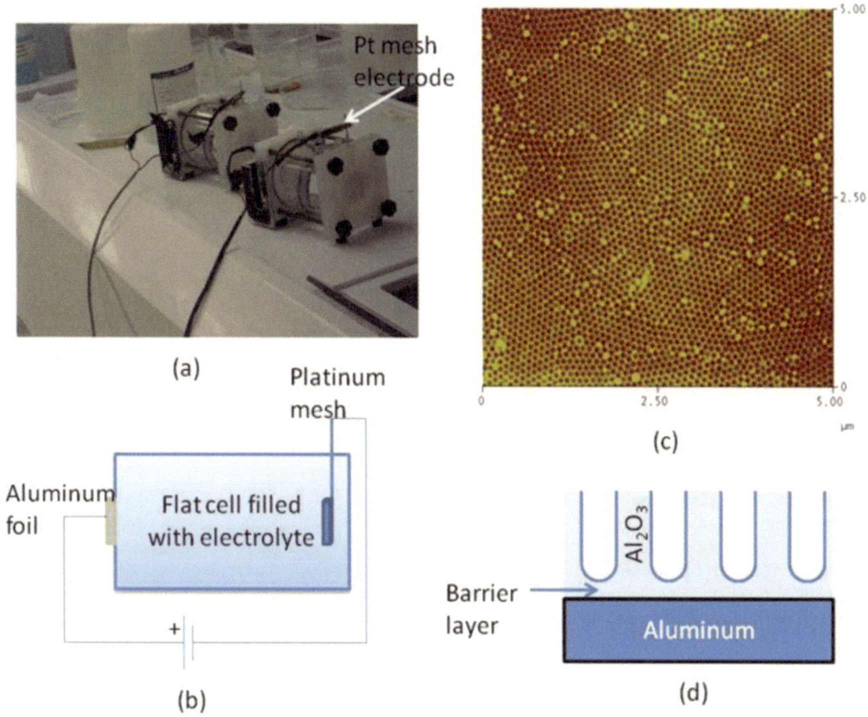

Figure 3.
(a) The anodizing flat cells, (b) the anodizing configuration where the aluminum foil is used as the anode and a platinum mesh as the cathode, (c) an atomic force micrograph of the nanoporous film formed after anodization. There is excellent ordering of the pores within domains of size ~1 µm and (d) cross-section of the pores showing the barrier layers which are later removed by soaking in phosphoric acid.

current to flow. The accumulated metal ions are deionized ($Co^{2+} + 2e^{-} = Co$) and are deposited within the pores. The spacer layer can also be electrodeposited by using appropriate solutions. For example, the procedure for depositing CdS is described in Refs. [13, 14]. Organics can be evaporated [3] or electrosprayed selectively within the pores [9]. In the end, we obtain a tri-layered nanowire of ferromagnet-spacer-ferromagnet as shown in **Figure 1**, constituting a nanowire spin valve. For electrical contacting, a metallic layer is deposited on top of the structure using electron beam evaporation. This layer and the bottom aluminum substrate are used as contacts and wires can be attached to these layers using silver paste.

2. Spin transport in nanowire spin valves formed in nanoporous alumina templates

2.1 Spin transport in organic nanowire spin valves

Organic nanowire spin valves were made of cobalt, Alq_3 (tris-(8-hydroxy-quinolinolato) aluminum) and nickel by first electrodepositing 500 nm of Ni within 50-nm diameter pores, then evaporating Alq_3 through a 1 mm^2 window in a mask at a base pressure of 10^{-7} Torr, and then evaporating Co on top without breaking vacuum. Alq_3 seeps into the pores by capillary action upon evaporation. The fact that it is a short stranded molecule with low molecular weight helps in transporting the molecules inside the pores.

Figure 4 shows a schematic of the structure formed, along with a transmission electron micrograph (TEM) of a single nanowire spin valve that is produced within

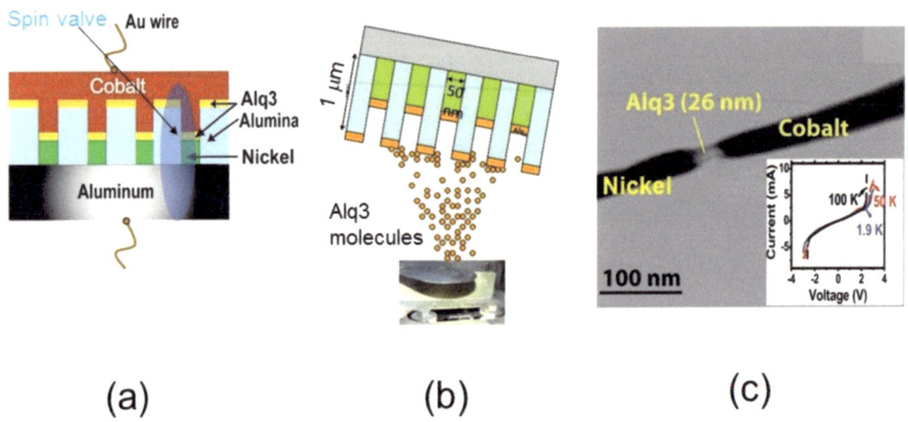

Figure 4.
(a) Schematic of the organic nanowire spin valves, (b) evaporation of Alq_3 is carried out by tilting the sample by 15° with respect to the source to get the best results, (c) transmission electron micrograph of a single nanowire spin valve. The inset shows the measured current-voltage characteristic of an array of spin valves at two different temperatures. The non-linear characteristic shows that the contacts between the ferromagnets and organic have a Schottky character, indicating that Schottky barriers form at these interfaces. Tunneling through the Schottky barriers aids spin injection and detection at the ferromagnetic contacts. Figures (a) and (c) are reproduced from Ref. [16].

a pore. The micrograph was obtained by releasing the spin valves from the alumina matrix by dissolving out the latter in hot chromic/phosphoric acid and recapturing the released wires on a TEM grid for imaging.

Longitudinal magnetoresistance traces of an ensemble of organic nanowire spin valves, measured at three different temperatures are shown in **Figure 5(a)**. From the heights of the peaks shown in the traces, the spin diffusion length can be estimated using the modified Julliere relation [17].

$$\frac{\Delta R}{R} = \frac{2P_1P_2e^{-d/L_s}}{1 - P_1P_2e^{-d/L_s}} \tag{1}$$

where ΔR is the height of the magnetoresistance peak measured from the baseline, R is the baseline resistance, P_1 and P_2 are the spin polarizations in the Co and Ni contacts, d is the length of the spacer layer (26 nm from **Figure 4**), and L_s is the spin diffusion length in Alq_3. This last quantity is plotted as a function of temperature in the top panel of **Figure 5(b)**.

From the spin diffusion length, we can estimate the spin relaxation time using the relation

$$\tau_s = \frac{q^2L_s^2}{kT\mu}, \tag{2}$$

where q is the electronic charge, kT is the thermal energy and μ is the mobility of carriers in Alq_3. Since the mobility in the organic is determined by carrier hopping, it is likely to be relatively independent of temperature. From the reported values of the mobility in Alq_3, we have found the extremum values of the spin relaxation time in the organic and these are plotted as a function of temperature in **Figure 5(b)**. These times are exceptionally long, perhaps among the longest reported in any solid above liquid nitrogen temperature (77 K). The reason why these times are so long is because Alq_3 is composed primarily of hydrocarbons which have low atomic number Z and hence weak spin-orbit interaction (the interaction strength is proportional to Z^4). The weak interaction preserves spin coherence over long durations, making organics a preferred platform for spintronics in some applications.

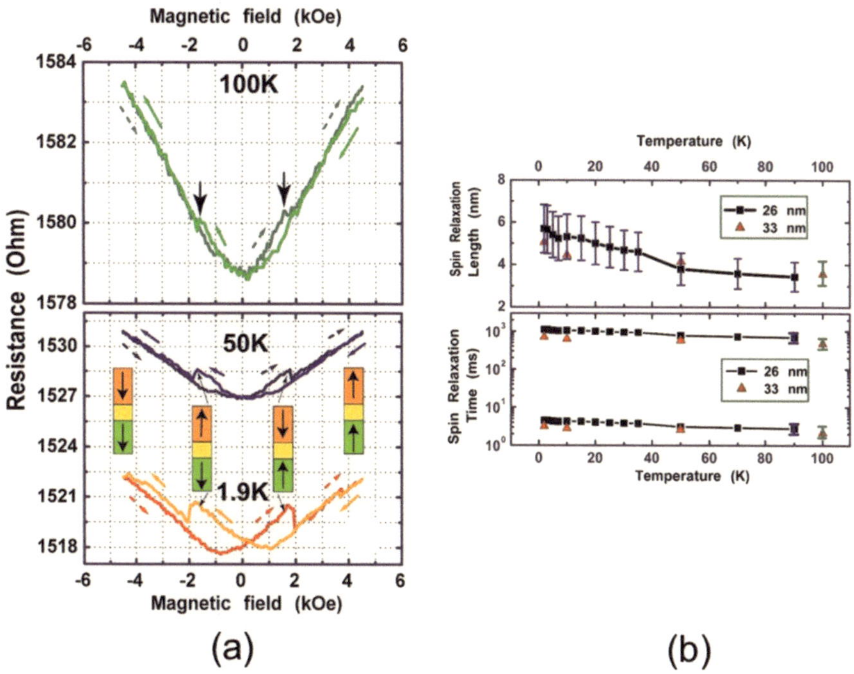

Figure 5.
(a) Longitudinal magnetoresistance traces of an ensemble of spin valves measured at three different temperatures, (b) the estimated spin relaxation lengths and times as a function of temperature for two different sets of samples with spacer layer thicknesses of 26 and 33 nm. In the lower panel, the extremum values are shown for the spin relaxation times. Reproduced from Ref. [16].

There is a bit of controversy regarding the primary mechanism responsible for spin relaxation in an organic like Alq_3. While Ref. [16] proposed that the main mechanism is the Elliott-Yafet spin relaxation [18], there is a strong differing opinion that it is hyperfine nuclear interaction [19]. The electric field dependence of the spin relaxation time (or spin diffusion length) observed in Ref. [3] speaks against hyperfine interaction and is more consistent with Elliott-Yafet relaxation. Recently, this view was concurred with in the context of spin transport in organic single crystal semiconductors [20].

2.2 Spin transport in semiconductor nanowire spin valves

Nanowire spin valves (diameter = 50 nm) with germanium as the spacer and Co, Ni as the ferromagnetic contacts have also been produced in our group using sequential electrodeposition of the constituent elements in 50 nm diameter pores in anodic alumina films. The Co and Ni were electrodeposited from aqueous solutions of $NiSO_4$ and $CuSO_4$. Ge was electrodeposited using the ionic liquid 1-butyl-3-methylimidazolium hexafluorophosphate (BMIM-PF_6), GeI_4 salts and dimethyl sulfoxide (DMSO) [21]. A gold top contact was deposited by electron beam evaporation and the Al foil was contacted from the back to electrically access an ensemble of spin valves as shown in **Figure 6(a)**. For characterization, the nanowire spin valves were released from their alumina hosts by dissolving out the alumina in hot chromic/phosphoric acid and then capturing the released spin vales in TEM grid for imaging. **Figure 6(b)** shows a transmission electron micrograph of a single spin valve that forms within a single pore.

Longitudinal magnetoresistance traces of these spin valve samples were obtained and from the spin valve peaks, the spin relaxation length (L_s) and time (τ_s)

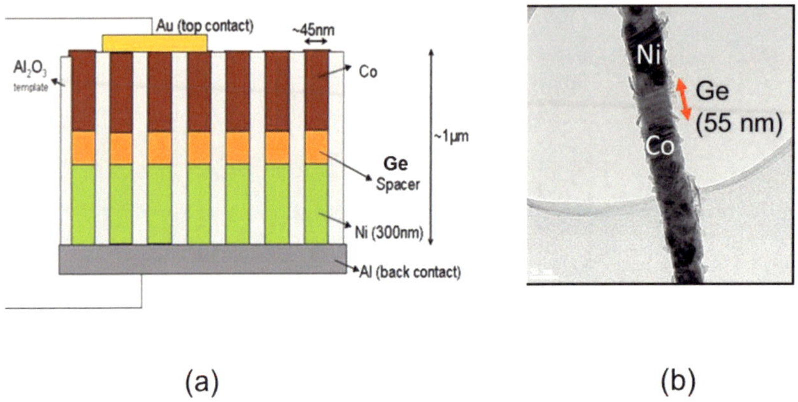

Figure 6.
(a) An array of vertically standing nanowire spin valves with Ge spacer hosted in an alumina matrix and (b) transmission electron micrograph of a single nanowire spin valve. Credit: Sridhar Patibandla.

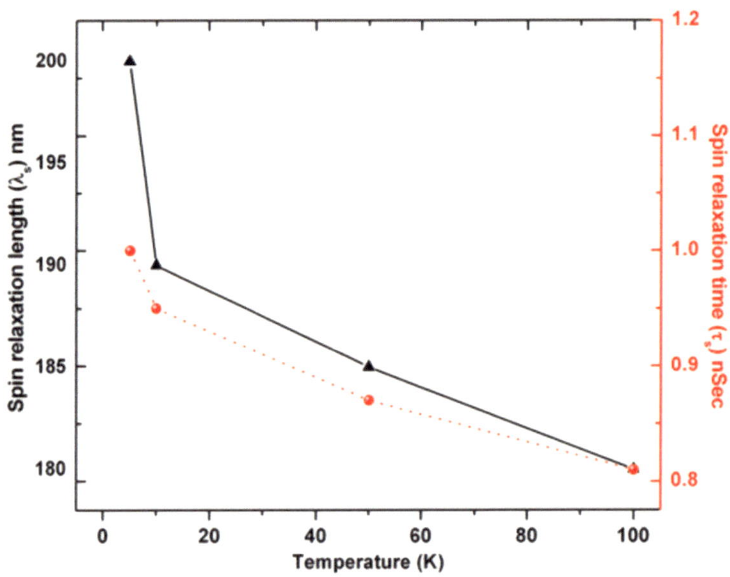

Figure 7.
Spin relaxation length and time in Ge nanowires as a function of temperature. Credit: Sridhar Patibandla.

were obtained following Eqs. (1) and (2). These quantities are plotted in **Figure 7** as a function of temperature.

Interestingly, the spin relaxation length at 4.2 K is ~200 nm which is longer than that reported in carbon nanotubes (130 nm) at that temperature [22]. Even at liquid nitrogen temperature (77 K), the spin relaxation length is over 180 nm, which makes Ge nanowires viable for spin-transistor type applications.

Germanium is an indirect gap semiconductor with the lowest conduction band valley at the L-point where the electron wavefunctions have some *p*-type character leading to non-zero intrinsic spin-orbit coupling. This should be the dominant spin relaxation mechanism in the Ge spacer layer since the Dresselhaus interaction [23] should be weak or non-existent owing to the fact that the Ge crystal is centrosymmetric. The Rashba interaction [24] should also be weak because of the absence of strong symmetry breaking by electric fields. The most common isotope of germanium ^{74}Ge has no net nuclear spin and hence hyperfine interaction with nuclear spins should also be weak in a Germanium nanowire. This leaves the intrinsic spin-orbit interaction as the likely dominant source of spin relaxation. The spin orbit interaction can cause both Elliott-Yafet [18] and D'yakonov-Perel' [25] spin relaxation in germanium nanowires, although in our samples, the poor mobility caused by interface scattering makes the Elliot-Yafet mechanism more likely to be dominant.

2.3 Spin transport in InSb nanowire spin valves at room temperature

Most spin transport experiments in nanowire spin valves have been conducted at low temperatures where the spin diffusion length is long enough to exhibit observable spin transport effects. Spin transport at room temperature is rarely, if ever, studied because the short spin relaxation lengths at room temperature obscure most spin effects. However, practical devices based on spin transport would require room temperature operation. This begs the question as to when spin transport effects may be observable at room temperature. To answer this question, one needs to understand what are the dominant sources of spin relaxation in semiconductor nanowires (in a given circumstance) and what could be done to suppress them in order to slow down the spin relaxation rate at room temperature.

In most semiconductors, the dominant spin relaxation mechanism at room temperature is the D'yakonov-Perel' mechanism [25]. This mechanism accrues from spin-orbit interaction that acts as an effective (velocity-dependent) magnetic field for electrons [26]. The electron spins precess about this effective magnetic field while transiting through a semiconductor. Collisions change the magnitude and direction of an electron's velocity and hence the direction and magnitude of the effective magnetic field. The spins therefore precess about randomly changing axes (direction of the effective magnetic field) with random angular frequencies since the frequency of precession is proportional to the magnetic field strength. This randomness causes ensemble spin relaxation.

In a strictly *one-dimensional* semiconductor structure (quantum wire), where only a *single* transverse subband is occupied by electrons, the D'yakonov-Perel' mechanism is completely suppressed as long as the axis of the nanowire is along certain principal crystallographic directions [26]. This would increase the spin relaxation time (and length) considerably, and may make spin effects observable at room temperature. In order to verify this prediction, we prepared InSb nanowire spin valves of 50 nm diameter in nanoporous alumina films. Since InSb has a very small electron effective mass, the energy spacing between subbands in a 50 nm diameter InSb nanowire is large enough that only a single subband could be occupied at room temperature. This would suppress the D'yaknov-Perel' mechanism in an InSb nanowire and might make spin effects observable at room temperature.

Vertically standing arrays of nanowire spin valves consisting of Co-InSb-Ni and encased in an alumina matrix were produced by sequentially electrodepositing these materials within 50-nm diameter nanopores using dc electrodeposition. The finished structure looks like that in **Figure 6(a)**, except the spacer layer is InSb instead of Ge. InSb is electrodeposited from a solution of 0.15 M $InSO_4$, 0.1 M $SbCl_3$, 0.17 M $Na_3C_6H_5O_7$ and 0.36 M $C_6H_8O_7$ (citric acid). The Co and InSb depositions were carried out at 3 V dc for 1 min each whereas the Ni electrodeposition was carried out at 5 V dc for 6 min to slightly overfill the pores and make the Ni spill out on the surface. These structures were contacted from the top and bottom with copper wires attached with silver paste.

We measured both the longitudinal and transverse magnetoresistance of the spin valve samples at room temperature using an electromagnet. The longitudinal magnetoresistance was measured with the magnetic field pointing in the direction of the nanowire axis (which is also the direction of current flow during

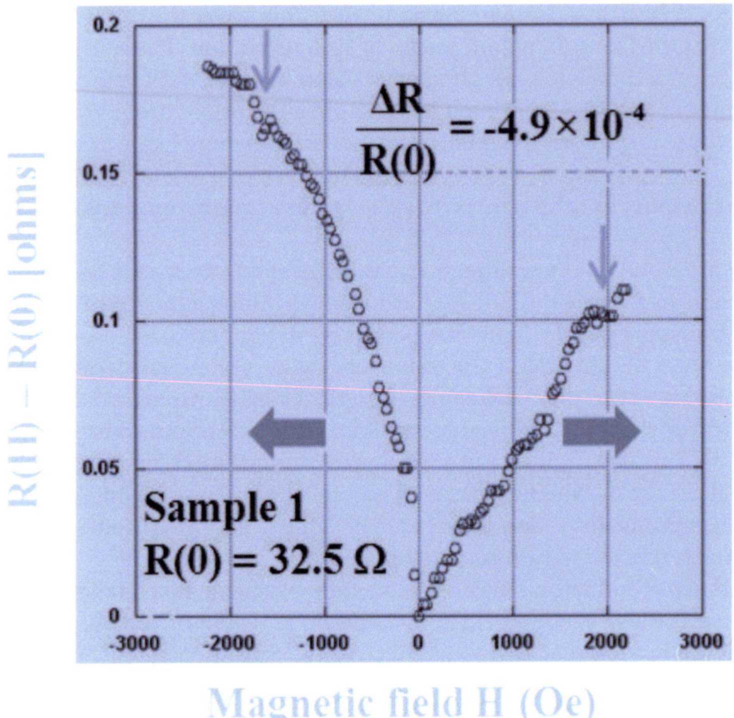

Figure 8.
Longitudinal magnetoresistance of InSb nanowire spin vales at room temperature. The spin valve troughs are indicated with vertical arrows. Reproduced from Ref. [28] with permission of Wiley.

magnetoresistance measurement). The transverse magnetoresistance was measured with the magnetic field perpendicular to the direction of current flow (and hence perpendicular to the nanowire axes).

Figure 8 shows the longitudinal magnetoresistance traces with the positive and negative field segments corresponding to forward and reverse scans of the magnetic field. There are two shallow "troughs" marked with vertical arrows. These are the tell-tale spin valve troughs. They are the counterparts of the spin valve peaks seen in **Figure 5**. Normally the spin valve effect would result in the magnetoresistance exhibiting peaks between the coercive fields of the two ferromagnetic contacts, but if there are defects close to either contact that electrons can resonantly tunnel through, then the spin polarization of that contact can be reversed, resulting in a trough as opposed to a peak in the longitudinal magnetoresistance [27]. From the measured depths of these troughs (ΔR), we can obtain the spin relaxation length L_s using Eq. (1) if we know what the spacer layer length L is. Transmission electron microscopy revealed the spacer layer length in the samples to be ~20 nm. The ratio L_s/L in the sample shown in **Figure 8** was found to be 0.29, indicating that the spin relaxation length in the InSb nanowire is 5.88 nm at room temperature under equilibrium.

We calculated that the energy spacing between the two lowest subbands in the InSb nanowires is 3.9 kT at room temperature (101 meV) which would force 96% of the electrons to reside in the lowest subband at room temperature [28]. This means that the samples are almost strictly one dimensional quantum wires where the D'yakonov-Perel relaxation will be strongly suppressed [28]. The experimental observations in Ref. [28] confirmed this view.

Because of the strict one-dimensionality of our nanowires, we can relate the spin relaxation length L_s and the spacer layer length L to the spin relaxation time (τ_s) and the transit time through the spacer (τ_t) according to

$$L = v_d \tau_t, \quad L_s = v_d \tau_s \tag{3}$$

where v_d is the drift velocity of electrons in the spacer.

In order to determine the transit time through the spacer layer, we measured the transverse magnetoresistance and observed Hanle oscillations in the magnetoresistance traces [28].

To understand the origin of Hanle oscillations in the magnetoresistance, one can refer to **Figure 9**. The magnetic field is directed perpendicular to the axis of the nanowire spin valve. The ferromagnetic contacts are initially magnetized along the axis with a longitudinal magnetic field. Depending on the nature of the ferromagnets, this will align the magnetizations of the two contacts either parallel, or antiparallel, to each other, as shown in **Figure 9**. The left contact, connected to the negative pole of the battery, injects electrons into the nanowire spacer with their spins pointing to the right. The transverse magnetic field makes the spins precess in the plane normal to the magnetic field. If the spins precess through an angle that is an odd multiple of π, then the spins arriving at the right ferromagnetic contact will be either antiparallel (in the case of parallel contacts) or parallel (in the case of antiparallel contacts) to the magnetization of the right contact. In the former case, the spins will be blocked by the right contact and the resistance of the spin valve will be high. In the latter case, the spins will transmit through the right contact and the resistance will be low. The reader can easily understand that the situation will be opposite when the spins precess through an angle that is an even multiple of π.

If we continuously increase the magnetic field, that will increase the precessional frequency (and hence the angle Φ through which the spins precess as they traverse the spacer layer) linearly with the field. Every time Φ reaches an odd integral value of π, the resistance either peaks or ebbs depending on the situation just described. That will lead to oscillations in the transverse magnetoresistance. These are the Hanle oscillations.

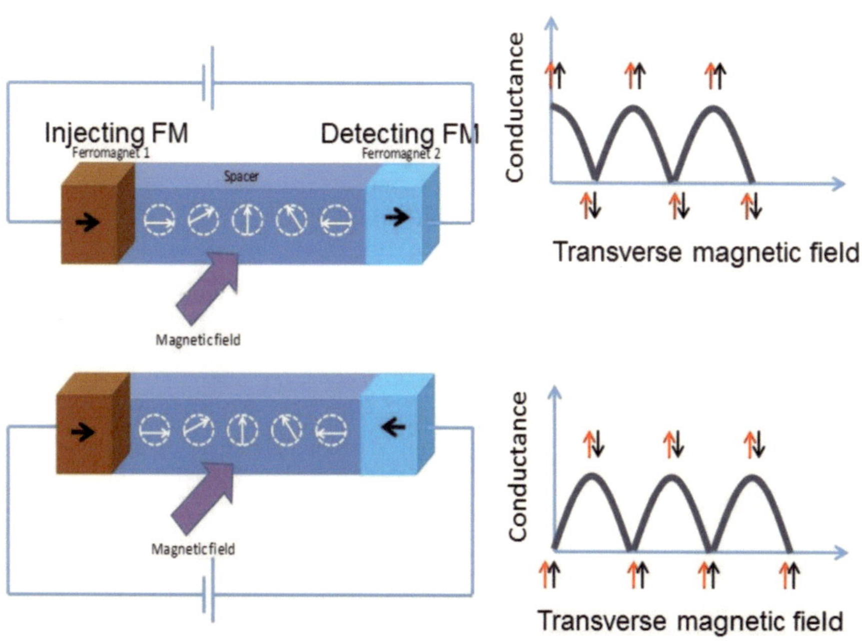

Figure 9.
Origin of Hanle oscillations in the magneto conductance. The left panel explains the origin of the effect and the right panel shows the Hanle oscillations in the magneto conductance.

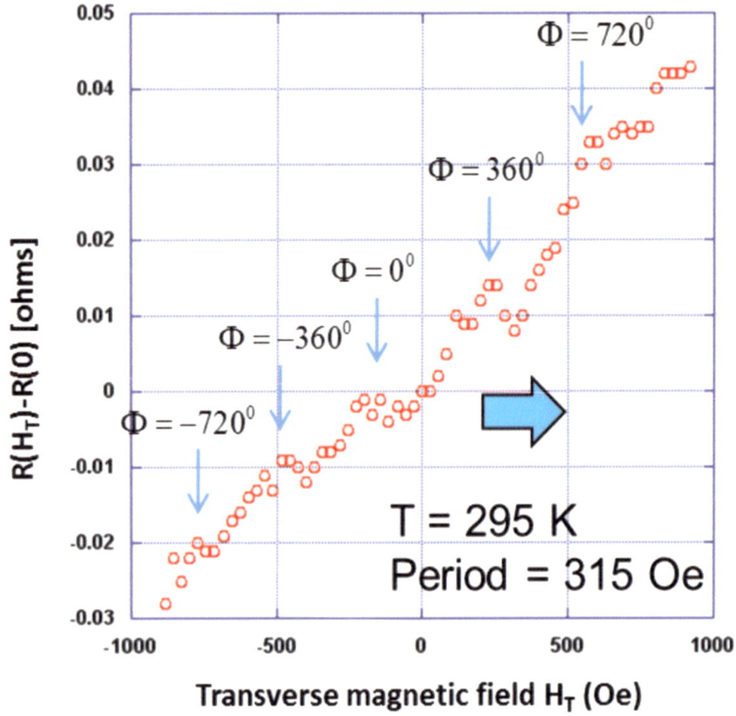

Figure 10.
Hanle oscillations in the transverse magnetoresistance traces of the InSb nanowire samples at room temperature. Reproduced from Ref. [28] with permission of Wiley.

In **Figure 10**, we show the observed Hanle oscillations at room temperature. The measured period of the oscillations in the transverse magnetic field is 315 Oe.

The transit time through the spacer layer is related to the oscillation period according to

$$\tau_t = \frac{h}{|g|\mu_B B_0}, \tag{4}$$

where h is the Planck constant, g is the Landé g-factor of InSb, μ_B is the Bohr magneton and B_0 is the period of the oscillation.

From Eq. (4), using B_0 = 315 Oe, we find that the average transit time through the spacer layers of the spin valves is 44 ps. Using this value in Eq. (3), we deduce that the drift velocity of electrons in the spacer layer is 454 m/s. From Eq. (3), we can then determine that that the spin relaxation time in the InSb spacer layer is 13 ps. This is approximately one order of magnitude larger than that reported in InSb or InAs quantum wells [29, 30].

3. Conclusions

In this chapter, I have described some spin transport experiments in nanowire spin vales produced by sequentially electrodepositing various materials (ferromagnet-spacer-ferromagnet) within nanopores of anodic alumina films. These experiments have revealed intriguing and important features of spin transport in restricted dimensionality systems. They have also demonstrated the immense capability afforded by electrochemical self-assembly methods.

Acknowledgements

The work described here has been supported by various grants through the US National Science Foundation at various times. Most of the work was carried out by my erstwhile Ph.D. students, post-doctoral mentees and high-school interns—Dr. Sridhar Patibandla (now at Intel Corporation, USA), Prof. Sandipan Pramanik (now at University of Alberta, Canada), Dr. Carmen-Gabriela Stefanita (now at IBM Data Science, USA), Dr. Iftekhar Hossain (now at Eastman Chemicals, USA), Dr. Hasnain Ahmad (now at IMEC, Belgium) and Mr. Saumil Bandyopadhyay (now at MIT, USA). I am also grateful to my collaborators Prof. Marc Cahay from University of Cincinnati, and Profs. Jayasimha Atulasimha and Gary C. Tepper from Virginia Commonwealth University.

The author declares no competing interest.

Author details

Supriyo Bandyopadhyay
Department of Electrical and Computer Engineering, Virginia Commonwealth University, Richmond, VA, USA

*Address all correspondence to: sbandy@vcu.edu

References

[1] Masuda H, Nishio K. Synthesis and applications of highly ordered anodic porous alumina. In: Adachi M, Lockwood DJ, editors. Self-Organized Nanoscale Materials, Nanostructure Science and Technology. New York: Springer; 2006. pp. 296-312

[2] Huber CA, Huber TE, Sadoqi M, Lubin JA, Manalis S, Prater CB. Nanowire array composites. Science. 1994;**263**:800-802

[3] Pramanik S, Bandyopadhyay S, Garre K, Cahay M. Normal and inverse spin-valve effect in organic semiconductor nanowires and the background monotonic magnetoresistance. Physical Review B. 1998;**74**:235329

[4] Ramanathan S, Patibandla S, Bandyopadhyay S, Edwards JD, Anderson J. Fluorescence and infrared spectroscopy of electrochemically self-assembled ZnO nanowires: Evidence of the quantum confined stark effect. Journal of Materials Science: Materials in Electronics. 2006;**17**:651-655

[5] Iwasaki T, Motoi T, Den T. Multi-walled carbon nanotubes growth in anodic alumina nanoholes. Applied Physics Letters. 1999;**75**:2044-2046

[6] Zeng H, Michalski S, Kirby RD, Sellmyer DJ, Menon L, Bandyopadhyay S. Effects of surface morphology on magnetic properties of Ni nanowire arrays in self-ordered porous alumina. Journal of Physics: Condensed Matter. 2002;**14**:715-721

[7] Kouklin N, Menon L, Wong AZ, Thompson DW, Woollam JA, Williams PF, et al. Giant photoresistivity and optically controlled switching in self-assembled nanowires. Applied Physics Letters. 2001;**79**:4423-4425

[8] Ramanathan S, Patibandla S, Bandyopadhyay S, Anderson J, Edwards JD. Fluorescence spectroscopy of electrochemical self-assembled ZnSe and Mn:ZnSe nanowires. Nanotechnology. 2008;**19**:195601

[9] Sarkar S, Kanchibotla B, Nelson JD, Edwards JD, Anderson J, Tepper GC, et al. Giant increase in the metal-enhanced fluorescence of organic molecules in nanoporous alumina templates and large molecule-specific red/blue shift of the fluorescence peak. Nano Letters. 2014;**14**:5973-5978

[10] Zeng H, Zheng M, Skomski R, Sellmyer DJ, Liu Y, Menon L, et al. Magnetic properties of self-assembled Co nanowires of varying length and diameter. Journal of Applied Physics. 2000;**87**:4718-4720

[11] Zeng H, Skomski R, Menon L, Liu Y, Bandyopadhyay S, Sellmyer DJ. Structure and magnetic properties of ferromagnetic nanowires self-assembled arrays. Physical Review B. 2002;**65**:134426

[12] Kouklin N, Bandyopadhyay S, Tereshin S, Varfolomeev A, Zarestky D. Electronic bistability in electrochemically self-assembled quantum dots: A potential non-volatile random access memory. Applied Physics Letters. 2000;**76**:460-462

[13] Kouklin N, Menon L, Bandyopadhyay S. Room-temperature single electron charging in electrochemically synthesized semiconductor quantum dot and wire array. Applied Physics Letters. 2002;**80**:1649-1651

[14] Varfolomeev A, Zarestky D, Pokalyakin V, Tereshin S, Pramanik S, Bandyopadhyay S. Admittance of CdS nanowires embedded in porous alumina template. Applied Physics Letters. 2006;**88**:113114

[15] Cahay M, Garre K, Fraser JW, Lockwood DJ, Semet V, Binh VT, et al. Characterization and field emission properties of lanthanum monosulfide nanoprotrusion arrays obtained by pulsed laser deposition on self-assembled nanoporous alumina templates. Journal of Vacuum Science and Technology B. 2007;**25**:594-603

[16] Pramanik S, Stefanita C-G, Patibandla S, Bandyopadhyay S, Garre K, Harth N, et al. Observation of extremely long spin relaxation times in an organic nanowire spin valve. Nature Nanotechnology. 2007;**2**:216-219

[17] Julliere M. Tunneling between ferromagnetic films. Physics Letters A. 1975;**54**:225-226

[18] Elliott RJ. Theory of the effect of spin-orbit coupling on magnetic resonance is some semiconductors. Physics Review. 1954;**96**:266-279

[19] Bobbert PA, Wagemans W, van Oost FWA, Koopmans B, Wohlgenannt M. Theory for spin diffusion in disordered organic semiconductors. Physical Review Letters. 2009;**102**:156604

[20] Tsurumi J, Matsui H, Kubo T, Häusermann R, Mitsui C, Okamoto T, et al. Coexistence of ultra-long spin relaxation time and coherent charge transport in organic single crystal semiconductors. Nature Physics. 2017;**13**:994-998

[21] Patibandla S, Pramanik S, Bandyopadhyay S, Tepper GC. Spin relaxation in a germanium nanowire. Journal of Applied Physics. 2006;**100**:044303

[22] Tsukogoshi K, Alphenaar BW, Ago H. Coherent transport of electron spin in a ferromagnetically contacted carbon nanotube. Nature. 1999;**401**:572

[23] Dresselhaus G. Spin-orbit coupling effects in zinc blende structures. Physics Review. 1955;**100**:580

[24] Bychkov YA, Rashba EI. Oscillatory effects and the magnetic susceptibility of carriers in inversion layers. Journal of Physics C: Solid State Physics. 1984;**17**:6039-6045

[25] D'yakonov MI, Perel' VI. Spin orientation of electrons associated with the interband absorption of light in semiconductors. Soviet Physics—JETP. 1971;**33**:1053

[26] Bandyopadhyay S, Cahay M. Introduction to Spintronics. 2nd ed. Boca Raton: CRC Press; 2015

[27] Tsymbal EY, Sokolov A, Sabirianov IF, Doudin B. Resonant inversion of tunneling magnetoresistance. Physical Review Letters. 2003;**90**:186602

[28] Bandyopadhyay S, Hossain MI, Ahmad H, Atulasimha J, Bandyopadhyay S. Coherent spin transport and suppression of spin relaxation in InSb nanowires with single subband occupancy at room temperature. Small. 2014;**10**:4379-4385

[29] Bhowmick M, Kini RN, Nontapot K, Goel N, Chung SJ, Mishima TD, et al. Probe of interband relaxations of photo-excited carriers and spins in InSb based quantum wells. Physics Procedia. 2010;**3**:1161-1165

[30] Olesberg JT, Lau WH, Flatté ME, Yu C, Altunkaya E, Shaw EM, et al. Interface contributions to spin relaxation in a short period InAs/GaSb superlattice. Physical Review B. 2001;**64**:201301(R)

Chapter 12

Concepts for Designing Tailored Thin Film Surfaces with Potential Biological Applications

Nicolás Eduardo Muzzio, Omar Azzaroni, Sergio E. Moya and Miguel Ángel Pasquale

Abstract

The tailoring of surfaces with nanostructured thin films, where interfacial properties can be controlled at the nanoscale, offers multiple possibilities for biological applications. The design of thin films with appropriate properties to induce desired biological responses at cell level requires the convergence of research from physics, chemistry, material science, biology, and medicine. Here, we will discuss the main surface properties that determine the behavior of isolated cells, cell colonies, and tissues interacting with a material. Surface roughness, morphological features, stiffness, wettability, chemical nature, and protein-surface interaction characteristics, as well as spatiotemporal heterogeneities, are expected to contribute to the desired biological performance of a material. A brief review in relation to thin films for biological applications will be presented. We will focus on examples in which basic rather simple processes play a key role in determining the triggering of a particular biological cell phenotype.

Keywords: tailored surfaces, biocompatible thin films, cell phenotype control, biocompatibility, multifunctionality

1. Introduction

The interfacial region of a real system interacting with the environment determines many of its properties [1]. The control of the structure, topography, composition, charge density and distribution, wettability, and the viscoelastic properties at the interface plays a key role in determining the ultimate characteristics of a material interacting with a biological system [2]. Thus, thin films and coatings can be used to obtain tailored interfaces with nanoscale controlled properties with the aim of generating systems with different complexity and specific functionalities of biological interest. This issue is of particular interest for engineering material surfaces with potential biomedical applications as implants or medical devices, or platforms for fundamental studies at the cell level [3–6].

Biological entities, such as proteins, prokaryote, or eukaryote cells, are expected to interact with a material when the material is in contact with biological fluids or tissues. In the case of the interaction of cells with an interface, the triggered biochemical signals that generate spatial and temporal changes in the interaction process should be considered [7]. Cells appear to be smart macromolecular systems that trigger cooperative phenomena upon interaction with a substrate with particular physicochemical and biochemical properties.

Depending on its specific applications, a proposed biocompatible material is required to enhance or inhibit protein and cell adhesion. For instance, interfacial properties of materials for implants should promote tissue-material integration [8] and avoid bacterial infections [9]. Moreover, for many medical devices to act as barriers, antiadherence and antifouling characteristics are needed [10, 11].

In recent years, the combination of new and improved techniques for patterning a substrate [12, 13], spatially controlling the deposition of a material [14], supramolecular synthesis [15], as well as new assembling procedures to fabricate tunable films at the nanoscale [16, 17], has allowed the introduction and control of spatial and temporal, physicochemical and biological cues. It has been possible to wisely design materials to (i) modulate and study cell behavior and fate [18–21], (ii) deliver, release and sense bioactive molecules [22–25], and (iii) increase the biocompatibility and functionalities of materials and devices [26, 27]. In many situations, the tailored system properties depend on both the characteristics of the pristine substrate and those of the coating. Typical examples (**Figure 1**) are presented to show some design strategies, without pretending to be complete, as a large number of systems are possible: (a) a thin film is coated onto a rigid substrate to improve biocompatibility; (b) the thin film can be used to modulate the viscoelastic properties of the system; (c) a rigid substrate is modified by a coating containing species that can be internalized by

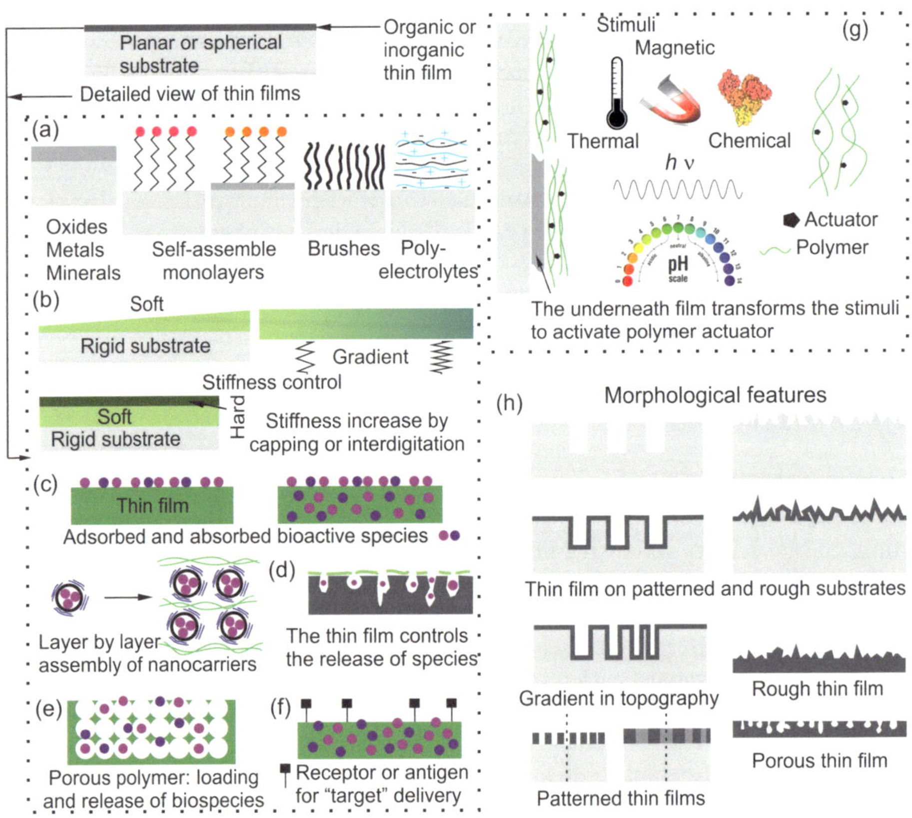

Figure 1.
Different strategies of thin film biological applications. (a) Organic and inorganic thin films that modify the interfacial properties toward cell behavior. (b) Stiffness modulation by organic thin films. (c) Thin film as cargo systems. (d) Thin film to control biomolecule release from a porous substrate. (e) Porous organic thin film for biomolecule release. (f) Thin film for "target" delivery. (g) Responsible thin films under either a direct or indirect stimulus. (h) Morphological features that can be incorporated by a thin film.

cells or are released to the medium; (d) the thin film controls the release of molecules from a solid substrate underneath; (e) the thin film is a porous material capable of releasing drugs encapsulated in the pores; (f) the thin film is tailored with species interacting specifically with a target; (g) either an underneath or an upper thin film bearing actuator moieties are responsive under external stimuli that trigger a change in the properties of the system; and (h) nanotopographic characteristics with spatial control can be introduced in the substrate surface. The topographic features can be combined with a thin film, and physicochemical properties such as specific moieties, stiffness, charge, and wettability can be tuned.

In the next sections, we will first give a brief comment about the desired biocompatibility of new proposed materials and survey some materials to fabricate thin films for biological applications. Then, we will depict the material properties that affect cell behavior and conclude with some examples of thin films to modulate cell behavior.

2. Thin films of biological interest

Thin films with potential applications in biology need to be biocompatible. There are many definitions for biocompatibility, but generally it is meant to refer to a material that is expected to carry out a suitable response in the host biological environment without producing any undesirable effects [28]. Different assays are performed to test the biocompatibility of a material, including in vitro experiments of cell proliferation and adhesion, interaction with biological fluids, and in vivo performance evaluation of the material function in model animals and its inflammatory response [29].

Many of the thin film materials presented in this chapter have potential applications but require further research for good biocompatibility.

2.1 Inorganic thin films

A wide variety of inorganic thin films and coatings have been proposed to be used in biology, including pure metals, alloys, metallic compounds, carbon-based thin films, oxide layers, and ceramic materials [3].

2.1.1 Metals and derived compounds

Titanium and its alloys are the most important materials for biomedical applications, particularly as bulk metal with proper surface modifications. Titanium oxide (TiO_2) and titanium nitride (TiN) coatings are very appealing as they form bioinert layers, protect substrates from corrosion and promote desirable biological functionalities, such as osteoblast adhesion and proliferation and apatite formation.

Mesoporous titania surfaces and nanotubular titania layers are appealing for biomedical applications since both titanium and titania exhibit good biocompatibility and are commonly utilized for biomedical devices and implants. Pores in the mesoporous titania can be designed for the encapsulation of drugs for delivery, growth factors, and antibiotics that can either facilitate cell material interaction or inhibit the presence of bacteria.

Gold and platinum metallic coatings are also used as microelectrodes to monitor signals from neurons and evaluate the electrophysiological activity of the neuronal network. Platinum-based thin films are also employed for pacemakers, implanted defibrillators, stents, and hearing assist devices, due to their good electrical conduction and biocompatibility.

2.1.2 Bioceramics and carbon-based materials

Bioceramic thin films are used for coating implants and prostheses, as they exhibit good corrosion resistance and better bioactivity than the metal substrate. Bioceramic thin films deposited on Ti prostheses yield better integration and bone growth in comparison with the bare metal. Some bioceramics utilized in patients are bioinert, for example, oxide and silicon ceramics. Other ceramics are resorbable, as is the case of calcium phosphates, and others are bioactive, such as bioglasses and hydroxyapatite.

Carbon-based materials, that is, diamond-like and carbon nitride thin films are potential materials for coating substrates for biological applications as biomedical devices. Diamond-like carbon shows lower platelet adhesion than carbon nitride. Despite this, carbon nitride thin films are able to yield harder coatings.

Unlike diamond-like and carbon nitride thin films that are amorphous carbon materials, nanocrystalline diamond thin films allow the tuning of their crystalline structure and can mimic the surface roughness of bone. These films also provide high chemical and corrosion resistance, making them very appealing for biological applications.

Carbon nanotubes and graphene sheets are promising materials for biomedical applications [30], presenting new opportunities due to their unique mechanical, electrical, and optical properties. Furthermore, carbon-based nanomaterials can mimic the microenvironment provided by the biological extracellular matrix [30]. These systems have been extensively studied as drug/gene delivery vehicles for anticancer treatments.

It has also been reported that graphene can favor the osteoblast linage of mesenchymal stem cells (MSCs) [31] and increases the rate of human MSCs osteogenic differentiation by a similar magnitude to bone morphogenic protein, BMP-2 [32].

2.2 Organic thin films and hybrid systems

Organic thin films, particularly those of natural origin, are very appealing for creating a biocompatible interface and are potentially suitable for biological applications. The drawback of these materials is their poor mechanical properties, as they resemble soft biological tissues. Soft materials are characterized by a limited cell adherence. Thus, they are often modified to fulfill the requirements of their intended application.

Examples of polymers used for thin film fabrication include the following:

1. Extracellular matrix components. For instance, collagen, fibronectin and laminin, growth factors and glycosaminoglycans, among others.

2. Zwitterionic polymers, which display an equal number of cationic and anionic groups making them hydrophilic and antifouling.

3. Polyethylene glycol-modified polymers (PEGylated polymers) and hyperbranched polymers with oligosaccharide surfactant hydrophilic moieties, which result in antifouling coatings.

4. Thermoresistive polymers derived from poly(N-isopropyl acrylamide) (PNIPAM) [33].

5. Polymers employed in the fabrication of devices (such as biosensors), used to prevent nonspecific adhesion of biological materials forming hydrophilic/

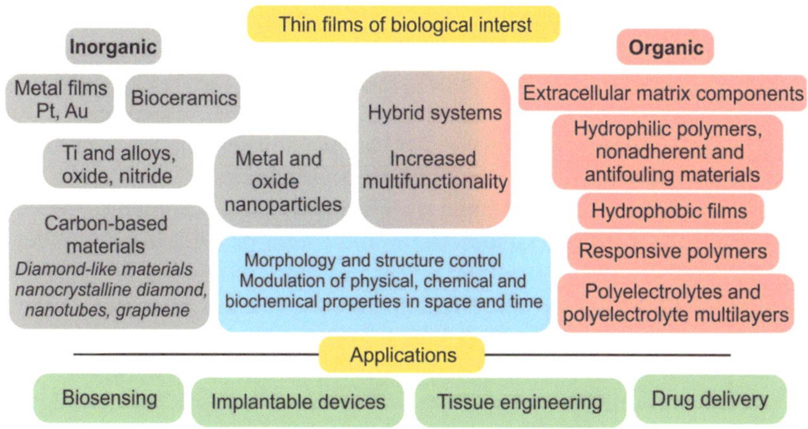

Figure 2.
Classification of materials for thin film coatings and applications.

hydrophobic layers, namely polydimethylsiloxane (PDMS), polyethylene (PE), perfluoropolyethylene (PFPE), polyimide (PI), parylene C, polyetheretherketone (PEEK), cyclic olefin/copolymer, poly(methyl methacrylate) (PMMA), polypropylene (PP), and polystyrene (PS) [34].

6. Polyelectrolyte multilayers (PEMs) of either synthetic or natural polyelectrolytes assembled by the versatile layer-by-layer (LbL) techniques driven by different molecular forces, that is, electrostatic or hydrogen bonding. The LbL technique allows the combination of different materials as building blocks and a wide variety of substrates can be coated.

Hybrid nature-inspired inorganic/organic materials have been fabricated, that is, layered systems can include metallic nanoparticles and nanotubes, mesoporous materials with organic moieties, hydrogels with in situ synthesized nanoparticles, etc. These systems can benefit from the multiple functionalities that can be carried by the nanomaterials to enhance gene delivery [35], imaging [36], and alternative disease treatments [37]. Furthermore, meso- and nanostructured titanium-based materials can be used to fabricate hybrid static and dynamic systems that include bioactive molecules as well as polymeric films to increase the material functionalities [21]. This approach is suitable to tackle the important issue of the miniaturization of medical devices.

A scheme of the classification of materials utilized to fabricate thin films of biological interest is depicted in **Figure 2**.

3. Thin film properties for tuning cell behavior

Living cells in an organism are sensitive to the environmental properties defined by neighboring cells and the extracellular matrix (ECM) [38]. Proteins play a key role in the cell sensing process of the microenvironment characteristics. For instance, cells are able to sense the viscoelastic and topographic properties of the environment through transmembrane proteins such as integrins [39]. Furthermore, growth factors, hormones, antibodies, vitamins, and other bioactive molecules can interact with specific cell receptors and induce a certain cell phenotype [40].

In general, cells in contact with a substrate interact with it through proteins deposited on the substrate. These proteins come from biological fluids or are produced by the cell itself upon interaction with the substrate [41]. Furthermore, in most of the experimental conditions employed for cell/material interaction studies, cells are cultured in medium supplemented with fetal bovine serum, and therefore, serum proteins interact with the substrate moieties. Among these proteins, bovine serum albumin (BSA) and fibronectin (FN) are usually considered as an example of nonadherent and adherent protein models, respectively.

Cell adhesion is the first step of many physiological and pathological processes such as wound healing, bacterial infections, and tumor progression. Eukaryote cell adhesion is possible if the intracellular forces exerted by stress fibers are balanced by the substrate stiffness [42]. The effect of the substrate stiffness is cell-type dependent, while adhesion of some cell lines is sensitive to changes in the substrate mechanical properties, other cells are less influenced [43]. Cells are able to sense the forces exerted by the environment and transduce them into biochemical signals; the study of this process is relatively recent and can profit from new actuator micro-devices such as microelectromechanical systems (MEMs) constructed with new biomaterials and methods to avoid toxicity [44].

On the other hand, cells can sense the stiffness of objects that are not in direct contact, depending on the compliant polymer thickness and its physicochemical properties [45]. Thus, cell adhesion can be modulated by depositing a compliant soft polymer upon a stiff support with controlled topography and by varying the thickness of the soft polymer [46]. Cells sense the stiffness that is the result of the combination of the stiffness of both the upper soft polymer and the rigid substrate underneath, namely "complex stiffness."

3.1 Substrate physicochemical properties

The cell adhesion characteristics and subsequent cell functionalities will depend strongly on the physicochemical properties of the biomaterial surface. In the next paragraphs, a brief survey of these properties will be given.

Wettability: Cell adhesion characteristics are optimal for surfaces with moderate wettability, that is, a contact angle of about 85°. It has been suggested that this behavior is due to the proper state of the adhesion-mediating ECM proteins that allow the interaction of cell receptors with specific protein motifs.

For highly hydrophobic surfaces, the amount of adsorbed protein is relatively large, but the intra-protein interaction and the protein/substrate interaction are high and may produce denaturing or hamper remodeling from cells.

On more polar hydrophilic surfaces, proteins adsorb in relatively low concentrations diminishing inter-protein interactions and preserving their native state. However, it has been observed that on highly hydrophilic surfaces, cell adhesion is limited and even completely unviable. In this case, the adhesion proteins are labile as relatively weak forces take place. This would cause the detachment of cells particularly when the monolayer density is high [2].

Surface charge: The charge of the coated substrate is an important factor for cell colony spreading. There are some research works that indicate that positively charged surfaces are more efficient toward cell adhesion than negatively charged surfaces. This fact is explained by the negative charge of most of the cell adhesion-mediating ECM species that would exhibit preferential adsorption at positively charged interfaces.

It has been reported that for UV-irradiated polytetrafluoroethylene in a NH3 atmosphere, the generated positive charge interacts synergistically with oxygen-containing groups, and the adhesion of mouse 3T3 fibroblast, human umbilical vein endothelial cells, and human embryonic kidney cells is stimulated [47–49].

Other works reported that the presence of carboxylate groups added to a copolymer of poly(DL-lactic acid) and poly(ethylene oxide) enhances rat aortic smooth muscle cells in comparison to a noncarboxylated surface [50]. The carboxylate groups may not only bring negative charge to the interface but they also increase its wettability to values suitable for cell adhesion [51].

It has been observed that bovine aortic endothelial cells adhere with a larger spreading area when the modified substrate bears a larger number of positively charged amine groups than negative carboxylate ones, both attached to self-assembled monolayers of alkanethiols [52]. In this work, the authors reported that the amount of osteopontin proteins adsorbed on both modified surfaces was very similar, thus they suggested that the orientation and geometrical conformation of osteopontin was more suitable for cell adhesion to the amino group-containing positive surface.

Substrate roughness and topography: In relation to these properties, it is convenient to distinguish three size scales: the macroroughness that involves features larger than 1 μm, the micro-/submicroroughness with features from 100 nm to 1 μm, and the nanoroughness with features below 100 nm.

Cell adhesion is insensitive to the macroroughness as cells are able to adhere to or between the surface irregularities that are larger than cells.

The micro-/submicroroughness significantly affects cell behavior as reviewed elsewhere [7]. Some studies have reported a positive influence of surface microroughness on cell adhesion, whereas in other studies, its influence has been shown to be negative. For instance, rat osteoblasts on microporous surface of titanium dental implants exhibited an increased average cell spreading area in comparison with flat surfaces [53]. Similarly, human osteoblast-like MG-63 cells cultured on Ti substrates with microroughness showed a more differentiated phenotype as a large expression of alkaline phosphatase (ALP), osteocalcin, and the faster production of an osteogenic environment were observed [54].

In contrast, a Ti-based alloy with microroughness exhibited poorer adhesion characteristics and decreased proliferation of MG-63 cells than flat surfaces [54].

These contradictory results are partially caused by the lack of systematic research work on the effect of micro-/submicroroughness, in part due to the ambiguity of the parameters used to characterize the surface roughness.

Finally, the nanoroughness of a biomaterial surface has been reported to have a beneficial influence on cell adhesion in a large number of papers. This fact occurs because nanoroughness resembles many aspects of the ECM morphological features. It appears that the proteins that induce cell adhesion adsorb on the biomaterial surface with the appropriate conformation for presenting specific sites, that is, Arg-Gly-Asp, to cell membrane receptors [2].

Substrate stiffness: The stiffness of the substrate plays a key role in cell function. For instance, by varying the elastic modulus of the polymer-coated substrate from about 0 to 500 kPa, the trigger of distinct cell functions has been observed and mechanistically described in relation to the forces exerted by cells [55]. Moreover, it has been reported that collagen cross-linking can modulate tissue stiffness to enhance integrin-mediated cell adhesion and induce the invasion by breast tumor cells [56].

Synthetic polymers such as polyacrylamide (PA) with modified stiffness have been utilized to study MSC differentiation. These cells differentiate into neurons on soft PA gels, myoblast on gels of intermediate stiffness and bone cells on stiffer gels [57].

PEM mechanical properties can be improved by adding layers of "hard" polyelectrolytes [58], properly changing the pH and the ionic strength during assembling [59], by cross-linking the layers [60], or incorporating nano-objects into the multilayer structure [20, 61]. In these cases, some of the reagents employed may be toxic and add additional synthesis steps.

3.1.1 Modulation of cell adhesion by thin film thermal annealing

In recent papers, we reported about the effect of thermal annealing of PEMs on the adhesion of different cell lines [62]. We employed polycationic terminated PEMs made up of 15 layers, $(\text{polycation/polyanion})_7$polycation: poly-L-lysine (PLL)/alginate (Alg); PLL/dextran sulfate (DEX) and chitosane (Chi)/hyaluronic acid (HA). By treating PLL/Alg and PLL/Dex PEMs at 37°C for 1–3 days, we reported an enhancement of cell adhesion. The cytoplasm spreading area of A549 and C2C12 cells on unannealed PLL/Alg films is poor, while cells seeded on annealed $(\text{PLL/Alg})_7$PLL PEMs present an extended cytoplasm with well-defined focal contacts and large actin fibers, like cells seeded on glass. In contrast, annealed Chi/HA PEMs exhibit a hampered adhesion of A549, C2C12, BHK, and MC-3T3 cells [26]. The advantage of this strategy is the absence of toxic substances and the simplicity of the procedure.

Recently, a well-characterized PEM based on poly(4-styrenesulfonic acid) (PSS) and poly(diallyl dimethyl ammonium chloride) (PDADMAC) has been employed to demonstrate the dynamic nature of the interface interacting with cells and the key role played by the strength of the interaction of proteins with the substrate [63]. Cells are able to sense the stiffness of the substrate provided they attach to adhesion proteins that are adsorbed on the interface with sufficiently high interaction energy. In this work, the authors measured an abrupt change in cell adhesion by varying the number of deposited layers (PEM thickness) rather slightly but maintaining its roughness constant, pointing out that the stiffness of the substrate was not an important property in the observed adhesion behavior, but rather the competition of proteins from the culture medium with preadsorbed ones. For these experiments, the authors used PEMs with variable surface charge and protein radiolabeling.

Thermal annealing results in the reorganization of PEMs, the polyelectrolytes rearrange in the multilayers from a stratified organization to molecular complexes of polycations and polyanions, where the interaction between oppositely charged polyelectrolytes is maximized [64]. For the case of PLL/Alg PEMs (**Figure 3a** and **c**), thermal annealing increases the stiffness of the film, increases the contact angle to about 90°, augments the negative charge, and renders a relatively flatter surface. Nevertheless, for both preadsorbed FN from a solution of FN alone and from a solution of FN and BSA, thermal annealing of PLL/Alg PEMs increases the stability of adsorbed FN as demonstrated by gamma counting of radiolabeled proteins and protein exchange assays. Furthermore, circular dichroism studies indicated that upon annealing, there is likely a larger number of arginine-glycine-aspartate (RGD) groups from FN that are able to interact with cell membrane integrins to favor cell adhesion [65].

In the case of Chi/HA PEMs (**Figure 3b** and **c**), the changes in physicochemical properties are significantly smaller after annealing than for unannealed PLL/Alg PEMs. The interface remains hydrophilic, the contact angle decreases from 30 to 21°, the charge becomes slightly more negative, and the stiffness remains unchanged. Contrarily, there is a significant change in the root mean square (RMS) roughness for annealed Chi/HA PEMs compared with unannealed Chi/HA PEMs: the RMS roughness decreases to half the value measured for unannealed films. This topographic change is expected to cause different interactions of the PEM surface with cells after annealing, increasing the antiadherent properties. For Chi/HA PEMs, we also conclude that thermal annealing reduces the interactions between adhesion proteins and the PEM interface, the reverse effect of that observed for PLL/Alg PEMs.

One of our goals in this work was to relate the changes in cell adhesion upon annealing of PEMs to the deposition and stability of adhesion and nonadhesion

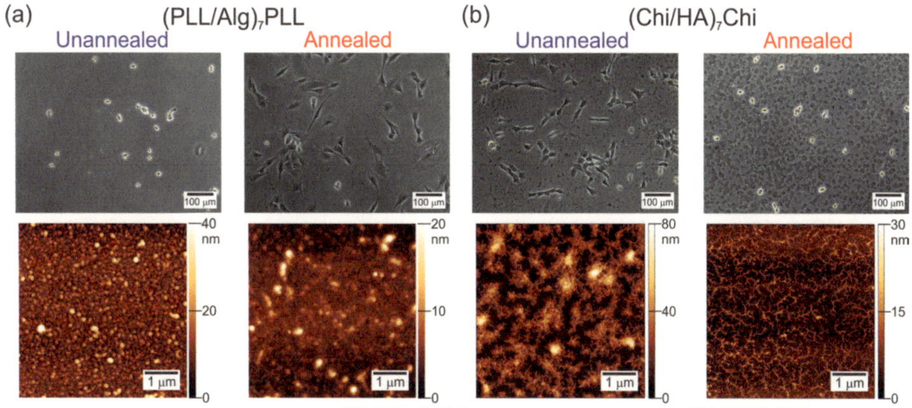

(c)

	(PLL/Alg)$_7$PLL		(Chi/HA)$_7$Chi	
	Unannealed	Annealed	Unannealed	Annealed
Roughness	4.3 nm	2.3 nm	9.3 nm	3.1 nm
Young's modulus	50 ± 5 MPa	450 ± 15 MPa	75 ± 8 MPa	73 ± 4 MPa
Zeta Potential	0 mV	-16 mV	-24 mV	-31 mV
Contact angle	36 ± 3°	92 ± 4°	30.2 ± 0.3°	20.6 ± 1.8°
FN exchangeability	Higher	Slower	High	Higher

Figure 3.
Cell adhesion modulation on unannealed and annealed (PLL/Alg)$_7$PLL and (Chi/HA)$_7$Chi PEMs. (a) Contrast phase images of A549 cells adhered to unannealed and annealed (PLL/Alg)$_7$PLL. The respective atomic force microimages are included. (b) Cell adhesion and atomic force images for unannealed and annealed (Chi/HA)$_7$Chi PEMs. (c) Physicochemical and fibronectin (FN) adsorption properties of unannealed and annealed (PLL/Alg)$_7$PLL and (Chi/HA)$_7$Chi PEMs.

proteins and rationalize the effect of the changes in the physicochemical properties of PEMs on cell adhesion.

Cells are able to expand due to proper forces applied to the substrate for increasing cell cytoplasm tension. Thus, cells sense the increase in the Young's modulus of PLL/Alg films upon annealing provided adhesion proteins interact tightly with the PEM surface (**Figure 4a**). Upon annealing, both BSA and FN exhibit an increased interaction with the substrate. Moreover, FN adsorbed on annealed PLL/Alg adopts an elongated tertiary structure, favoring the exposure of RGD adhesive groups that enhance cell adhesion. For Chi/HA PEMs, results indicate a combined effect of a larger deposition of BSA (nonadhesive protein), and lower deposition of FN, the latter with an increased exchangeability, which renders the PEM surface unfavorable for stable interactions with integrins from the cell membrane (**Figure 4b**).

3.1.2 Modulation of cell adhesion by biologically inspired materials

Biologically inspired materials play a key role in the design of realistic platforms to modulate cell function. We proposed a novel biologically inspired supramolecular coating generated via one-step dip coating of the substrate in an aqueous solution of polyallylamine and phosphate anions [66]. We found selective cell adhesion, following the order in adhesion C2C12 myoblast > HeLa cells > MC-3 T3 preosteoblast, by varying the deposition time, that is, the thickness of the film from 20 to 120 nm (**Figure 5**) due to changes in the "complex stiffness". The proposed platform supports a cell-type dependent exponential proliferation rate, with viability tested during a month.

This phosphate-polyamine film can be used to coat either adherent or nonadherent substrates and perhaps even more interesting, relatively thick films (≈100 nm) that exhibit poor cell adhesion properties, irrespective of the cell line, recover their adhesive properties toward eukaryote cells upon annealing at 37° for 2–3 days

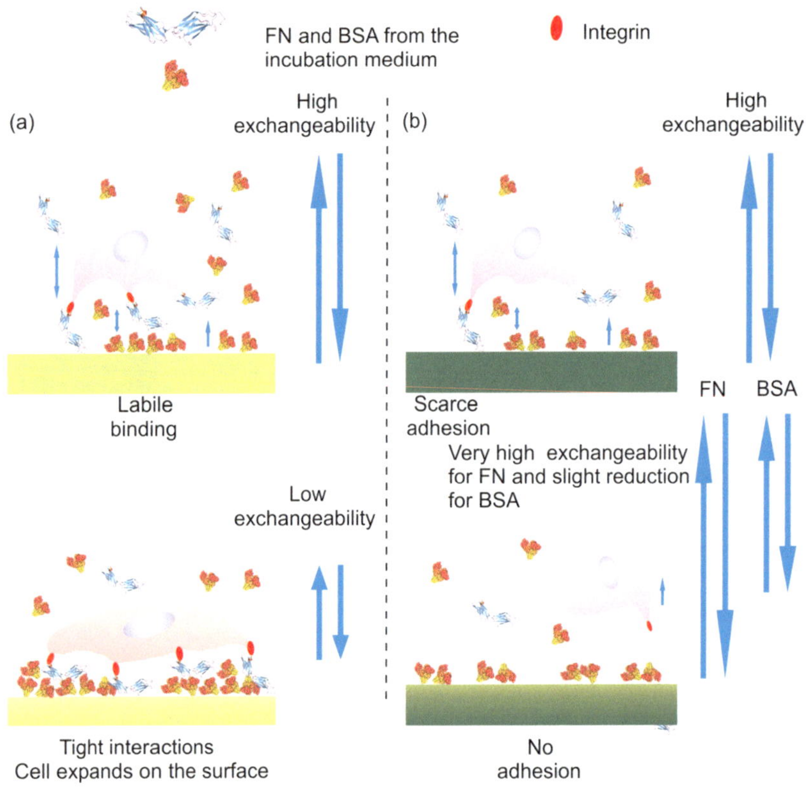

Figure 4.
Scheme of cell interaction with protein adsorbed on unannealed (top) and annealed (bottom) $(PLL/Alg)_7PLL$ (a) and $(Chi/HA)_7Chi$ PEMs (b).

(**Figure 5**). This effect was mainly explained by the increase in the film stiffness due to partial irreversible dehydration accompanied with an increase in the film roughness. Upon thermal annealing, the contact angle increases approaching about 70°, the surface charge remains negative, and the mass of FN adsorbed accessed by quartz crystal microbalance remains constant (**Figure 5d**).

3.2 Polymeric films with gradients

Cell behavior is markedly affected by the spatial distribution of mechanical and biochemical cues at the cell microenvironment. The overall cooperative result impacts on either physiological or pathological processes.

The fabrication of films with a gradual change in their swelling behavior has been obtained by gradual immersion of assembled PSS/PDADMAC multilayers into NaCl solutions [67]. In this work, the migration of smooth muscle cells (SMCs) at an appropriate cell density, in the direction of lowering hydration (relatively low swelling) has been observed. Furthermore, by time-controlled immersion in a solution of natural cross-linkers, a stable stiffness gradient on either free standing PEMs or assembled on silica substrates presenting differential cell adhesion has been reported [68].

A multilayer composed of poly(acrylic acid) (PAA) and poly(allylamine hydrochloride) (PAH) grafted with photosensitive benzophenone was employed to generate stiffness gradient (55–140 MPa) by illuminating the assembled films with a gradient density filter that regulates the amount of UV light deposited onto the films [69]. Here, SMCs and MC-3T3 cells adhere and spread better on stiffer regions.

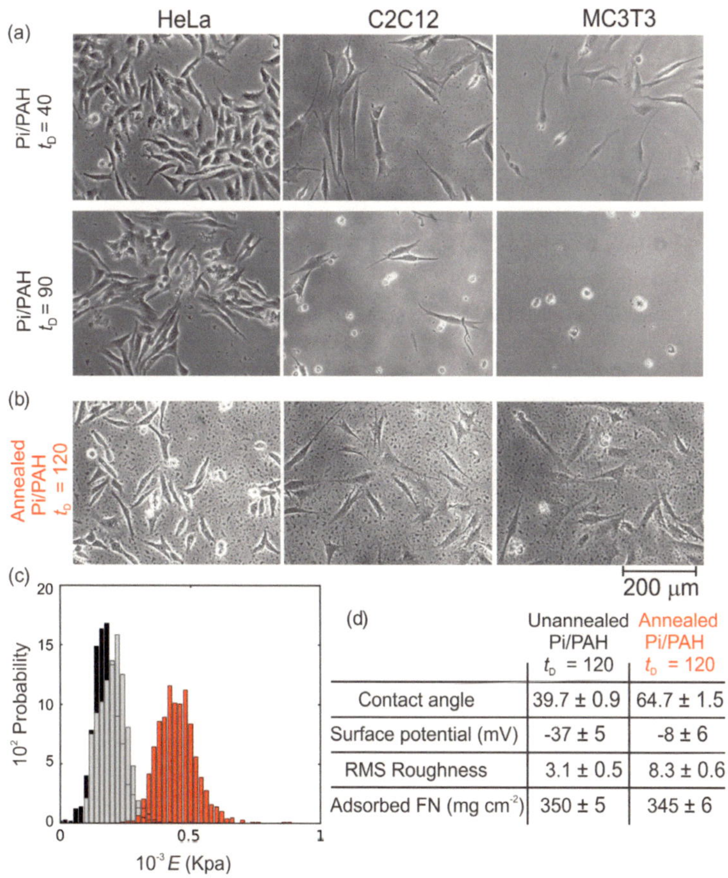

Figure 5.
Selective cell adhesion modulation on Pi/PAH-coated coverslips and the effect of thermal annealing at 37°C for 2–3 days. (a) Phase contrast images of HeLa, C2C12, and MC-3T3 cells seeded on unannealed Pi/PAH-coated substrates with t_D = 40 and 90 min, as indicated in the figure. (b) Contrast phase images of cells adhered to annealed Pi/PAH-coated substrates with t_D = 120 min. For C2C12 myoblasts and MC-3T3 preosteoblasts, cell adhesion characteristics improve in comparison with unannealed samples. (c) Stiffness (distribution of Young's modulus, E) for unannealed Pi/PAH samples with t_D = 40 min (black bars), t_D = 120 min (gray bars) and annealed Pi/PAH samples (red bars). (d) Some physicochemical properties of unannealed and annealed Pi/PAH-coated coverslips.

Other systems that include stiffness gradients [70] and concentration gradients of cell adhesive peptides or growth factors have been reported [71]. These concentration gradients are usually developed by microfluidic devices, allowing a spatially controlled surface arrangement of biological cues that are combined with physical ones to modulate cell fate.

We have recently reported on the post-assembly treatment of PLL/Alg PEMs with a temperature gradient between 10 and 50°C in samples 2 cm in length [19]. The resulting temperature gradient appears to be linear with the distance between sources at fixed temperature. A continuous change in the physicochemical properties appears to set in along the substrate upon the application of a thermal gradient. For the substrate region held at the highest temperature, C2C12 cell adhesion was roughly close to that observed on glass. In contrast, at the coldest region, cell adhesion was very poor. The most significant changes in the average cell spreading area were observed between 30 and 22°C (**Figure 6**). The application of a thermal gradient to locally modify the physicochemical properties of films as well as the adhesion characteristics toward different cell lines can be extended to other polymeric systems to increase the versatility of new materials for rapid testing of cell functionalities.

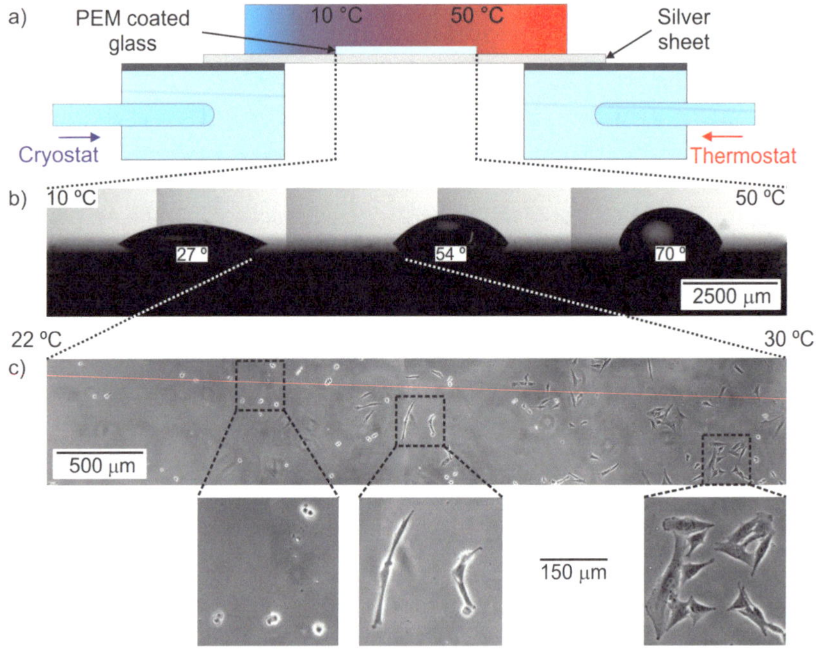

Figure 6.
Thermal gradient applied to (PLL/Alg)$_7$PLL PEM. (a) Scheme of the experimental setup for applying a thermal gradient. (b) Contact angle variation in the sample. (c) Spatial variation of cell adhesion characteristics.

4. Some hybrid multifunctional systems

The design of hybrid systems is motivated by the aim of profiting from the properties of the different materials composing them. For example, metal- and metallic oxide-derived thin films combined with polymeric layers that eventually contain bioactive molecules are useful to obtain multifunctionality.

Ti, its oxides, and alloys are widely used materials for biomedical applications. A detailed knowledge of the surface characteristics of Ti-based materials is needed for understanding the relation with their reactivity and biological performance [72]. Titanium oxide surfaces can be treated by chemical etching, sonochemical modifications or electrochemical treatment, to obtain hierarchically micro-/nanostructured porous surfaces [21]. These treatments enhance the photoactivity of the oxide layer. And, besides the inherent effect of the modified micro-/nanotopography on cell functions, for instance, the appropriate size of topographic features to efficiently promote integrin clustering, the porous materials can be loaded with bioactive molecules that promote prokaryote cell-material interaction [73].

By combining metallic surfaces with stimuli-responsive polyelectrolyte/polymer multilayers and supramolecular interacting systems, eukaryote and prokaryote cell behavior modulation has been achieved. A review of the application of these concepts is given in [21].

An example in relation to cell modulation behavior by a hybrid multifunctional system is provided in this section. The system is built up with a photocatalytically active TiO_2 film sonochemically nanostructured. Upon illumination, a photohole and a photoelectron are formed on the oxide surface as well as both H^+ and HO^-, one species prevailing over the other according to the composition of the interface. Thus, upon illumination for a certain time, a change in the pH at the interface is obtained. The system also includes a pH-sensitive high-amplitude actuating polymer, a linear triblock terpolymer consisting of polybutadiene (B), poly(methacrylic acid) (MAA), and quaternized poly[2-(dimethylamino)ethyl methacrylate] (Dq) [74]. In aqueous solution, the system (BMAADq) forms self-assembled micelles (BCM) with a hydrophobic core and a pH-sensitive MAA shell, and a strong Dq corona. BCMs are positively charged and are assembled with PAA to fabricate the pH sensitive coating. The resulting TiO_2/BCM/PAA/BCM/PAA architecture is switched in response to the local changes in pH induced by the photocatalytic reactions on the oxide surface, resulting in a layer of variable thickness and stiffness. Thus, the authors achieve an efficient and controllable strategy to convert energy of electromagnetic illumination into a local pH variation that triggers the reversible response of the soft polymer. The same authors reported that the MC-3T3 cell migration is modulated in the proposed system. By local illumination of the film, cells are induced to migrate from softer to stiffer regions [75].

The strategy briefly commented above shows the capability of hybrid smart systems to modulate cell behavior by spatial and temporal control of physicochemical cues.

5. Conclusions

In this chapter, a number of materials and strategies employed to obtain inorganic, organic, and hybrid thin film functional systems with actual and potential applications in biosciences have been presented. We have depicted some materials of natural and synthetic origin employed to fabricate thin films to be used as platforms for fundamental cellular studies and potential applications. The system properties that affect cell functions and modulate its behavior and fate in a context approaching real situations have been also pointed out.

Furthermore, we briefly comment on some of our research works in relation to biocompatible materials based on polyelectrolyte assembly and post-assembly treatments to change the performance toward cell adhesion and proliferation.

The biocompatibility of these materials determines the ultimate success of a developed material. Despite the large number of wisely designed systems with tunable spatial and temporal properties, even at the nanoscale, the challenge to bring these advances to the clinic is a central issue and requires the convergence of basic and applied sciences as well as commercial and social considerations.

Acknowledgements

S.E.M. thanks the MAT2017-88752-R and the CTQ2017-87637-R Retos projects from the Spanish Ministry of Science. N.E.M., M.A.P., and O.A. thank the Consejo Nacional de Investigaciones Científicas y Técnicas (CONICET, Argentina) (Grant No. PIP 0602).

Author details

Nicolás Eduardo Muzzio[1], Omar Azzaroni[2], Sergio E. Moya[3] and Miguel Ángel Pasquale[2]*

1 Institute of Physics of Buenos Aires (IFIBA, UBA-CONICET), CABA, Argentina

2 Institute of Theoretical and Applied Physical Chemistry (INIFTA, CONICET), Chemistry Department, National University of La Plata, La Plata, Argentina

3 Soft Matter Nanotechnology Group, CIC biomaGUNE, San Sebastián, Guipúzcoa, Spain

*Address all correspondence to: miguelp@inifta.unlp.edu.ar

References

[1] Adamson AW, Gast AP. Physical Chemistry of Surfaces. 6th ed. New York: Wiley; 1997

[2] Bacakova L, Filova E, Parizek M, et al. Modulation of cell adhesion, proliferation and differentiation on materials designed for body implants. Biotechnology Advances. 2011;**29**:739-767

[3] Nazarpour S, editor. Thin Films and Coatings in Biology. Dordrecht: Springer; 2013. DOI: 10.1007/978-94-007-2592-8. [Epub ahead of print]

[4] Monge C, Almodóvar J, Boudou T, et al. Spatio-temporal control of LbL films for biomedical applications: From 2D to 3D. Advanced Healthcare Materials. 2015;**4**:811-830

[5] Mano JF, Choi IS, Khademhosseini A. Biomimetic interfaces in biomedical devices. Advanced Healthcare Materials. 2017;**6**:1700761

[6] Beri P, Matte BF, Fattet L, et al. Biomaterials to model and measure epithelial cancers. Nature Reviews Materials. 2018;**3**:418-430

[7] Stevens MM, George JH. Exploring and engineering the cell surface interface. Science. 2005;**310**:1135-1138

[8] Moroni L, Elisseeff JH. Biomaterials engineered for integration. Materials Today. 2008;**11**:44-51

[9] Lentino JR. Prosthetic joint infections: Bane of orthopedists, challenge for infectious disease specialists. Clinical Infectious Diseases. 2003;**36**:1157-1161

[10] George Pados AM, Tarlatzis B. Endoscopy. Rijeka: InTech; 30 April 2013. DOI: 10.5772/50355. [Epub ahead of print]

[11] Posaric-Bauden M, Isaksson K, Åkerberg D, et al. Novel anti-adhesive barrier biobarrier reduces growth of colon cancer cells. The Journal of Surgical Research. 2014;**191**:196-202

[12] Falconnet D, Csucs G, Grandin HM, et al. Surface engineering approaches to micropattern surfaces for cell-based assays. Biomaterials. 2006;**27**:3044-3063

[13] Gangnaik AS, Georgiev YM, Holmes JD. New generation electron beam resists: A review. Chemistry of Materials. 2017;**29**:1898-1917

[14] Arnold M, Schwieder M, Blümmel J, et al. Cell interactions with hierarchically structured nano-patterned adhesive surfaces. Soft Matter. 2009;**5**:72-77

[15] Lehn JM. Supramolecular chemistry: Where from? Where to? Chemical Society Reviews. 2017;**46**:2378-2379

[16] Richardson JJ, Cui J, Björnmalm M, et al. Innovation in layer-by-layer assembly. Chemical Reviews. 2016;**116**: 14828-14867

[17] Guo HC, Ye E, Li Z, et al. Recent progress of atomic layer deposition on polymeric materials. Materials Science and Engineering: C. 2017;**70**:1182-1191

[18] Muzzio NE, Pasquale MA, Gregurec D, et al. Polyelectrolytes multilayers to modulate cell adhesion: A study of the influence of film composition and polyelectrolyte interdigitation on the adhesion of the A549 cell line. Macromolecular Bioscience. 2016;**16**:482-495

[19] Muzzio NE, Pasquale MA, Moya SE, et al. Tailored polyelectrolyte thin film multilayers to modulate cell adhesion. Biointerphases. 2017;**12**:04E403

[20] Gribova V, Auzely-Velty R, Picart C. Polyelectrolyte multilayer assemblies on materials surfaces:

From cell adhesion to tissue engineering. Chemistry of Materials. 2012;**24**:854-869

[21] Zhukova Y, Skorb EV. Cell guidance on nanostructured metal based surfaces. Advanced Healthcare Materials. 2017;**6**:1600914

[22] Simovic S, Losic D, Vasilev K. Controlled drug release from porous materials by plasma polymer deposition. Chemical Communications. 2010;**46**:1317-1319

[23] Park S, Han U, Choi D, et al. Layer-by-layer assembled polymeric thin films as prospective drug delivery carriers: Design and applications. Biomaterials Research. 2018;**22**:29

[24] Wei M, Gao Y, Li X, et al. Stimuli-responsive polymers and their applications. Polymer Chemistry. 2017;**8**:127-143

[25] Liu D, Yang F, Xiong F, et al. The smart drug delivery system and its clinical potential. Theranostics. 2016;**6**:1306-1323

[26] Muzzio NE, Pasquale MA, Diamanti E, et al. Enhanced antiadhesive properties of chitosan/hyaluronic acid polyelectrolyte multilayers driven by thermal annealing: Low adherence for mammalian cells and selective decrease in adhesion for gram-positive bacteria. Materials Science and Engineering: C. 2017;**80**:677-687

[27] Liu L, Li W, Liu Q. Recent development of antifouling polymers: Structure, evaluation, and biomedical applications in nano/micro-structures. Wiley Interdisciplinary Reviews: Nanomedicine and Nanobiotechnology. 2014;**6**:599-614

[28] Williams DF. Definitions in Biomaterials: Proceedings of a Consensus Conference of the European Society for Biomaterials; March 3-5, 1986; Chester, England. Elsevier; 1987

[29] Ratner BD. Biomaterials: Been there, done that, and evolving into the future. Annual Review of Biomedical Engineering. 2019;**21**:171-191

[30] Ku SH, Lee M, Park CB. Carbon-based nanomaterials for tissue engineering. Advanced Healthcare Materials. 2013;**2**:244-260

[31] Kalbacova M, Broz A, Kong J, et al. Graphene substrates promote adherence of human osteoblasts and mesenchymal stromal cells. Carbon N Y. 2010;**48**:4323-4329

[32] Nayak TR, Andersen H, Makam VS, et al. Graphene for controlled and accelerated osteogenic differentiation of human mesenchymal stem cells. ACS Nano. 2011;**5**:4670-4678

[33] Glinel K, Déjugnat C, Prevot M, et al. Responsive polyelectrolyte multilayers. Colloids and Surfaces A: Physicochemical and Engineering Aspects. 2007;**303**:3-13

[34] Becker H, Gärtner C. Polymer microfabrication technologies for microfluidic systems. Analytical and Bioanalytical Chemistry. 2008;**390**:89-111

[35] Dou Q, Fang X, Jiang S, et al. Multi-functional fluorescent carbon dots with antibacterial and gene delivery properties. RSC Advances. 2015;**5**:46817-46822

[36] McGrath AJ, Dolan C, Cheong S, et al. Stability of polyelectrolyte-coated iron nanoparticles for T2 -weighted magnetic resonance imaging. Journal of Magnetism and Magnetic Materials. 2017;**439**:251-258

[37] Ye E, Loh XJ. Polymeric hydrogels and nanoparticles: A merging and

emerging field. Australian Journal of Chemistry. 2013;**66**:997

[38] Rianna C, Radmacher M. Influence of microenvironment topography and stiffness on the mechanics and motility of normal and cancer renal cells. Nanoscale. 2017;**9**:11222-11230

[39] Ingber D. Integrins as mechanochemical transducers. Current Opinion in Cell Biology. 1991;**3**:841-848

[40] Muzzio NE, Carballido M, Pasquale MA, et al. Morphology and dynamics of tumor cell colonies propagating in epidermal growth factor supplemented media. Physical Biology. 2018;**15**:046001

[41] Anselme K, Ploux L, Ponche A. Cell/material interfaces: Influence of surface chemistry and surface topography on cell adhesion. Journal of Adhesion Science and Technology. 2010;**24**:831-852

[42] Mih JD, Marinkovic A, Liu F, et al. Matrix stiffness reverses the effect of actomyosin tension on cell proliferation. Journal of Cell Science. 2013;**125**:5974

[43] Tilghman RW, Cowan CR, Mih JD, et al. Matrix rigidity regulates cancer cell growth and cellular phenotype. PLoS One. 2010;**5**:e12905

[44] Chorsi MT, Curry EJ, Chorsi HT, et al. Piezoelectric biomaterials for sensors and actuators. Advanced Materials. 2019;**31**:1802084

[45] Buxboim A, Ivanovska IL, Discher DE. Matrix elasticity, cytoskeletal forces and physics of the nucleus: How deeply do cells 'feel' outside and in? Journal of Cell Science. 2010;**123**:297-308

[46] Chang H, Zhang H, Hu M, et al. Surface modulation of complex stiffness via layer-by-layer assembly as a facile strategy for selective cell adhesion. Biomaterials Science. 2014;**3**:352-360

[47] Heitz J, Švorčík V, Bačáková L, et al. Cell adhesion on polytetrafluoroethylene modified by UV-irradiation in an ammonia atmosphere. Journal of Biomedial Materials Research Part A. 2003;**67A**:130-137

[48] Mikulikova R, Moritz S, Gumpenberger T, et al. Cell microarrays on photochemically modified polytetrafluoroethylene. Biomaterials. 2005;**26**:5572-5580

[49] Parizek M, Kasalkova N, Bacakova L, et al. Improved adhesion, growth and maturation of vascular smooth muscle cells on polyethylene grafted with bioactive molecules and carbon particles. International Journal of Molecular Sciences. 2009;**10**:4352-4374

[50] Bačáková L, Lapčíková M, Kubies D, et al. Adhesion and Growth of Rat Aortic Smooth Muscle Cells on Lactide-Based Polymers. Boston, MA: Springer. pp. 179-189

[51] Bet MR, Goissis G, Vargas S, et al. Cell adhesion and cytotoxicity studies over polyanionic collagen surfaces with variable negative charge and wettability. Biomaterials. 2003;**24**:131-137

[52] Liu L, Chen S, Giachelli CM, et al. Controlling osteopontin orientation on surfaces to modulate endothelial cell adhesion. Journal of Biomedical Materials Research Part A. 2005;**74**:23-31

[53] Sammons RL, Lumbikanonda N, Gross M, et al. Comparison of osteoblast spreading on microstructured dental implant surfaces and cell behaviour in an explant model of osseointegration. Clinical Oral Implants Research. 2005;**16**:657-666

[54] Zhao G, Schwartz Z, Wieland M, et al. High surface energy enhances cell response to titanium substrate microstructure. Journal of Biomedical Materials Research Part A. 2005;**74**:49-58

[55] Kocgozlu L, Lavalle P, Koenig G, et al. Selective and uncoupled role of substrate elasticity in the regulation of replication and transcription in epithelial cells. Journal of Cell Science. 2010;**123**:29-39

[56] Levental KR, Yu H, Kass L, et al. Matrix crosslinking forces tumor progression by enhancing integrin signaling. Cell. 2009;**139**:891-906

[57] Engler AJ, Sen S, Sweeney HL, et al. Matrix elasticity directs stem cell lineage specification. Cell. 2006;**126**:677-689

[58] Hemmerle J, Ball V, Lavalle P, et al. Stiffening of soft polyelectrolyte architectures by multilayer capping evidenced by viscoelastic analysis of AFM indentation measurements. Journal of Physical Chemistry C. 2007;**111**:8299-8306

[59] Thompson MT, Berg MC, Tobias IS, et al. Tuning compliance of nanoscale polyelectrolyte multilayers to modulate cell adhesion. Biomaterials. 2005;**26**:6836-6845

[60] Silva JM, García JR, Reis RL, et al. Tuning cell adhesive properties via layer-by-layer assembly of chitosan and alginate. Acta Biomaterialia. 2017;**51**:1-15

[61] Schmidt S, Madaboosi N, Uhlig K, et al. Control of cell adhesion by mechanical reinforcement of soft polyelectrolyte films with nanoparticles. Langmuir. 2012;**28**:7249-7257

[62] Muzzio NE, Gregurec D, Diamanti E, et al. Thermal annealing of polyelectrolyte multilayers: An effective approach for the enhancement of cell adhesion. Advanced Materials Interfaces. 2017;**4**:1600126

[63] Arias CJ, Surmaitis RL, Schlenoff JB. Cell adhesion and proliferation on the 'living' surface of a polyelectrolyte multilayer. Langmuir. 2016;**32**:5412-5421

[64] Diamanti E, Muzzio N, Gregurec D, et al. Impact of thermal annealing on wettability and antifouling characteristics of alginate poly-L-lysine polyelectrolyte multilayer films. Colloids and Surfaces B: Biointerfaces. 2016;**145**:328-337

[65] Muzzio NE, Pasquale MA, Rios X, et al. Adsorption and exchangeability of fibronectin and serum albumin protein corona on annealed polyelectrolyte multilayers and their consequences on cell adhesion. Advanced Materials Interfaces. 2019;**6**:1900008

[66] Muzzio NE, Pasquale MA, Marmisollé WA, et al. Self-assembled phosphate-polyamine networks as biocompatible supramolecular platforms to modulate cell adhesion. Biomaterials Science. 2018;**6**:2230-2247

[67] Han L, Mao Z, Wu J, et al. Directional cell migration through cell–cell interaction on polyelectrolyte multilayers with swelling gradients. Biomaterials. 2013;**34**:975-984

[68] Silva JM, Caridade SG, Oliveira NM, et al. Chitosan-alginate multilayered films with gradients of physicochemical cues. Journal of Materials Chemistry B. 2015;**3**:4555-4568

[69] Martinez JS, Lehaf AM, Schlenoff JB, et al. Cell durotaxis on polyelectrolyte multilayers with photogenerated gradients of modulus. Biomacromolecules. 2013;**14**:1311-1320

[70] Almodóvar J, Crouzier T, Selimović Š, et al. Gradients of physical

and biochemical cues on polyelectrolyte multilayer films generated via microfluidics. Lab on a Chip. 2013;**13**:1562-1570

[71] Almodóvar J, Guillot R, Monge C, et al. Spatial patterning of BMP-2 and BMP-7 on biopolymeric films and the guidance of muscle cell fate. Biomaterials. 2014;**35**:3975-3985

[72] Berger O, Fischer W-J. Photo-induced switchable TiO_2 thin films for biological applications. Journal Bioinspired, Biomimetic and Nanobiomaterials. 2013;**2**:100-116

[73] Escobar A, Muzzio N, Coy E, et al. Antibacterial mesoporous titania films with embedded gentamicin and surface modified with bone morphogenetic protein 2 to promote osseointegration in bone implants. Advanced Materials Interfaces. 2019;**6**:1801648

[74] Ulasevich SA, Brezesinski G, Möhwald H, et al. Light-induced water splitting causes high-amplitude oscillation of pH-sensitive layer-by-layer assemblies on TiO_2. Angewandte Chemie International Edition. 2016;**55**:13001-13004

[75] Ulasevich SA, Brezhneva N, Zhukova Y, et al. Switching the stiffness of polyelectrolyte assembly by light to control behavior of supported cells. Macromolecular Bioscience. 2016;**16**:1422-1431

Chapter 13

Multilayer Thin Films on Fine Particles

Sajjad Habibzadeh, Ehsan Rahmani, Mohammad Reza Saeb, Mohammad Reza Ganjali and Jamal Chaouki

Abstract

The tunable construction of multilayer thin-film-based particulate has opened up new horizons in materials science and led to exciting new developments in many scientific areas during the past two decades. Indeed, to utilize the synergistic properties of thin film coatings and the core particles, the thin film immobilized on fine particles can be a promising approach. The interaction between the thin films and the core fine particles results in adjustable properties of the coated particles. Therefore, such coated systems have been considered as an important class of emerging powder technology for a wide range of applications. Namely, multilayer structural features can lead to designing a highly active and selective catalytic systems. In addition, multilayer-coated nano/micro particles (NMPs) can be employed in the development of many new properties, ease of functionalization, conjugation of biomolecules, etc. Such structure with multilayer coatings can also revolutionize the energy storage and conversion systems.

Keywords: fine (nano/micro) particles, powder technology, multilayer coatings, thin films

1. Introduction

Single-and/or multilayer deposition of films or coatings of particular interest, without altering their bulk properties can be employed to tune the surface properties of fine particles in different fields such as catalysis, energy production, microelectronics, optoelectronics, etc. [1–3]. In particular, multilayer deposition on particles offers a broad range of nanostructured thin films suitable for multiple applications, including photocatalysis [4], advanced energy storage systems [5], drug delivery [6, 7], etc.

Individual fine particles (nano/micro particles, NMPs) were previously studied due to their much superior properties as compared to the agglomerated ones. Later, it was found that composite or coated semiconductor particles could render more enhanced performance than their corresponding individual particles where they might even develop the particles with novel features. Lately, multilayer semiconductor nanoparticles have been synthesized towards improving properties of such semiconductor materials [8]. Furthermore, the demands of modern technology pushed the research activity more forward to synthesize more advanced materials. This demand accompanied by the development of characterization techniques has also greatly helped to establish the structures of multicomponent nano/micro

IntechOpen

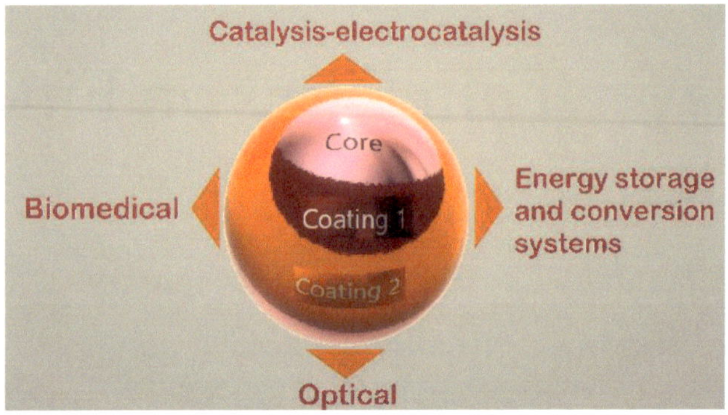

Figure 1.
Applications of multilayer coated-NMPs in advanced materials.

particles. Therefore, multilayer coated-NMPs structured have become one of the most popular research topics in recent years.

Nevertheless, synthesis of multilayer coated-NMPs has faced with concerns in terms of conformality and uniformity of coatings (shells). In one hand, the core particles might not only be considered as the spherical shape but also a variety of other shapes such as cube, prism, hexagon, octahedron, disk, wire, rod, tube, etc. can be employed. In the other side, fine particles are liable to clustering either by agglomeration (reversible) or aggregation (irreversible) processes due to the interparticle attractive force (IPF)-to-particle weight ratio (i.e., granular Bonding number) [9]. Therefore, to address such concerns, coating of fine particles as the synthesis of advanced materials needs a multidisciplinary understanding of powder technology and surface engineering.

In the present chapter, several multilayer coated-NMPs with different morphologies and compositions are discussed. The corresponding coated NMPs in various applications including catalytic, biomedical, energy, optical systems are reviewed (see **Figure 1**). Before digging into multilayer coated-NMPs, we found it quite necessary to make an applied understanding of particulate technology and classification. This would help to design the multilayered coating systems based on the corresponding powder characteristics. Indeed, the role of powder engineering cannot be ignored while particles are used as the substrate in a deposition system.

2. Particulate technology and classification

Powder technology is critically employed in the whole domains of particle processing and applications. Besides, a process involving solid particles cannot be conducted without the essential characterization of the particles of interest. Such characterizations encompass not only the intrinsic static parameters such as size, density, shape, and morphology) but also their dynamic properties attributed to the fluid flow, e.g., drag coefficient and terminal velocity. There are many techniques that can be employed to characterize particles, either simple or complicated. Namely, sieve analysis, imaging technique, dynamic light scattering (DLS) are used to directly characterize the particle size and shape. In addition, the physical gas adsorption technique based on the well-known BET (Brunauer–Emmett–Teller) method on monolayer coverage of adsorptives such as nitrogen is employed to measure the powder surface area and the pore size.

For a particle moving in a fluid (e.g., fluidization), the force acting on the surface of a particle depends only on the flow of the fluid in its immediate vicinity.

A widely used classification of particulate materials based on their fluidization properties has been long proposed by Geldart [10]. The empirical observations on fluidized particles in terms of particle size versus the relative density difference between the fluid phase and the solid particles are shown in **Figure 2** (classical Geldart's diagram). Such classification is in accordance with the behavior of powders as they are fluidized by dry air at ambient pressure and temperature.

Analogous to the Geldart classification typically used in the fluidized particles context, coated-particles can also be categorized based on the particle density and size. A useful nondimensional number to interpret the Geldart's diagram is the granular Bond number Bo_g, defined as the ratio of interparticle attractive force F_0 to particle weight W_p [11]. Fine cohesive particles ($Bo_g > > 1$, typically $d_p \leq 20$ μm), possess the strong attractive forces existing between the particles when compared to their weight, are so-called Geldart C particles. For slightly cohesive particles ($Bo_g \sim 1$, 20 μm $\leq d_p \leq$ 100 μm, i.e., Geldart A particles), the particles are smoothly flowable. Therefore, Geldart A particles seem quite efficient for industrial applications. This might lead to an important criteria in terms of particles size and density when the coated-particles are needed to be utilized in the various industrial sectors. In addition, powders within the limit of noncohesive particles ($Bo_g \leq 1$, typically $d_p \geq$ 100 μm for dry particles,) are deemed in the group of Geldart B particles (i.e., granular materials).

2.1 Importance of multilayer deposition on particles

To utilize a synergistic properties of thin film coatings and the core particles, the thin film immobilized on fine particles can be a promising approach. The interaction between the thin films and the core fine particles result in adjustable properties of the coated particles. Therefore, such coated systems have been considered as an important class of emerging powder technology for a wide range of applications as it was mentioned above.

Coating the core materials successively with similar or different types of materials can result in so-called "core–multishell NMPs", which have been

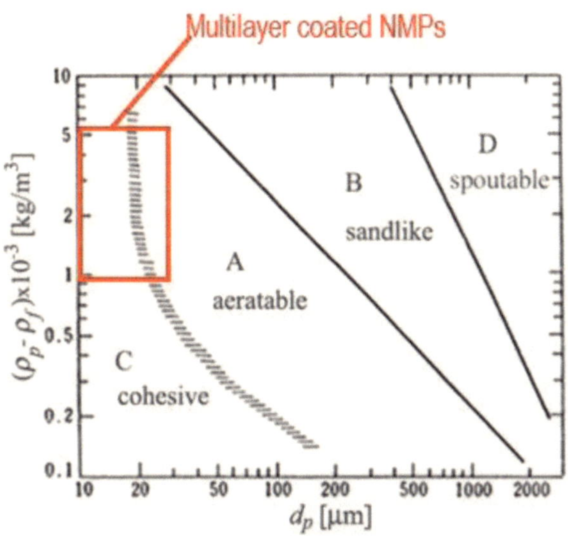

Figure 2.
Classical Geldart's diagram for the types of behavior of powders fluidized by dry air at ambient conditions based on the difference between particle and gas density (vertical axis) and particle size (horizontal axis) [10].

explored recently for a number of applications ranging from catalysis to biology. The tunable construction of multilayer thin-film-based particulate has opened up new horizons in materials science and led to exciting new developments in many scientific areas during the past two decades. The multilayer coated particles benefit from the intrinsic potential for the combination of diverse building blocks through complementary interactions, i.e.: electrostatic interactions, hydrogen bonding, etc., to create thin films displaying functional groups and chemical entities at controlled sites in nanoscale arrangements [12]. In addition, a stepwise procedure is employed to utilize different functionalities of the multishell rather than just to protect the core materials. Furthermore, shielding the core from incompatible outer layers of such NMPs might be performed by the inner layers [13]. It should be noted that, traditionally, composite NMPs composed of inner material (core) and outer layer material (shells) are often defined as core-multishell NMPs (the notation of "@" represents the core-multishell structure, core @ shell_1 @ shell_2 @ ... @ shell_n).

Multilayer coating of powders can be applied by either wet chemical processes, such as sol–gel and impregnation [14, 15] or by dry techniques (i.e., the use of a reactive gas phase), including pyrolysis and chemical vapor deposition (CVD) [16]; however, most attention in the literature has been paid to the deposition on fine particles using wet chemical processes. This includes synthesis of core–shell particles or layer-by-layer (LbL) assembly [4]. Indeed, LbL technique initially employed the alternate deposition of oppositely charged polyelectrolytes to produce nanoscale films. Further, the LbL method was successfully applied to other building blocks. A broad variety of multilayer composite films have been constructed by replacing one or both polyelectrolyte counterparts with other charged building blocks such as proteins, dendrimers, lipids and colloidal nanoparticles [12].

2.1.1 Multilayer-coated particles in catalytic applications

Indeed, any advancement in the technology of powders can be employed to augment the catalytic reaction performance while working as a process engineer. In other words, multilayer structural features lead to designing a highly active and selective, low-cost and eco-friendly catalytic systems. This is due to the fact that the multilayer coated particles, i.e., catalysts, can synergistically take the advantages of the core (i.e., support) and the coated layers. For instance, a superior catalytic activity can be obtained by tuning the surface properties (surface area, porosity, etc.) of the core particles as the coating layers are properly engineered. In addition, combination of the properties of the core and shell might be utilized towards improved/combined applications (e.g. magnetically separable nanocatalysts with the possibility of the repetitive use without the loss of the catalytic efficiency).

Fe_3O_4@SiO_2@ TiO_2 coated-particles were developed through a hierarchical structure. The first layer of SiO_2 on the superparamagnetic Fe_3O_4 core was coated by a modified Stöber process. Subsequently, TiO_2 was deposited on top of the silica layer to obtain Fe_3O_4@SiO_2@ TiO_2 composite. According to the x-ray diffraction (XRD) results, a uniform and well-defined morphologies with anatase crystalline TiO_2 was detected on the surface. However, the confirmation of the multilayer coatings of Fe_3O_4@SiO_2@TiO_2 with techniques other than microscopy has been yet a crucial challenge in the coated-NMPs context. In this study, the Fourier transform infrared spectroscopy (FTIR) spectrum was used to detect Si-O-Ti at the wave number around 940–960 cm^{-1}. Moreover, decomposition of RhB under UV irradiation surpassed that of P25 composites, indicating good photocatalytic activity (see **Figure 3(d)**). Further, little drop in efficiency over 18 cycles together with the magnetic recoverability and good cyclability were reported [17].

Habibzadeh et al. deposited a multilayer film of $TiO_2/SiO_2/TiO_2$ on the surface of fine soda lime glass (SLG) particles by a fluidized bed chemical vapor deposition (FB CVD) technique (see **Figure 4(a)**). The optimal values of the gas velocity for different temperatures of the powder bed were obtained from the differential pressure measurements. Titanium dioxide was deposited at 300°C whereas the silicon dioxide deposition was implemented at room temperature. Focused ion beam

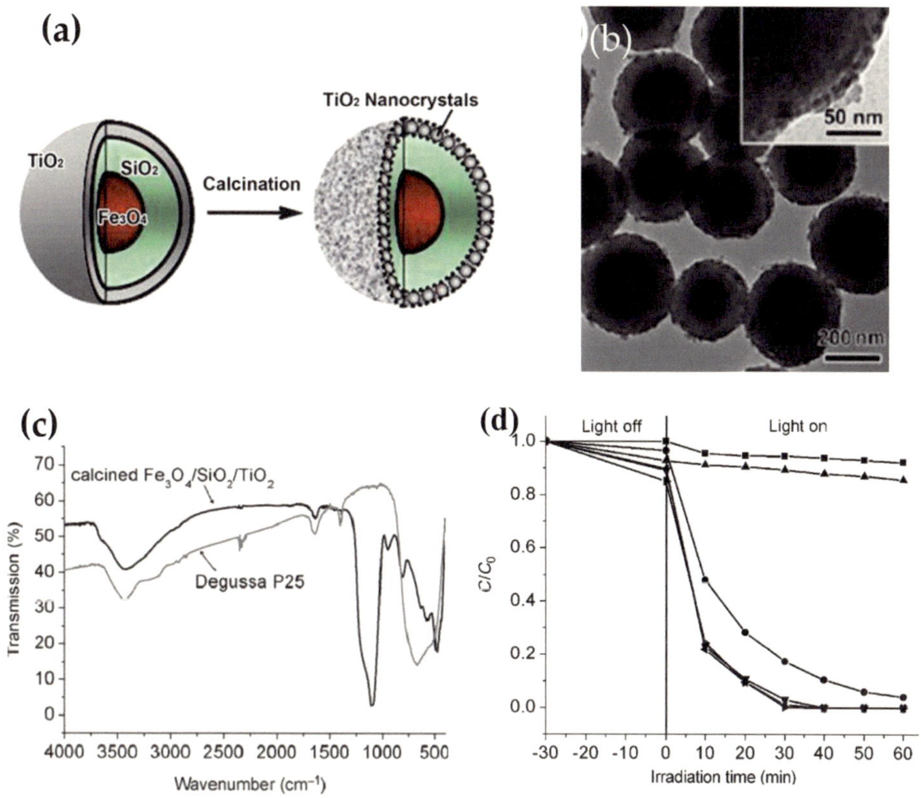

Figure 3.
(a) Scheme of as-synthesized composite structure, (b) TEM images of calcined $Fe_3O_4/SiO_2/TiO_2$ particles (inset in TEM image is the magnified ones of portions of the composite particles showing the morphological change in the TiO_2 shell due to calcination), (c) FTIR spectra of the calcined $Fe_3O_4/SiO_2/TiO_2$ structures and Degussa P25 (commercial TiO_2) and (d) Photocatalytic performance of various samples of catalysts (■ = UV, • = TiO_2 spheres, Δ = Fe_3O_4/SiO_2, ∇ = SiO_2/TiO_2, > $Fe_3O_4/SiO_2/TiO_2$, < P25) [17].

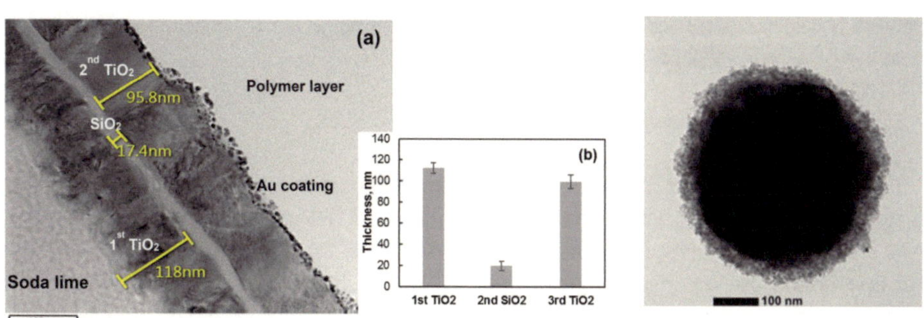

Figure 4.
(a) FIB cross-sectional TEM image and (b) the average film thicknesses of the $TiO_2/SiO_2/TiO_2$ system on the surface of SLG fine particles. The gold and polymer layers are added during the preparation of the TEM sample (c) HRTEM images of CdS/PDA/TiO_2 core/shell hybrids [4, 18].

(FIB) cross-sectional transition electron microscopy (TEM) images (**Figure 4(a)**) showed a thickness of ~110 and ~20 nm for titanium and silicon dioxide deposited layers, respectively.

X-ray photoelectron spectroscopy (XPS) results confirmed pinhole-free and uniform multilayer coatings on the surface of silica microspheres. Moreover, the anatase crystalline and amorphous structures of TiO_2 and SiO_2, respectively, were detected by XRD of the multilayer-coated particles. Uniform distribution of Ti and Si were also observed by energy-dispersive x-ray spectroscopy (EDS) elemental mapping analysis. In addition, it was found that the multilayer coatings developed in this study can be effectively employed as a photocatalyst in wastewater treatments [4].

The photocatalytic activity of the multilayer coatings, CdS/polydopamine/TiO_2 (CdS/PDA/TiO_2-**Figure 4(b)**) was determined. The ternary coated-particles were synthesized where dopamine self-polymerization and titanium (IV) butoxide hydrolysis were conducted on CdS nanospheres through hydrothermal processes. A considerable increase in the reaction rate constants of such ternary hybrids (more than three times) towards the degradation of RhB, methylene blue (MB) and phenol as the pollutants were seen as compared to CdS/TiO_2 hybrids. The amended photocatalytic degradation was credited to the greater light absorption and charge carrier separation efficiency together with the introduction of PDA [18].

Xiong et al. [1] showed an enhanced catalytic activity of 3D ordered meso-macroporous $Ce_{0.3}Zr_{0.7}O_2$-supported PdAu@CeO_2 nanoparticles for soot oxidation. The multilayered structure catalyst found to improve the activation property for gaseous reactants. In addition, the synergetic effect of Pd-Au-CeO_2 could promote the rate determining step. Yi et al. [19] reported the synthesis of a nanocomposite of Pd/SiO_2/Fe_2O_3 catalyst system utilized in hydrogenation of nitrobenzene. First, monodisperse silica-coated Fe_2O_3 nanoparticles (SiO_2/Fe_2O_3) were obtained in water-in-cyclohexane reverse microemulsion. Then, they were functionalized with mercapto and amine functionalities, which have been known to have strong affinity with transition metal nanoclusters, such as Pd, Fe_3O_4, and Au. The multilayer coated catalyst resulted in five times higher conversion rate as compared to the commercial Pd/C.

Moreover, Pd@CdS@ZnO multilayer nanorods (NRs) showed a superior catalytic activity for the degradation of toxic organic pollutants due to the effective separation of electrons-hole pairs. The excitation-wavelength-dependent catalytic performance of Pd@CdS@ZnO NRs showed significant enhancement at the wavelengths equivalent to the CdS layer. The recyclability results revealed a strong ability with favorable reusable photocatalytic efficiency. Besides, electrochemical impedance and photocurrent detection analysis further confirmed a reduced in the charge transfer resistance of Pd@CdS@ZnO NRs owing to embedded Pd NPs and coating of a CdS layer on ZnO NRs [20].

2.1.2 Multilayer-coated particles in biomedical applications

The smaller size and high surface to volume ratio of fine particles are the key features which render them useful in the biomedical fields. In addition, multilayer-coated NMPs can be employed in the development of many new properties, ease of functionalization, conjugation of biomolecules, etc. In particular, the application of novel nanomaterials in the biomedical engineering has directed to the development of the new contrast agent and drug delivery vehicle, which are of great importance in the area of health care [7]. The discrepancy of nanoparticles has been applied in several researches attributed to disease diagnostics, early detection studies, and better contrast agents for enhanced imaging techniques.

Owing to greater bio- and cyto-compatibility as well as increased circulation time of new drug delivery vehicles the load of the drugs along with the efficacy of the drug in the system have been reduced. Thus, the appearance of nanoparticles has revolutionized all the matters related to the medical biotechnology and biomedical engineering. This results in improving the conventional techniques together with the experimentation of new and advanced techniques for drug delivery and its monitoring.

In biological applications multilayer-coated NMPs have major advantages over individual particles leading to the improvement of properties such as less cytotoxicity, more dispersibility, bio- and cyto-compatibility, better conjugation with other bioactive molecules, greater thermal and chemical stability and so on [21]. In other words, when the particles of interest are toxic which may cause plenty of issues to the host tissues and organs, the multilayer coatings of benign materials on top of the core render the nanoparticles much less toxic and bio-compatible.

Sometimes the coating layers not only act as nontoxic layers, but also improve the core material property. Moreover, hydrophilicity of particles is quite important to disperse them in biological systems (aqueous). The increase in biodispersivity, bio- and cyto- compatibility makes it a useful alternative to conventional drug delivery vehicle. In addition, many bio-applications are dependent upon the conjugation of biomolecules on the surface of particles. In particular, coating of a suitable bio-compatible material can help to conjugate a particular type of biomolecules with the surface of particles. Further, coating of an inert material generally enhances the stability of core particles when the core materials are susceptible to chemical or thermal change during exposure to surrounding environment. Thus, coated NMPs are more appropriate for biological applications than single nanoparticles.

In an in vivo experiment, dual imaging contrast agent for MRI and optical imaging purpose Gadolinium tetraazacyclododecanetetraacetic acid (Gd-DOTA) is covalently linked with silanized nanoparticle to form Gd-DOTA, and then attached to SiO_2 coated CdS/ZnS quantum dots (QDs) [22]. In fact, excitons (electron and hole) in a QDs (i.e., semiconductor nanoparticles) can restrict the movement of conduction band electrons, valence band holes and excitons in all three spatial directions. However, The main problem associated with QDs is its high toxicity, because of the presence of heavy metals in its core, which causes cell death due to leaching out to the external environment. The deposition of outer layers or employing another metal in place of toxic cadmium, from the respective QDs might lead to addressing the problems caused by cadmium. Kim et al. [23] performed cell labeling in Hela cells, a human cervical cancer cell line by CTAB to prepare CdSe/CdS/ZnS (core/shell/shell) QDs. Apart from CdSe@CdS@ZnS QDs which has been widely applied in cell labeling, multilayer coated-QDs such as CdSe@ZnSe@ZnS [24] and CdSe@$Zn_{0.5}Cd_{0.5}$Se@ZnSe@ZnS [25] have been used for imaging applications. In addition, CdSe@ZnS@SiO_2 nanoparticles [26] which gave stronger emissions with consistent fluorescence intensity than bare CdSe–ZnS nanoparticles alone, leading to improve imaging applications. Besides, the silica shell renders it suitable for conjugating with bioactive agents, needed for targeted delivery to the concerned imaging site [27].

The other aspect of coated-NMPs research revolves around targeted drug delivery. Namely, the cancer therapy and treatment of static tumors due to a low therapeutic index (ratio of therapeutic efficacy to side effects) of medication causes severe deterioration in health [28]. Therefore, the effects of these drugs should be target-oriented where the disorder of vulnerable cells is critical. Next, diabetes and cardiovascular diseases are regarded after cancer therapy considering medical importance; indeed, malfunctioning of certain tissue regions, or cells result in such diseases. For instance, malfunctioning of pancreatic β cells leads to diabetes. Hence, a precise and permanent cure by targeting these cells particularly for drug therapy may eliminate the problem once and for all. Chitosan/B-lactoglobulin core/shell particles were successfully synthesized as nutraceutical carriers for nutrient delivery increasing permeability of the molecules, increasing gastric residence time. This subsequently provides the environmental stability (that it lacked in normal food processing). This encapsuled nutrient system rendered the oral administration safe [29].

As mentioned above, most particles-based drug delivery research is focused on cancer therapy. The material comprising the particles plays a significant role on the working mechanism. In the category of particle-based drug delivery carriers, two types are the most important; first, magnetic materials (i.e., iron and its oxides and sulfides), and secondly, noble metals such as gold and silver. The former materials, i.e. the magnetic, are mostly used as core materials since the corresponding surface exposure because is not required to exploit their properties. However, Au owing to its efficient surface functionalization of different ligands, might be used as a shell material.

In addition, most inorganic materials are toxic to the living system. Other metal oxide particles, i.e., TiO_2, SiO_2, ZnO have also been studied for drug delivery. Gold nanorods/mesoporous silica/hydroxyapatite (Au/SiO_2/HAP) hybrid nanoparticles with AuNR core and SiO_2/HAP hybrid inorganic shell for multi-responsive drug delivery had been prepared. According to the in vitro results of the drug release, Au/SiO_2/HAP nanoparticles showed high drug loading efficiency, superb near-infrared (NIR)- and pH-responsive drug release properties. In addition, Au/SiO_2/HAP nanoparticles displayed a higher drug release of 37.62% upon NIR irradiation at pH 4.5 as compared to the drug release of Au/SiO_2 nanoparticles over 12 h (about 6.35%). This is due to the NIR-responsiveness of AuNRs and the pH-responsiveness of HAP in acid media. Noticeable biocompatibility of the Au/SiO_2/HAP nanoparticles was shown based on the cell viability results [30].

Furthermore, applying an external magnetic field facilitates conducting the drug towards specific target cells while employing magnetic nanoparticles such as ferrites because of their biocompatibility and superparamagnetic behavior [31]. Although such magnetic core particles have been employed as the potential drug carriers through in vitro systems (i.e., simulating human physiological systems), one must take into caution the toxicity of the bare Fe particles. This can be attributed to the thrombosis formation (aggregation of Fe particles), induction of free radicals in bloodstream and their facile oxidizability. However, encapsulating of these ferrite particles with a noble metal layer, leading to better surface chemistry between the drug and the carrier and thus the corresponding structural stabilization. Namely, a shell of gold has been used in anticancer drugs such as doxorubicin. Such layer can effectively adsorb amine-groups, thus reducing particle aggregation by steric hindrance [32]. The adsorbed drug can be released to the target sites in response to ionic, pH stimuli or externally controlled mechanism (magnetic or thermal). Moreover, magnetic core/mesoporous silica shell structures have been studied for in-vitro drug carrier application [33]. While Fe/Au core/shell nanocomposites [34], Fe_3O_4/$CaCO_3$/PMMA [35], Au/poly(L-aspartate-doxorubicin)-b-poly(ethylene glycol) copolymer [36] may also find similar biomedical applications in the near future.

2.1.3 Multilayer-coated particles in energy and other applications

Developing low-cost, eco-friendly systems with large energy conversion and storage efficiency have been recently addressed by emerging intricate nanocomposites. Such structure with multilayer coatings can revolutionize the energy storage systems, i.e., supercapacitors, li-ions batteries and hydrogen storage as well as the corresponding energy conversion technologies including quantum dot solar cells, dye-sensitized solar cells, silicon/organic solar cells and fuel cell.

Supercapacitors (SCs) are typical energy storage devices encompassing electrochemical double layer capacitors (EDLCs), pseudocapacitors and asymmetric capacitors. The electrode materials of EDLCs can be used in mobile and stationary systems while high power pulses are required. EDLCs based on carbon materials have attracted much attention due to its high power density and long cycling life [37], which is attributed to the good electronic conductivity of carbon materials together with their high chemical stability and large specific surface areas. However, simultaneous fast charging/discharging rate, high energy density, and long cycle performance cannot be practically obtained through a single nanostructured electrode material owing to some inherent limitations of individual composition. For instance, carbon materials have excellent cyclical stability and high rate capability while faced with low specific capacitance. In general, the ideal multilayer coating structured nanomaterials in supercapacitors should meet a few critical requirements for coating materials. First, such coatings maintain structural integrity and limit volume expansion. It can protect the core particle from being affected by outside environmental and causes a rapid transport of electrons and ions.

As the second category of SCs, pseudocapacitors which store charges through a Faradaic process possess higher energy density compared to EDLCs [38]. Metal oxides and conducting polymers are employed as the promising materials for pseudocapacitors. Namely, metal oxides such as, ruthenium, manganese, nickel and iron oxides with two or more oxidation states and good conductivity as well as conducting polymers, e.g., polythiophene (PTh), polypyrrole (PPy), polyaniline (PANI) and their derivatives with high storage capacity/porosity and high potential window are considered as outstanding candidates for pseudocapacitors.

A hierarchical $ZnO@MnO_2@PPy$ multilayer arrays (see **Figure 5(a–c)**) fabricated through layer-by-layer process as an active material for energy storage showed an enhanced electrochemical performance. Specific capacitance of 1281 Fg^{-1} at a current density of 2.5 A g^{-1} associated with a significant areal capacitance of 1.793 F cm^{-2} at a current density of 3.5 mA cm^{-2} were resulted from such electrodes. Furthermore, excellent cycling stability of the $ZnO@MnO_2@PPy$ ternary core–shell electrode (**Figure 5(d)**) with a capacitance retention of 90% after 5000 charge–discharge cycles at 5 A g^{-1} was obtained [39].

Apart from the application of multilayer coatings in different types of supercapacitors, such coatings can be well employed as the electrode materials for lithium ion batteries (LIBs). Indeed, many impediments, such as the slow Li^+ diffusion and high resistance at the electrolyte/electrode interface can be addressed by improving the reactivity and conductivity of Li^+ ion/electron, accelerating charge across the electrolyte/electrode interface and shortening the Li^+ extraction/insertion pathway. One of the most significant roles of the coatings on the cathodic and anodic LIBs' active materials is attributed to the electronic conductivity enhancement. This has typically been conducted by employing carbon material such as graphite or metal/metal oxides as well as conductive polymers. Zhang et al. [40] synthesized graphene@Fe_3O_4@C core–shell nanosheets (**Figure 6(a, b)**) have been rationally designed and fabricated by a solvothermal method, which have been demonstrated as a high-performance LIB anode material with a capacitance of 1200 mAh g^{-1} after almost 100 cycles, indicating the great stability and usability of this modern structure. The obtained graphene@Fe_3O_4@C nanosheets contains Fe_3O_4 nanoparticles

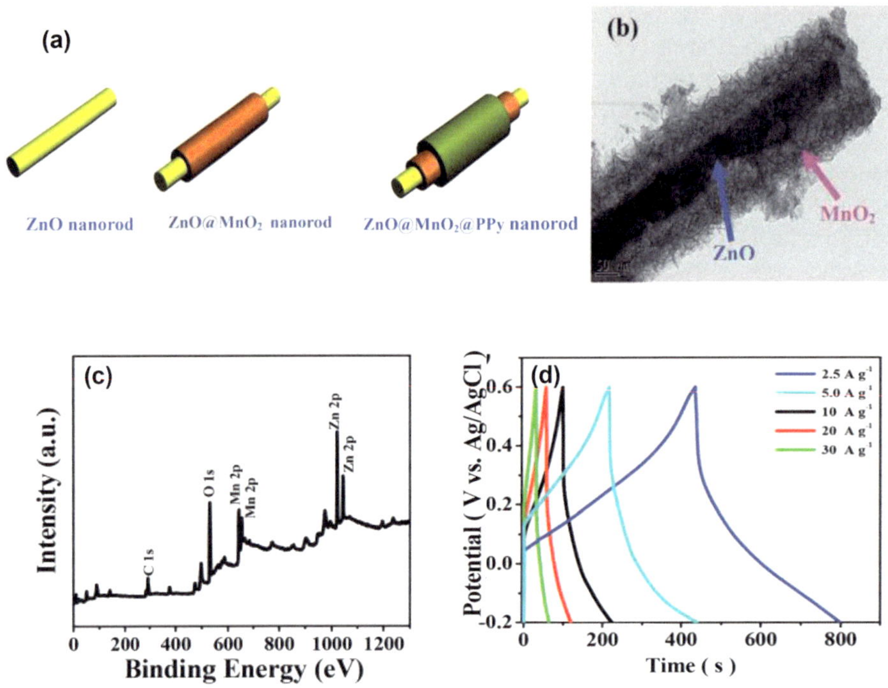

Figure 5.
(a) Schematic illustration of the formation process for ZnO@MnO_2@PPy ternary core–shell nanorod arrays, (b) typical TEM images of ZnO@MnO_2@PPy nanorod arrays, (c) XPS survey spectrum of the ZnO@MnO_2 nanorod arrays, (d) the galvanostatic charge–discharge (GCD) plots at various current densities of ZnO@ MnO_2 (c) and ZnO@MnO_2@PPy [39].

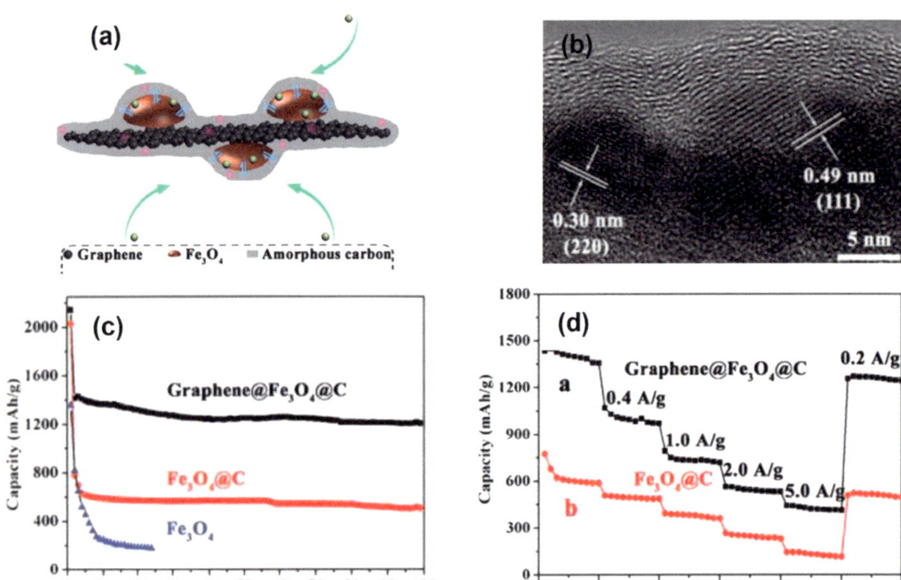

Figure 6.
(a) Schematic representation of the electrochemical reaction path on the graphene@Fe_3O_4@C core–shell nanosheet electrode (b) HRTEM image of the graphene@Fe_3O_4@C core–shell nanosheet. (c) Cycling performance of the graphene@Fe_3O_4@C core–shell nanosheet, Fe_3O_4@C core–shell nanosphere, and Fe_3O_4 nanosphere electrodes at a constant current density of 200 mA g^{-1}, (d) cycling performance of the graphene@ Fe_3O_4@C core–shell nanosheet (a) and Fe_3O_4@C core–shell nanospheres (b) at different current densities [40].

(6 nm) on the graphene sheets and coated by a uniform flexible carbon shell, as well as show a high surface area of 136 $m^2 g^{-1}$.

The three-dimensional graphene foams encapsulated hollow SnO_2@Co_3O_4 were synthesized using 300 nm spherical SiO_2 particles as a template. The multilayer coated spheres fabricated through self-assembly in hydrothermal process from graphene oxide nanosheets and metal oxide hollow spheres. The encapsulated architectures could considerably enhance the capacity, cycling stability and rate capability of hollow SnO_2@Co_3O_4 spheres electrodes as the supercapacitor. This was because of the highly conductive networks and flexible buffering matrix [41]. Furthermore, the 3D h-SnO2@Co3O4@GF electrode was also evaluated as LIBs (**Figure 7(a–c)**). The lithium storage properties were conducted by using the as-synthesized products as anode materials and Li metal as the cathode. **Figure 7(d)** shows the discharge capacities of the different electrodes at current rates of 200, 500, 800 and 1000 mA g^{-1}, then returns to 200 mA g^{-1}. Obviously, the 3D h-SnO_2@Co_3O_4@GF electrodes reveal superbly high rate capability relative to the other electrodes. For example, at a large current density of 1000 mA g^{-1}, 3D h-SnO_2@Co_3O_4@GF electrode still delivers favorable capacity of about 380 mAh g^{-1}, while h-SnO_2@Co_3O_4 electrode only exhibits capacity of about 120 mAh g^{-1}, 3D h-SnO_2@GF electrode retains 160.7 mAh g^{-1} and h-SnO_2 electrode retains only 56.4 mA g^{-1}. When the current rate is returned to 200 mA g^{-1}, the stable capacity of 3D h-SnO_2@Co_3O_4@GF is 676.6 mAh g^{-1}, which is 86.9% of the initial reversible capacity at 200 mA g^{-1} (**Figure 7(e)**). The values are much larger than these of 3D h-SnO_2@GF (362 mAh g^{-1} and 57% retention), h-SnO_2@Co_3O_4 (287.9 mAh g^{-1} and 32.5% retention) and h-SnO_2 (165.2 mAh g^{-1} and 16.5% retention) [41].

Figure 7.
(a) Digital photograph of 3D h-SnO_2@Co_3O_4@GF, (b) 3D graphene foams encapsulated hollow SnO_2@Co_3O_4 spheres, (c) HR-TEM of 3D h-SnO_2@Co_3O_4@GF structure (d) specific capacitance of 3D graphene foams, h-SnO_2@Co_3O_4 and 3D h-SnO_2@Co_3O_4@GF electrodes, (e) cycling performance at various rates (from 200, 500, 800, 1000 to 200 mA g^{-1}) for the h-SnO_2, 3D h-SnO_2@GF, h-SnO_2@Co_3O_4 and 3D h-SnO_2@Co_3O_4@GF [41].

Xie et al. reported a core@double-shell structure to synthesize barium titanate (BT)-based high performance polymer nanocomposites. The first layer of hyperbranched aromatic polyamide (HBP) and the second one of poly (methyl methacrylate) (PMMA) were deposited on top of the core particles. The synergistic effect of both polymer shells, resulting in superior dielectric property. This cannot be obtained in nanocomposites prepared by the conventional blending methods. The energy densities of BT@HBP@PMMA nanocomposites were higher than that of BT/PMMA nanocomposites accordingly. The analyses of the dielectric response of the nanocomposites together with the mechanisms resulting in the higher dielectric constant and lower dielectric loss in BT@HBP@PMMA nanocomposites. As a result, the core@double-shell structure polymer nanocomposites can be considered a newly-designed architecture with desirable dielectric properties [42].

The role of multilayer coatings NMPs in the optical application is mostly dependent upon the shell layers' material and thickness. By varying the core and shell thicknesses, the color of such shells can be varied across a broad range of the optical spectrum spanning the visible and the near infrared spectral regions. The use of semiconductors either as the core or shells (multilayer coatings) have been much raised. The typical band gap for semiconductor particles is normally less than 4 eV, which are those normally found in insulating materials. In comparison with other fluorescent probes (organic dye or proteins), semiconductor nanoparticles or quantum dots as a new class of fluorescently labeled compounds render exceptional properties. In addition, since the fluorescent emission spectra of these particles can be tuned by changing the particle size, such semiconductor materials are preferably used in photoluminescence (PL) applications [43].

It should be mentioned that the main advantages of multilayer semiconductor NMPs are ascribed to the corresponding higher quantum yield, higher photoluminescence efficiency, improved optical properties, increased half-life times. In addition, facile detection of emission spectra of the semiconductor coated-NMPs is owing to their higher wavelength shift in the visible range, photo-oxidation stability, improved appropriate electronic properties (band gap, band alignment), and finally better structural (lattice mismatch) properties than unlayered fine particles. Moreover, lattice mismatch between the core and shell material can improve the luminescence properties no matter how significant is the shell thickness. Lattice mismatch refers to "the situation where two materials having different lattice constants are brought together by deposition of one material on top of another". It should be noted that the band gap as well as the lattice structure of the material are employed to form the semiconductor core particles and the multilayer coatings. Namely, CdSe/HgTe/CdTe [44], CdS/HgS/CdS [45] CdSe/ZnTe/ZnS [46] and etc. are the common multilayered coating structures applied.

Kim et al. reported a multilayered core-shell composite particles with a high reflective efficiency and durability. The multilayer SiO_2 and TiO_2 were deposited onto the Fe/Ag microspheres by the corresponding precursors of tetraethyl orthosilicate (TEOS) and titanium n-butoxide (TBOT), respectively. The coating of shell layers were conducted through a sol–gel process. It was shown that the SiO_2 and TiO_2 multilayered shells render the Fe/Ag microspheres high efficient for preserving the reflectivity and increasing the durability of the core particles [47].

3. Synthesis and characterization of multilayer-coated particles

Conformal and uniform multilayer coatings NMPs are considered as an enormous challenge in the fine particles deposition. In particular, the situation would be more severe when the coating layer thickness must be controlled. Some of the various synthetic methods for multilayer coated- NMPs used by different research groups are categorized into dry techniques (i.e., the use of a reactive gas phase), such as pyrolysis, chemical vapor deposition (CVD) [16] and atomic layer deposition (ALD) [48] onto particles or wet chemical techniques including precipitation, polymerization, microemulsion, impregnation and sol–gel condensation [8, 14, 15], layer by layer adsorption techniques [12], etc.

In order to conduct the synthesis of multilayer coated-NMPs, the techniques used to synthesize one of the most applied multilayer coatings, i.e., metal oxides, will be next discussed. Namely, the sol–gel technique is widely used as a wet chemical process when metal oxides are employed as the multilayer coatings. Indeed, this synthesis process encompasses three different routes. The reaction pathway might go through the gelation of solutions of a colloidal powder. The other pathway might be the hydrolysis Eq. (1) and polycondensation of metal alkoxides or metal salt precursors Eq. (2) followed by hypercritical drying of the gels. Apart from such two pathways, the commonly-used route for the synthesis of multilayer metal oxide coatings has been implemented by hydrolysis and polycondensation of metal alkoxide precursors followed by aging and drying under ambient atmosphere. It would be worth mentioning that the sol–gel process has preferentially been employed to synthesize the multilayer of silicon and titanium oxides.

$$M(OR)_n + n\,H_2O \rightarrow M(OH)_n + n\text{ROH Hydrolysis} \tag{1}$$

$$M(OH)_n + M(OH)_n \rightarrow (OH)_{n\text{-}1}M - O - M - (OH)_{n\text{-}1} + H_2O\text{ Condensation} \tag{2}$$

Furthermore, most of the various types of oxides can be synthesized through the coprecipitation techniques. This is particularly applied for the metals which form the hydroxide precipitate in alkaline pH. In fact, a precipitation reaction results in at least one water insoluble salt which precipitates from the precursor media (i.e., two or more water-soluble salts). Such a precipitation reaction is significantly dependent upon the solubility product of the precipitated compound. Namely, the particle formation would begin as the concentration of the product goes beyond the solubility product value of the respective compound in the reaction media. The mechanism of multilayer coating formation via precipitation would follow the process of nucleation and growth (i.e., molecular addition, Ostwald ripening and aggregation) [8, 49, 50].

The characterization of multilayer coated-nanoparticles is not as straightforward as that of multilayer coatings on the common substrates with macro-dimension, e.g., silicon, steel, plastic, etc. In addition, it would be somewhat troublesome to distinguish the various coated layers on the fine particles; therefore, a reliable characterization technique is always necessary for the core and multilayer coatings. A similar approach for characterization of single particles can be applied for multilayer coated-NMPs; however one technique may not be sufficient. The characterization techniques for the measurement of size, shell thickness, elemental and surface analysis, optical properties, and thermal stability play more important role for multilayer coated-nanoparticles. Therefore, dynamic light scattering (DLS), scanning electron microscopy (SEM) or much higher magnification - FESEM (field-emission SEM) SEM, transmission electron microscopy (TEM), thermal gravimetric analysis (TGA), X-ray photoelectron spectroscopy (XPS), photoluminescence (PL), and UV–vis spectroscopy are the ones most often used [4, 8, 51–55].

Among the electron microscopy techniques (SEM, TEM), TEM provides much more important information. Namely, TEM can result in corroboration of core/multilayer coatings through contrast difference, overall particle size, core size, shell thickness, uniform or nonuniform shell coating, lattice fringes of the coating material, etc. Besides, the size and morphology of the particles can be determined from the contrast difference of the core and shell materials. In order to reach the resolution even at the molecular level, much higher magnification, HR-TEM (high-resolution TEM) is used. Images can release valuable outcomes on crystallinity, lattice fringes, and even the d-spacing of the multilayer coated-NMPs. Nevertheless, two-dimensional images of the surface cannot be obtained by these electron microscopy techniques (SEM, TEM); thus, it is difficult to understand the roughness of the surface. Therefore, additional information can be found through the scanning probe microscopy (STM, AFM, etc.) techniques. Besides, the diameter and height of the particles can be obtained using such microscopy techniques.

More importantly, XPS is a useful spectroscopic technique to reveal surface information including elemental composition, chemical status, and electronic state or binding modes of surface ligands, and depth analysis or atomic composition with depth. The kinetic energy (KE) and/or binding energy (BE) together with the number of electrons escaping from a 1–10 nm thick layer of the surface is possible to measure using this technique. The main disadvantages of XPS are the maintenance and operating costs. Namely, it requires an ultrahigh vacuum (UHV) chamber and characterization is only possible to a depth 10 nm from the particle surface [4].

4. Conclusions

This chapter reviewed a newly-emerged topic in the coating research area. Multilayer coated NMPs have been showed to render the powder technology more room to grow. The new properties of the coated particles can be employed to develop multicomponent nano/micro particles in the various applications. Particulate technology together with the surface engineering concepts was properly discussed to address the current advancement and challenges in the particle deposition technology.

Author details

Sajjad Habibzadeh[1,5]*, Ehsan Rahmani[1], Mohammad Reza Saeb[2], Mohammad Reza Ganjali[3,4] and Jamal Chaouki[5]

1 Department of Chemical Engineering, Amirkabir University of Technology (Tehran Polytechnic), Tehran, Iran

2 Department of Resin and Additives, Institute for Color Science and Technology, Tehran, Iran

3 School of Chemistry, College of Science, Center of Excellence in Electrochemistry, University of Tehran, Tehran, Iran

4 Biosensor Research Center, Endocrinology and Metabolism Molecular-Cellular Sciences Institute, Tehran University of Medical Sciences, Tehran, Iran

5 Department of Chemical Engineering, École Polytechnique de Montréal, Montréal, QC, Canada

*Address all correspondence to: sajjad.habibzadeh@mail.mcgill.ca

References

[1] Xiong J, Mei X, Liu J, Wei Y, Zhao Z, Xie Z, et al. Efficiently multifunctional catalysts of 3D ordered meso-macroporous Ce0.3Zr0.7O2-supported PdAu@CeO2 core-shell nanoparticles for soot oxidation: Synergetic effect of Pd-Au-CeO2 ternary components. Applied Catalysis B: Environmental. 2019;**251**:247-260

[2] Habibzadeh S, Li L, Omanovic S, Shum-Tim D, Davis EC. Biocompatibility of Ir/Ti-oxide coatings: Interaction with platelets, endothelial and smooth muscle cells. Applied Surface Science. 2014;**301**:530-538

[3] Caruso RA, Susha A, Caruso F. Multilayered titania, silica, and laponite nanoparticle coatings on polystyrene colloidal templates and resulting inorganic hollow spheres. Chemistry of Materials. 2001;**13**(2):400-409

[4] Habibzadeh S, Zabeida O, Argoitia A, Sargent R, Klemberg-Sapieha J, Chaouki J, et al. Conformal multilayer photocatalytic thin films on fine particles by atmospheric pressure fluidized bed chemical vapor deposition. Industrial and Engineering Chemistry Research. 2018;**57**(31):10345-10353

[5] Zhou T, Zheng Y, Gao H, Min S, Li S, Liu HK, et al. Surface engineering and design strategy for surface-amorphized TiO_2@graphene hybrids for high power Li-ion battery electrodes. Advanced Science. 2015;**2**(9):1500027

[6] Chatterjee K, Sarkar S, Jagajjanani Rao K, Paria S. Core/shell nanoparticles in biomedical applications. Advances in Colloid and Interface Science. 2014;**209**:8-39

[7] Santos AC, Caldas M, Pattekari P, Fontes Ribeiro C, Ribeiro AJ, Lvov Y, et al. Chapter 16—Layer-by-layer coated drug-core nanoparticles as versatile delivery platforms. In: Grumezescu AM, editor. Design and Development of New Nanocarriers. Elseveir: William Andrew Publishing; 2018. pp. 595-635. Available from: https://www.sciencedirect.com/science/article/pii/B9780128136270000168

[8] Ghosh Chaudhuri R, Paria S. Core/shell nanoparticles: Classes, properties, synthesis mechanisms, characterization, and applications. Chemical Reviews. 2012;**112**(4):2373-2433

[9] Valverde JM, Castellanos A. Types of gas fluidization of cohesive granular materials. Physical Review E. 2007;**75**(3):031306

[10] Geldart D. Types of gas fluidization. Powder Technology. 1973;**7**(5):285-292

[11] Valverde JM, Castellanos A, Mills P, Quintanilla MA. Effect of particle size and interparticle force on the fluidization behavior of gas-fluidized beds. Physical Review. E, Statistical, Nonlinear, and Soft Matter Physics. 2003;**67**(5 Pt 1):051305

[12] Coustet M, Irigoyen J, Garcia TA, Murray RA, Romero G, Susana Cortizo M, et al. Layer-by-layer assembly of polymersomes and polyelectrolytes on planar surfaces and microsized colloidal particles. Journal of Colloid and Interface Science. 2014;**421**:132-140

[13] Gawande MB, Goswami A, Asefa T, Guo H, Biradar AV, Peng D-L, et al. Core–shell nanoparticles: Synthesis and applications in catalysis and electrocatalysis. Chemical Society Reviews. 2015;**44**(21):7540-7590

[14] Habibzadeh S, Shum-Tim D, Omanovic S. Surface and electrochemical characterization of IrTi-oxide coatings: Towards the improvement of radiopacity

for coronary stent applications. International Journal of Electrochemical Science. 2013;**8**(5):6291-6310

[15] Dahl M, Liu Y, Yin Y. Composite titanium dioxide nanomaterials. Chemical Reviews. 2014;**114**(19): 9853-9889

[16] Serp P, Kalck P, Feurer R. Chemical vapor deposition methods for the controlled preparation of supported catalytic materials. Chemical Reviews. 2002;**102**(9):3085-3128

[17] Ye M, Zhang Q, Hu Y, Ge J, Lu Z, He L, et al. Magnetically recoverable core–shell nanocomposites with enhanced photocatalytic activity. Chemistry—A European Journal. 2010;**16**(21):6243-6250

[18] Wang M, Cui Z, Yang M, Lin L, Chen X, Wang M, et al. Core/shell structured CdS/polydopamine/TiO_2 ternary hybrids as highly active visible-light photocatalysis. Journal of Colloid and Interfacc Science. 2019;**544**:1-7

[19] Yi DK, Lee SS, Ying JY. Synthesis and applications of magnetic nanocomposite catalysts. Chemistry of Materials. 2006;**18**(10):2459-2461

[20] Oh JY, Yu J-M, Chowdhury SR, Lee TI, Misra M. Significant impact of Pd nanoparticle and CdS nanolayer of Pd@CdS@ZnO core-shell nanorods on enhancing catalytic, photoelectrochemical and photocurrent generation activity. Electrochimica Acta. 2019;**298**:694-703

[21] Law W-C, Yong K-T, Roy I, Xu G, Ding H, Bergey EJ, et al. Optically and magnetically doped organically modified silica nanoparticles as efficient magnetically guided biomarkers for two-photon imaging of live cancer cells. The Journal of Physical Chemistry C. 2008;**112**(21):7972-7977

[22] Gerion D, Herberg J, Bok R, Gjersing E, Ramon E, Maxwell R, et al. Paramagnetic silica-coated nanocrystals as an advanced MRI contrast agent. The Journal of Physical Chemistry C. 2007;**111**(34):12542-12551

[23] Kim JH, Park J, Won N, Chung H, Kim S. A highly effective and facile way to prepare cellular labelling quantum dots with cetyltrimethylammonium bromide. Journal of Experimental Nanoscience. 2009;**4**(2):105-112

[24] Sun YH, Liu YS, Vernier PT, Liang CH, Chong SY, Marcu L, et al. Photostability and pH sensitivity of CdSe/ZnSe/ZnS quantum dots in living cells. Nanotechnology. 2006;**17**(17):4469-4476

[25] Zhang H, Chen D, Zhang J, Wang Z, Cui Y, Shen L. Effect of shell layers on luminescence of colloidal CdSe/Zn0.5Cd0.5Se/ZnSe/ZnS Core/multishell quantum dots. Journal of Nanoscience and Nanotechnology. 2010;**10**(11):7587-7591

[26] Won Y-H, Jang HS, Chung D-W, Stanciu LA. Multifunctional calcium carbonate microparticles: Synthesis and biological applications. Journal of Materials Chemistry. 2010;**20**(36):7728-7733

[27] Zhelev Z, Ohba H, Bakalova R. Single quantum dot-micelles coated with silica shell as potentially non-cytotoxic fluorescent cell tracers. Journal of the American Chemical Society. 2006;**128**(19):6324-6325

[28] Kateb B, Chiu K, Black KL, Yamamoto V, Khalsa B, Ljubimova JY, et al. Nanoplatforms for constructing new approaches to cancer treatment, imaging, and drug delivery: What should be the policy? NeuroImage. 2011;**54**:S106-SS24

[29] Chen L, Subirade M. Chitosan/β-lactoglobulin core–shell nanoparticles as nutraceutical carriers. Biomaterials. 2005;**26**(30):6041-6053

[30] Song Z, Liu Y, Shi J, Ma T, Zhang Z, Ma H, et al. Hydroxyapatite/mesoporous silica coated gold nanorods with improved degradability as a multi-responsive drug delivery platform. Materials Science and Engineering: C. 2018;**83**:90-98

[31] Sahiner N, Pekel N, Akkas P, Güven O. Amidoximation and characterization of new complexing hydrogels prepared from N-vinyl 2-pyrrolidone/acrylonitrile systems. Journal of Macromolecular Science, Part A. 2000;**37**(10):1159-1172

[32] Kayal S, Ramanujan RV. Anti-cancer drug loaded iron–gold core–shell nanoparticles (Fe@Au) for magnetic drug targeting. Journal of Nanoscience and Nanotechnology. 2010;**10**(9):5527-5539

[33] Zhao W, Gu J, Zhang L, Chen H, Shi J. Fabrication of uniform magnetic nanocomposite spheres with a magnetic core/mesoporous silica shell structure. Journal of the American Chemical Society. 2005;**127**(25):8916-8917

[34] Jafari T, Simchi A, Khakpash N. Synthesis and cytotoxicity assessment of superparamagnetic iron–gold core–shell nanoparticles coated with polyglycerol. Journal of Colloid and Interface Science. 2010;**345**(1):64-71

[35] Wang C, Yan J, Cui X, Cong D, Wang H. Preparation and characterization of magnetic hollow PMMA nanospheres via in situ emulsion polymerization. Colloids and Surfaces A: Physicochemical and Engineering Aspects. 2010;**363**(1):71-77

[36] Prabaharan M, Grailer JJ, Pilla S, Steeber DA, Gong S. Gold nanoparticles with a monolayer of doxorubicin-conjugated amphiphilic block copolymer for tumor-targeted drug delivery. Biomaterials. 2009;**30**(30):6065-6075

[37] Fedorov MV, Kornyshev AA. Ionic liquids at electrified interfaces. Chemical Reviews. 2014;**114**(5):2978-3036

[38] Zhang J, Fu J, Zhang J, Ma H, He Y, Li F, et al. Co@Co3O4 core–shell three-dimensional nano-network for high-performance electrochemical energy storage. Small. 2014;**10**(13):2618-2624

[39] Ma W, Shi Q, Nan H, Hu Q, Zheng X, Geng B, et al. Hierarchical ZnO@MnO2@PPy ternary core–shell nanorod arrays: An efficient integration of active materials for energy storage. RSC Advances. 2015;**5**(50):39864-39869

[40] Zhang Z, Wang F, An Q, Li W, Wu P. Synthesis of graphene@Fe_3O_4@C core–shell nanosheets for high-performance lithium ion batteries. Journal of Materials Chemistry A. 2015;**3**(13):7036-7043

[41] Zhao B, Huang S-Y, Wang T, Zhang K, Yuen MMF, Xu J-B, et al. Hollow SnO2@Co3O4 core–shell spheres encapsulated in three-dimensional graphene foams for high performance supercapacitors and lithium-ion batteries. Journal of Power Sources. 2015;**298**:83-91

[42] Xie L, Huang X, Huang Y, Yang K, Jiang P. Core@double-Shell structured BaTiO3–polymer nanocomposites with high dielectric constant and low dielectric loss for energy storage application. The Journal of Physical Chemistry C. 2013;**117**(44):22525-22537

[43] Feng H-P, Tang L, Zeng G-M, Zhou Y, Deng Y-C, Ren X, et al. Core-shell nanomaterials: Applications in energy storage and conversion. Advances in Colloid and Interface Science. 2019;**267**:26-46

[44] Kim D-W, Cho K, Kim H, Park B, Young Sung M, Kim S. Optoelectronic characteristics of CdTe/HgTe/

CdTe quantum-dot quantum-well nanoparticles. Solid State Communications. 2006;**140**(5):215-218

[45] Schooss D, Mews A, Eychmüller A, Weller H. Quantum-dot quantum well CdS/HgS/CdS: Theory and experiment. Physical Review B. 1994;**49**(24):17072-17078

[46] Cheng C-T, Chen C-Y, Lai C-W, Liu W-H, Pu S-C, Chou P-T, et al. Syntheses and photophysical properties of type-II CdSe/ZnTe/ZnS (core/shell/shell) quantum dots. Journal of Materials Chemistry. 2005;**15**(33):3409-3414

[47] Kim S-D, Choe W-G, Choi J, Jeong J-R. High reflective efficiency and durability of multilayered core-shell composite particles with controlled shell thickness. Dyes and Pigments. 2018;**153**:53-60

[48] Herrmann CF, Fabreguette FH, Finch DS, Geiss R, George SM. Multilayer and functional coatings on carbon nanotubes using atomic layer deposition. Applied Physics Letters. 2005;**87**(12):123110

[49] Thanh NTK, Maclean N, Mahiddine S. Mechanisms of nucleation and growth of nanoparticles in solution. Chemical Reviews. 2014;**114**(15):7610-7630

[50] Sokolov SV, Tschulik K, Batchelor-McAuley C, Jurkschat K, Compton RG. Reversible or not? Distinguishing agglomeration and aggregation at the nanoscale. Analytical Chemistry. 2015;**87**(19):10033-10039

[51] Habibzadeh S, Kazemi-Beydokhti A, Khodadadi AA, Mortazavi Y, Omanovic S, Shariat-Niassar M. Stability and thermal conductivity of nanofluids of tin dioxide synthesized via microwave-induced combustion route. Chemical Engineering Journal. 2010;**156**(2):471-478

[52] Habibzadeh S, Li L, Shum-Tim D, Davis EC, Omanovic S. Electrochemical polishing as a 316L stainless steel surface treatment method: Towards the improvement of biocompatibility. Corrosion Science. 2014;**87**:89-100

[53] Attia M, Farag S, Habibzadeh S, Hamzehlouia S, Chaouki J. Fast pyrolysis of lignocellulosic biomass for the production of energy and chemicals: A critical review. Current Organic Chemistry. 2016;**20**(23):2458-2479

[54] Shahrezaei M, Babaluo AA, Habibzadeh S, Haghighi M. Photocatalytic properties of 1D TiO_2 nanostructures prepared from polyacrylamide gel–TiO_2 nanopowders by hydrothermal synthesis. European Journal of Inorganic Chemistry. 2017;**2017**(3):694-703

[55] Shahrezaei M, Habibzadeh S, Babaluo AA, Hosseinkhani H, Haghighi M, Hasanzadeh A, et al. Study of synthesis parameters and photocatalytic activity of TiO_2 nanostructures. Journal of Experimental Nanoscience. 2017;**12**(1):45-61